Edited by
Adriano Zecchina, Silvia Bordiga,
and Elena Groppo

Selective Nanocatalysts and Nanoscience

Further Reading

Zhou, Q.-L. (Ed.)

Privileged Chiral Ligands and Catalysts

2011
Hardcover
ISBN: 978-3-527-32704-1

Guo, J. (Ed.)

X-Rays in Nanoscience

Spectroscopy, Spectromicroscopy, and Scattering Techniques

2011
Hardcover
ISBN: 978-3-527-32288-6

Cybulski, A., Moulijn, J. A., Stankiewicz, A. (Eds.)

Novel Concepts in Catalysis and Chemical Reactors

Improving the Efficiency for the Future

2010
Hardcover
ISBN: 978-3-527-32469-9

Cejka, J., Corma, A., Zones, S. (Eds.)

Zeolites and Catalysis

Synthesis, Reactions and Applications

2010
Hardcover
ISBN: 978-3-527-32514-6

de Jong, K. P. (Ed.)

Synthesis of Solid Catalysts

2009
Hardcover
ISBN: 978-3-527-32040-0

Astruc, D. (Ed.)

Nanoparticles and Catalysis

2008
Hardcover
ISBN: 978-3-527-31572-7

Ertl, G., Knözinger, H., Schüth, F., Weitkamp, J. (Eds.)

Handbook of Heterogeneous Catalysis

8 Volumes
Second completely revised and enlarged edition

2008
Hardcover
ISBN: 978-3-527-31241-2

Edited by Adriano Zecchina, Silvia Bordiga, and Elena Groppo

Selective Nanocatalysts and Nanoscience

Concepts for Heterogeneous and Homogeneous Catalysis

WILEY-VCH Verlag GmbH & Co. KGaA

The Editors

Prof. Adriano Zecchina
Dipto. di Chimica IFM
Università di Torino
Via Pietro Giuria 7
10125 Torino
Italy

Prof. Dr. Silvia Bordiga
University of Turin
NIS Centre of Excellence
via P. Giuria 7
10125 Torino
Italy

Dr. Elena Groppo
University of Turin
NIS Centre of Excellence
via P. Giuria 7
10125 Torino
Italy

Library of Congress Card No.: applied for

British Library Cataloguing-in-Publication Data
A catalogue record for this book is available from the British Library.

Bibliographic information published by the Deutsche Nationalbibliothek
The Deutsche Nationalbibliothek lists this publication in the Deutsche Nationalbibliografie; detailed bibliographic data are available on the Internet at <http://dnb.d-nb.de>.

Cover Design Adam Design, Weinheim
Typesetting Toppan Best-set Premedia Limited, Hong Kong
Printing and Binding Fabulous Printers Pte Ltd, Singapore

Printed on acid-free paper

ISBN: 978-3-527-32271-8
ePDF ISBN: 978-3-527-63570-2
ePub ISBN: 978-3-527-63569-6
Mobi ISBN: 978-3-527-63571-9
oBook ISBN: 978-3-527-63568-9

Contents

Preface

Catalysis and selective catalysis are at the core of synthetic chemistry. In the last century, the development of catalysis has followed two distinct paths, heterogeneous and homogeneous, initially separated and then intimately interconnected. In the first part of the 20th century, heterogeneous catalysis obtained important results, as evidenced by the six Nobel Prizes awarded in just 50 years, beginning with W. Ostwald (1909), followed by P. Sabatier (1912), F. Haber (1918), K. Bosh (1931), up to K. Ziegler, and G. Natta (1956). These results led quickly to important industrial applications, ranging from nitric acid production from ammonia oxidation, to ammonia synthesis, to hydrogenation reactions, and finally to olefin polymerization. Although the majority of these reactions, occurring on transition metal surfaces or on isolated transition metal sites, are relatively simple, at that time little was known about the reaction mechanism and only the development of surface science and computational methods, with the contribution of the Nobel laureates I. Langmuir (1932), C. Inshelwood (1933), and G. Ertl (2007), led to a progressively accurate understanding of the surface structures involved in the catalytic events. The research in heterogeneous catalysis gradually stimulated both the synthesis and the study of finely divided materials (metal oxides, metals, and supported metals), exhibiting a high surface area. These studies certainly contributed to open the era of nanoscience. Similarly, the need of surface characterization has stimulated the development of increasingly surface-sensitive methods.

Although the problem of selectivity in heterogeneous catalysis had not been neglected until then, it is certainly true that this started to become critical with the advent of Ziegler–Natta catalysts and the related synthesis of isotactic polypropylene and since then, the problem of selectivity has started to attract the attention of an increasing number of researchers involved in the construction and characterization of catalytic centers, having the desired selectivity properties. The results in this research area were remarkable, although prevalently obtained through an empirical approach, more than as a result of a rational *ab initio* design. However, an overall achievement has emerged from such studies, namely the selectivity is the result of a complex design of surface active sites, through the fine tuning of the ligands.

Approximately at the same time, chemists started to develop homogeneous catalysts showing increasingly better defined structures; the list of Nobel Prizes awarded in this field, starting with G. Wilkinson (1973), and followed by W.S.

Kowles, R. Noyori, and B. Sharpless (2001), and Y. Chauvin, R. Grubbs, and R. Schrock (2005), fully testifies to these contributions. One of the most remarkable examples of construction of a class of homogeneous catalysts based on a rational design of the active centers is that of Zr-based metallocenes for selective olefin polymerization, for which the steroselective properties were obtained by appropriate design of the ligands sphere.

After about a century since the first Nobel Prize was awarded to catalysis, we can state that both heterogeneous and homogeneous approaches lead to the same general conclusion: a selective catalyst can be considered a nanomachine obtained through a precise control of the structure of the active sites, of the three-dimensional environment and of their relationship. For homogeneous and heterogeneous selective catalysts, the three-dimensional environment around the active sites resembles the tunable structure of enzymes, which are the most efficient catalysts optimized by nature over billions of years. In this regard, a point that merits a specific comment is the fact that, while in the past heterogeneous and homogeneous catalysis mainly followed separate development dynamics, today it is becoming increasingly clear that they are strongly interconnected and that the achievements obtained in one area have influence on the other one. In other words, selective catalysis is a single chapter of science, whatever it is, homogeneous, heterogeneous, or even enzymatic. The chapters of this book, devoted to both heterogeneous and homogeneous catalysts, have been selected following this basic approach:

1. *The Structure and Reactivity of Single and Multiple Sites on Heterogeneous and Homogeneous Catalysts: Analogies, Differences, and Challenges for Characterization Methods* by A. Zecchina, S. Bordiga, and E. Groppo.

2. *Supported Nanoparticles and Selective Catalysis: A Surface Science Approach* by W. Zhang.

3. *When Does Catalysis with Transition Metal Complexes Turn into Catalysis by Nanoparticles?* by J. DeVries.

4. *Capsules and Cavitands: Synthetic Catalysts of Nanometric Dimension* by G. Borsato, J. Rebek Jr., and A. Scarso.

5. *Photocatalysts: Nanostructured Photocatalytic Materials for Solar Energy Conversion* by K. Domen.

6. *Chiral Catalysts* by J.M. Fraile, J.I. García, and J.A. Mayoral.

7. *Selective Catalysts for Petrochemical Industry: Shape Selectivity in Microporous Materials* by S. Svelle and M. Bjørgen.

8. *Crystal Engineering of Metal-Organic Frameworks (MOFs) for Heterogeneous Catalysis* by Chuan-De Wu.

9. *Mechanism of Stereospecific Propene Polymerization Promoted by Metallocene and Nonmetallocene Catalysts* by A. Correa and L. Cavallo.

From the above-mentioned titles, the effort to mix both homogeneous and heterogeneous catalysts in a single book is evident.

Elena Groppo
Silvia Bordiga
Adriano Zecchina

List of Contributors

Morten Bjørgen
Norwegian University of Science and Technology
Department of Chemistry
Trondheim N-7491
Norway

Silvia Bordiga
Department of Inorganic, Physical and Material Chemistry (IFM)
NIS Centre of Excellence
Nanostructured Interfaces and Surfaces
University of Torino
Via P. Giuria 7
10125 Torino
Italy

Giuseppe Borsato
Università Ca' Foscari di Venezia
Dipartimento di Scienze Molecolari e Nanosistemi
Calle Larga S. Marta 2137
30123 Venice
Italy

Luigi Cavallo
Università di Salerno
Dipartimento di Chimica
Via ponte don Melillo
84084 Fisciano, SA
Italy

Andrea Correa
Università di Salerno
Dipartimento di Chimica
Via ponte don Melillo
84084 Fisciano, SA
Italy

Kazunari Domen
The University of Tokyo
School of Engineering
Department of Chemical System Engineering
Tokyo 113-8656
Japan

José M. Fraile
Instituto de Síntesis Química y Catálisis Homogénea (ISQCH)
Departamento de Química Orgánica
Facultad de Ciencias, Universidad de Zaragoza-C.S.I.C.
C/ Pedro Cerbuna s/n
50009 Zaragoza
Spain

José I. García
Instituto de Síntesis Química y Catálisis Homogénea (ISQCH)
Departamento de Química Orgánica
Facultad de Ciencias, Universidad de Zaragoza-C.S.I.C.
C/ Pedro Cerbuna s/n
50009 Zaragoza
Spain

Elena Groppo
Department of Inorganic, Physical and Material Chemistry (IFM)
NIS Centre of Excellence
Nanostructured Interfaces and Surfaces
University of Torino
Via P. Giuria 7
10125 Torino
Italy

José A. Mayoral
Instituto de Síntesis Química y Cattálisis Homogénea (ISQCH)
Departamento de Química Orgtánica
Facultad de Ciencias, Universidad de Zaragoza-C.S.I.C.
C/ Pedro Cerbuna s/n
50009 Zaragoza
Spain

Julius Rebek Jr.
The Scripps Research Institute
The Skaggs Institute for Chemical Biology
La Jolla, CA 92037
USA

Alessandro Scarso
Università Ca' Foscari di Venezia
Dipartimento di Scienze Molecolari e Nanosistema
Calle Larga S. Marta 2137
30123 Venice
Italy

Stian Svelle
University of Oslo
Innovative Natural Gas Processes and Products (inGAP), Department of Chemistry
Oslo N-0315
Norway

Johannes G. de Vries
DSM Innovative Synthesis BV
A Unit of DSM Pharma Chemicals
P.O. Box 18
6160 MD Geleen
The Netherlands

Da Wang
Nankai University
Institute of Polymer Chemistry, Key Laboratory of Functional Polymer Materials of Ministry of Education
Tianjin 300071
China

Chuan-De Wu
Zhejiang University
Department of Chemistry
Hangzhou 310027
China

Rui Yan
Nankai University
Institute of Polymer Chemistry, Key Laboratory of Functional Polymer Materials of Ministry of Education
Tianjin 300071
China

Adriano Zecchina
Department of Inorganic, Physical and Material Chemistry (IFM)
NIS Centre of Excellence
Nanostructured Interfaces and Surfaces
University of Torino
Via P. Giuria 7
10125 Torino
Italy

Wangqing Zhang
Nankai University
Institute of Polymer Chemistry, Key Laboratory of Functional Polymer Materials of Ministry of Education
Tianjin 300071
China

1
The Structure and Reactivity of Single and Multiple Sites on Heterogeneous and Homogeneous Catalysts: Analogies, Differences, and Challenges for Characterization Methods

Adriano Zecchina, Silvia Bordiga, and Elena Groppo

1.1
Introduction

The content of this book is specifically devoted to a description of the complexity of the catalytic centers (both homogeneous and heterogeneous) viewed as nanomachines for molecular assembling. Although the word "nano" is nowadays somewhat abused, its use for catalysts science (as nanoscience) is fully justified. It is a matter of fact that (i) to perform any specific catalytic action, the selective catalyst must necessarily possess sophisticated structure where substrates bonds are broken and formed along a specific path and (ii) the relevant part of this structure, usually constituted by a metal center or metal cluster surrounded by a sphere of ligands or by a solid framework or by a portion of functionalized surface, often reaches the nanometric dimension. As it will emerge from the various chapters, this vision is valid for many types of selective catalysts including catalysts for hydrogenation, polymerization, olygomerization, partial oxidation, and photocatalytic solar energy conversion. Five chapters are devoted to the above-mentioned reactions. From the point of view of the general definition, homogeneous and heterogeneous selective catalysts can be treated in the same way. As homogeneous selective catalysts are concerned, the tridimensional structure surrounding the metal center can be organized with cavitand shape, while for heterogeneous catalysts the selectivity is the result of an accurate design and synthesis of the framework structures (often microporous and crystalline) where the sites are anchored. This is the case of zeolitic and metallorganic materials. The structure of catalytic centers in capsules and cavitands is discussed specifically in a single chapter, while the zeolitic and metallorganic catalyst are treated in two chapters. One of the intriguing aspects of catalysis science is represented by the wide recognition that the chemists usually design and prepare precursor structures and that the really "working" centers are formed after an induction period in the presence of reactants. For this reason, it is not sufficient to know as much as possible about the precursor structures and *in situ* characterization methods under operando conditions are beneficial. Catalyst characterization is consequently a relevant aspect of catalysis science and two chapters are devoted partially or totally to this problem.

Selective Nanocatalysts and Nanoscience, First Edition. Edited by Adriano Zecchina, Silvia Bordiga, Elena Groppo.

1.2 Definition of *Multiple-* and *Single-Site* Centers in Homogeneous and Heterogeneous Catalysis

1.2.1 Single-Site Homogeneous Catalysts: Prototype Examples Taken from the Literature

The usual definition of single-site molecular catalyst is a catalyst which contains a single metal center. The metal atom usually has an open-side, or an easily replaceable group, where it binds substrates and where bonds are broken and formed. A prototype well describing this definition is represented by the classic Wilkinson homogeneous catalysts for hydrogenation reactions, whose structure is represented in Figure 1.1a (shaded area). This is definitely the first example of single-site homogeneous catalyst having activity similar to that of heterogeneous counterparts [1]. It must be noted, however, that the reported structure is that of the *precursor species*. In fact, the really reactive center is universally thought to derive from the previous structure by the initial dissociation of one or two triphenylphosphine ligands to give 14- or 12-electron complexes, respectively. The elimination of the phosphine ligand generates a coordinatively unsaturated species able to

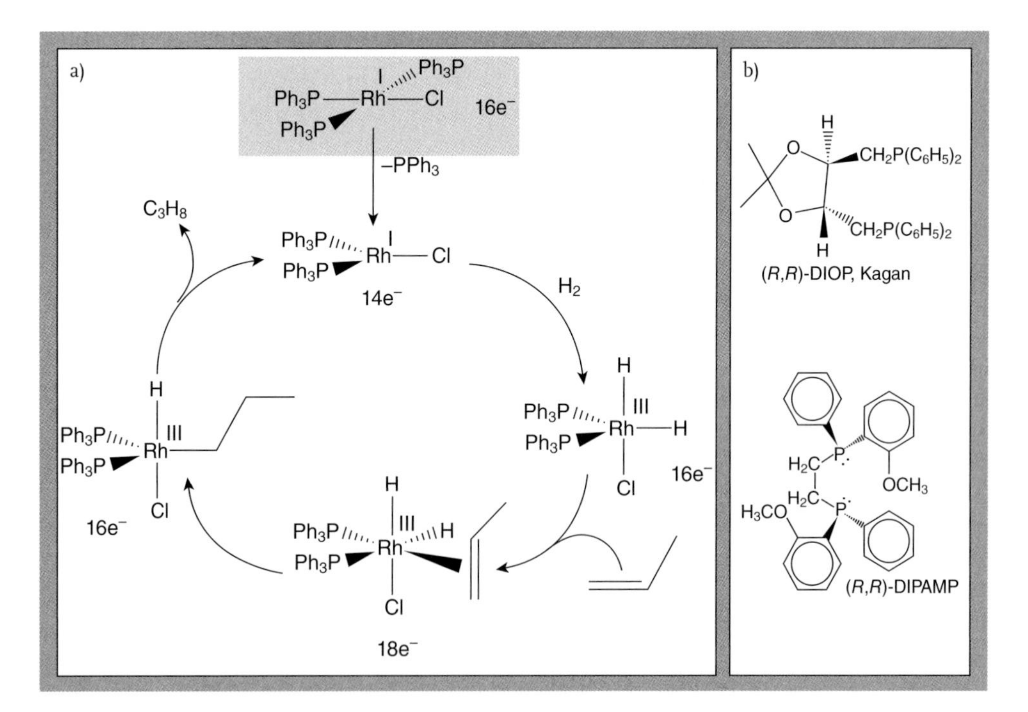

Figure 1.1 (a) Wilkinson homogeneous catalysts for hydrogenation reactions, and catalytic cycle during hydrogenation of propene. (b) Two examples of chiral ligands used instead of the phosphine groups to modify the selectivity of the Wilkinson catalyst.

dissociate the hydrogen molecule (oxidative addition of H_2 to the metal). Subsequent π-complexation of alkene, intramolecular hydride transfer (olefin insertion), and reductive elimination result in the release of the alkane product as depicted in the scheme reported in Figure 1.1a.

When the Wilkinson catalysts cycle is described, the attention is normally focused on the change in coordination and valence state of the Rh center, that is, on its electronic properties. The role of the bulky ancillary ligands (triphenylphosphine groups) is usually less considered, although it certainly plays a role in determining the selectivity in the chemospecific hydrogenation of terminal alkenes in presence of internal alkenes or other easily reducible groups [2]. That the role of ligands is central in determining the catalyst selectivity is demonstrated by the well-known fact that the use of chiral ligands instead of phosphine groups (like the chelating DIOP or DIPAMP ligands represented in Figure 1.1b) [1] transforms the Wilkinson complex into an efficient asymmetric catalyst characterized by outstanding enantioselectivity.

The role of properly tuned ancillary ligands structures covalently bonded to the metal site is not limited to the Wilkinson hydrogenation catalysts. In fact, the well-known enantioselectivity of homogeneous Ziegler type, Osborn–Schrock, and Crabtree type hydrogenation systems [3, 4] is based on the same principle. As for the Osborn–Schrock and Crabtree cationic catalysts, further role is known to be played also by the solvent molecules and by the counteranionic species, which can act as external ancillary ligands (coordinatively or electrostatically bonded) influencing both reactivity and selectivity [5, 6]. The same is valid for the role of anionic counterparts on the activity and selectivity of cationic (Wilkinson type) chiral Rh(I) catalysts [7].

A preliminary conclusion that can be derived directly from the previous considerations is that, as already underlined [6], in order to develop selectivity, the single metal center forming the core of the catalysts (where the substrates bonds are broken and formed) must be surrounded by a complex ligands framework (including the solvent molecules directly interacting with the metal center). This makes the whole structure similar to a functional nanomachine for molecular assembling. On the basis of these considerations, we can slightly modify the classical definition of selectivity [8] from "the selectivity of a catalyst is its ability to direct conversion of reactants along one specific pathway" to "the selectivity of a catalyst is its ability to direct conversion of reactants along one specific pathway also with the intervention of covalently and coordinatively bonded ancillary structures."

The considerations derived from the hydrogenation catalysts can be easily extended to other single-site catalysts. Among all, an outstanding example of single-site homogeneous catalysts is represented by the metallocene and postmetallocene family of catalysts for olefin polymerization. The first example of catalysts belonging to this family, discovered by Kaminsly [9, 10], is represented in Figure 1.2a, where M is Ti, Zr, and Hf. In the following decades, new catalysts were synthesized and tested showing a greatly improved selectivity in propene polymerization, with an increasingly sophisticated character of the ligand sphere (see, e.g., structures displayed in Figure 1.2b and c). The new structures differ from the

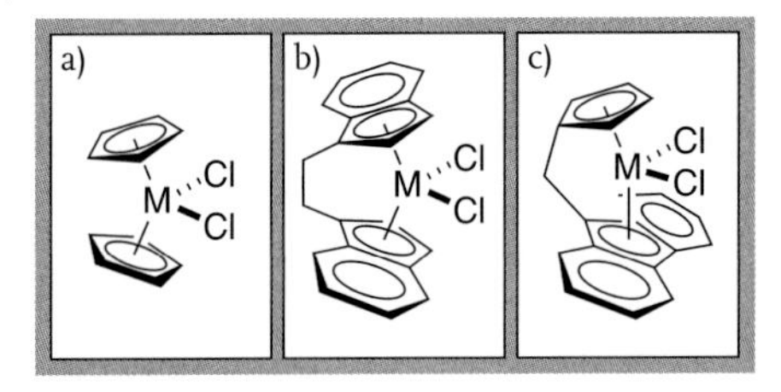

Figure 1.2 Examples of metallocene catalysts characterized by an increasingly sophisticated character of the ligand sphere (M = Ti, Zr, and Hf).

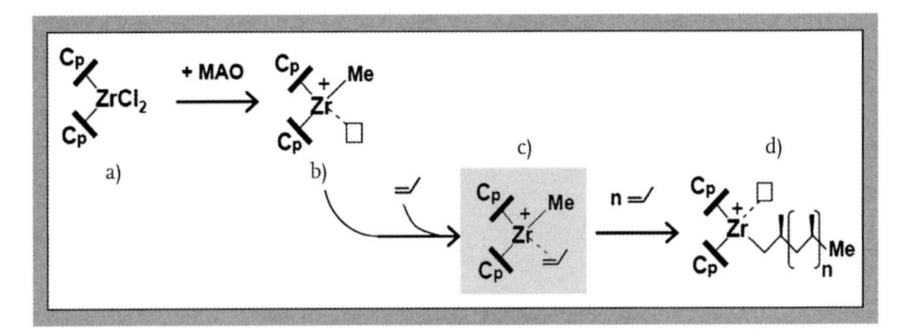

Figure 1.3 Accepted mechanism explaining the formation of the active site in case of metallocene catalysts by interaction of the metallocene precursor (a) with an activator (MAO), creation of a coordination vacancy (b), subsequent insertion of the olefinic monomer (c), followed by growth of the polymeric chain (d).

original one for the presence of an engineered ligands sphere, which confers to the catalytic center an increased rigidity and appropriately designed. The whole subject concerning the evolution of the catalysts design represents one of the most exciting pieces of research of the second half of the last century and has been excellently reviewed by Resconi *et al.* [11].

It must also be underlined that in the case of metallocene catalysts, *the represented structures are the precursors of the really active species,* which derive from them by interaction with an activator (usually MAO) following the well-known reaction reported in Figure 1.3. The resulting coordinatively unsaturated cationic structure coordinates a propene molecule and then originates the polymeric chain by a subsequent insertion reaction. It must be noticed that the complex with propene (activated complex or intermediate, shaded area in Figure 1.3) has never been experimentally observed, and consequently it must be considered as a result of mechanistic and quantum computational approaches [11].

So far, the presence of the anionic counterion has been neglected. When this is considered, the resulting ion-pair character of the catalyst is immediately emerging. As the interaction between the anion and cation forming the pair is critically dependent upon external factors such as the solvating power of the anion and the solvent polarity, it comes out that the activity and selectivity of the catalyst are not only influenced by the structure of the covalently bonded ancillary ligand sphere, but also by external factors [5, 11, 12]. In particular, it is expected that the anion can greatly influence the insertion of the monomer molecule in the growing chain [13]. When all the involved factors are simultaneously taken into consideration, the sophisticated character of this molecular assembling nanomachine is clearly emerging.

1.2.2
Single-Site Heterogeneous Catalysts

The usual definition of single-site heterogeneous catalyst is *a catalyst constituted by a metal atom, ion, or small cluster of atoms, held by surface ligands to a rigid framework*. These sites are isolated inside the hosting structure. As the supporting solid (or rigid framework) usually exposes different faces and hence different anchoring situations, the structure of the anchored sites is usually not well known. Following this definition, the substantial difference between homogeneous and heterogeneous single sites is represented by the ligands sphere (which is accurately engineered in the first case and less defined in the second case). In the following, we will illustrate these concepts starting from cases where the anchoring structure is characterized by a definition comparable to that of the homogeneous sites, and moving to situations of increasing complexity. As previously discussed for homogenous catalysts, it will be illustrated that by appropriate surface and hosting structures engineering the selectivity of the catalyzed reactions can be tuned up to levels comparable with those of the homogeneous systems.

1.2.2.1 TS-1 and the Shape Selectivity
Among the solid frameworks which can anchor or host the catalytically active centers, zeolites and zeolitic materials occupy an outstanding position, because they are crystalline and the resulting materials have large practical applications. One of the catalytic systems where the structural situation of the metal center (Ti^{IV}) is better known is titanium silicalite (TS-1, Figure 1.4a), a catalyst which has found wide practical applications in oxidation reactions with hydrogen peroxide (Figure 1.4b) [14–18].

The precursor structure of TS-1 (i.e., the structure of the catalyst before contact with reagents) is illustrated in Figure 1.4c. It has been fully demonstrated that in the virgin samples, the Ti^{IV} centers occupy regular tetrahedral positions of the framework, because they substitute the silicon atoms. When immersed in the hydrogen peroxide water solutions, the coordination sphere of the Ti^{IV} centers expands to an octahedral situation. This is the result of the formation of the "active single-site structure" reported in Figure 1.4c, which is formed in solution by hydrolysis of ≡Si–O–Ti≡ bridges and interaction with water and hydrogen peroxide molecules. The structure of the active site under conditions approaching the reaction situation has been the subject of many investigations with physical methods such as UV-Vis, XANES, EXAFS, Raman, and FTIR spectroscopies [14–20]. When

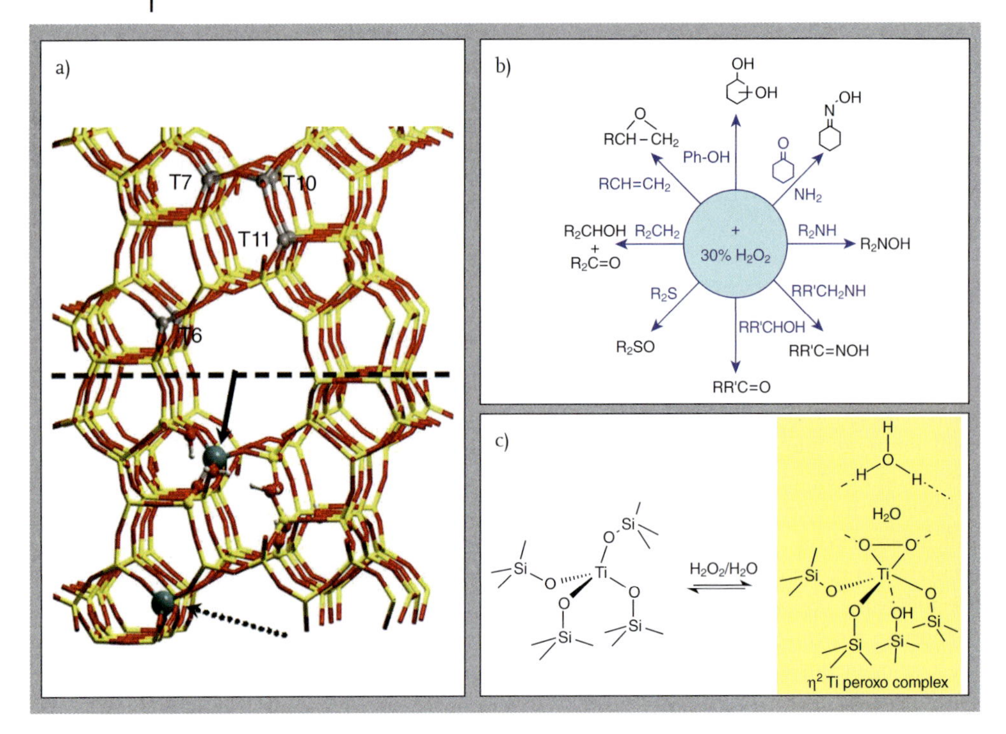

Figure 1.4 (a) Schematic representation of the preferential location of Ti atoms and Si vacancies in the TS-1 framework (upper part) and their interplay (lower part). Yellow and red sticks represent Si and O of the regular MFI lattice; green balls refer to Ti, while red and white ones to O and H of defective internal OH groups. Dotted and full arrows evidence regular $[Ti(OSi)_4]$ and defective $[Ti(OSi)_3OH]$ sites, respectively. (b) Role of TS-1 in oxidation in chemical reactions. (c) Formation of the active single-site structure by hydrolysis of ≡Si–O–Ti≡ bridges and interaction with water and hydrogen peroxide molecules.

a substrate molecule (olefin, phenol, etc.) is approaching the active center, a hydrogen-bonding interaction with the Ti–OOH group is likely to occur, followed by an oxygen transfer. In this case, the hydrogen-bonded species is one of the reaction intermediates.

The TS-1 system is interesting not only because it is one of the heterogeneous catalysts where the structure of the active sites is better known, but also because it allows to introduce the concept of shape selectivity. TS-1 catalyzes the oxidation of phenol (PhOH) to diphenol (Figure 1.4b). Not all the isomers are, however, formed: in fact, the formation of the *m*-isomer is inhibited. This is due to the shape selective character of the catalyst characterized by channels with shape not allowing the formation and diffusion of the more spatially demanding *m*-isomer. We are here in the presence of catalytic events in confined spaces, where selection of products is made on the basis of the shape of the reactants. Following the considerations already made for homogeneous single-site catalysts, the structure of the channels can be considered as a special and new type of external ligands sphere, having no precedent in the cases illustrated in the previous section.

The shape selectivity is usually considered as belonging to the heterogeneous catalyst domain; however it is worth mentioning that, although not relevant in the previously discussed cases, the role of confinement effects is well known also in homogeneous catalysis. It is sufficient to cite the enzyme action mechanism, where the active center (often constituted by a single metal atom or by a dimer where the substrates bonds are broken and formed) is confined in a pocket or channel of the fluxional protein structure [6, 21, 22]. Furthermore, more and more examples of homogeneous catalytic systems where the active center is located in a suitably engineered pocket are known.

1.2.2.2 **H-ZSM-5: A Popular Example of Protonic Zeolite**

H-ZSM5 is a zeolitic material with the same MFI structure of TS-1 described above, the substantial difference being represented by the presence of Al instead of Ti in the framework. To compensate the negative charge of the framework, in correspondence of the substitutional Al atoms, a positive center must be present (alkaline ion or proton). The protonic sites protruding into the channel in correspondence of each Al atom present in the framework are fully available for interaction with shape-selected molecules penetrating the channels (Figure 1.5). As the Si/Al ratio is never less than 15, the protonic sites are isolated and consequently can behave as single-site catalytic centers in a variety of Brønsted-catalyzed reactions.

From the point of view of the isolation and structural definition, the H-ZSM5 is one of the most defined heterogenous single-site catalysts reported in the literature. The interaction of the Brønsted sites with molecules of increasing basicity has been thoroughly studied [23–25]. When the proton affinity of the base is low, the result is the formation of a hydrogen-bonded adduct. In the case of olefin/ Brønsted site interaction, the formation of the hydrogen-bonded species can be followed by protonation and formation of oligomeric species with carbocationic character following the reaction path illustrated below:

$$H^+ + C_2H_4 \rightarrow (H^+ \cdots C_2H_4) \rightarrow C_2H_5^+ \tag{1.1}$$

$$C_2H_5^+ + C_2H_4 \rightarrow (C_2H_5^+ \cdots C_2H_4) \rightarrow C_4H_9^+ \tag{1.2}$$

The formation of these carbocationic species is only the initial stage of the reaction, whose final products can be more complex linear and branched olefins (like butanes), or branched and aromatic molecules. The formation step of the protonated species is thought to occur via a hydrogen-bonded intermediate. Whether this intermediate is sufficiently stable and long lived to be observed can only be decide by means of an operando experiment. This will be discussed in the following paragraph. The protonic site can also interact with methanol and ethanol giving H_2O and olefins and more complex molecules following a path similar to that illustrated above. It is thought that under reaction conditions, the real catalyst is a pool.

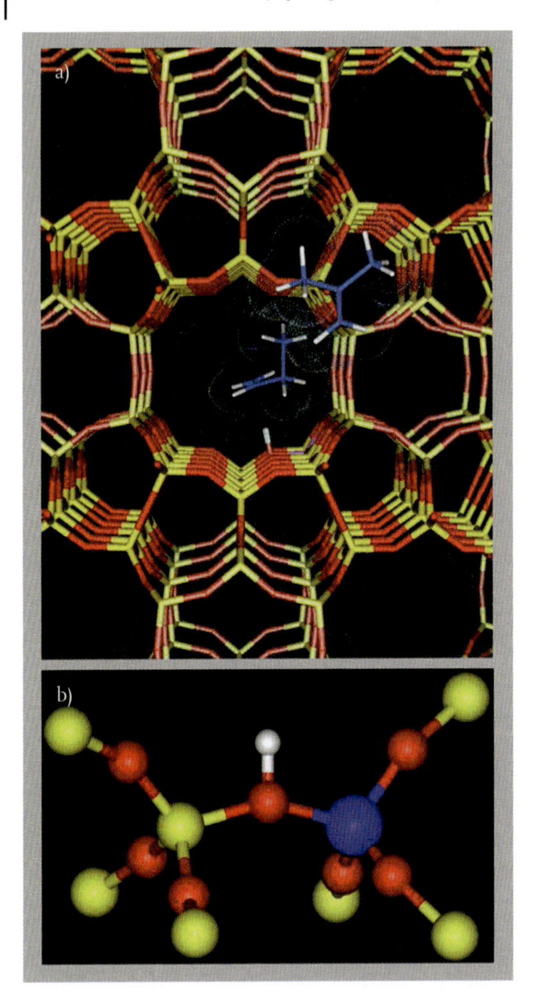

Figure 1.5 (a) Schematic representation of an MFI structure with a protonic site protruding into the main channel and interacting with shape-selected molecules penetrating the channel. (b) It shows an enlargement of the protonic site. Si, O, Al, and H atoms are represented in yellow, red, blue, and white, respectively.

In some other cases, protons cannot be considered as active sites as they are only the starting point needed for the formation of more complex carbocationic species that act as catalytic center. This is the case of methanol-to-hydrocarbons (MTH) reaction, for which the complex mechanism has been subject of numerous studies during the past 30 years [26–28]. In MTH, it has been observed a rapid conversion of methanol to dimethyl ether (DME) and water, followed by olefin formation and finally by concurrent formation of aromatics and paraffins, in agreement with thermodynamic equilibrium under the chosen conditions. Today, it is generally accepted that the MTH reaction proceeds over a hybrid site consisting of the zeotype/zeolite lattice including the active and acidic site and an adsorbed hydrocarbon intermediate, which reacts with methanol or dimethyl ether to form a larger hydrocarbon entity, and subsequently splits off olefins. This mechanistic scheme is referred to as the so-called hydrocarbon pool mechanism [29–31]. It has been claimed that the number of methyl groups on the main methylbenzene and their protonated analogues change in consideration of the catalysts topologies (H-SAPO-34, H-ZSM-5, H-beta) [32–38]. Wide-pore H-beta zeolite stabilizes the higher methyl benzenes and gives a high selectivity to propene and butene, while the lower methyl benzene analogues in H-ZSM-5 give a high selectivity to ethene and propene [37–40]. A relevant fact that complicates MTH reaction are parallel reactions, especially olefin methylation reactions [39, 40], and fast deactivation by coke formation. In order to obtain a higher single product selectivity and coking resistance, several studies have been devoted to look for new zeolitic topologies and to consider the effect of acid strength and site density.

1.2.2.3 The Ziegler–Natta Polymerization Catalyst

Another single-site heterogeneous catalyst is undoubtedly the Ziegler–Natta system for olefin polymerization. After the initial discovery of the activity of the$TiCl_3/Al(C_2H_5)_2Cl$ system in olefin polymerization [41, 42], a decisive advancement in the industrial utilization of Ti^{III}-based system was obtained when Ti centers were supported on high surface area microcrystalline $MgCl_2$ [43]. Following Busico *et al.* [44] " . . . half a century after the discovery and in spite of 20 years competition with metallocenes, the classical Ziegler–Natta catalysts still dominate the industrial production of isotactic polypropylene. According to recent estimates, over 98% of installed capacity is based on $Ti/MgCl_2$ system promoted by Al-trialkyls and by electron donors adsorbed on the surface." The first step of the preparation is the adsorption of $TiCl_4$ on $MgCl_2$ surface. A common assumption is that this adsorption has epitactic character and that (100) and (110) surfaces of α-$MgCl_2$ phase are involved, resulting in the structures reported in Figure 1.6. However, the exact location of the Ti sites on the $MgCl_2$ crystal faces, edges, and corners has been matter of a debate which is still alive in the literature.

The adsorbed $TiCl_4$ is then reduced to Ti[3+] by Al-trialkyl [45]. One of the external Cl atoms bonded to Ti[3+] is then exchanged with an alkyl group R leading to the formation of the real active center where olefin polymerization is occurring. The whole process is displayed in Figure 1.7.

As for the polymerization of the prochiral propene molecule, for which the stereoselectivity is the major issue [46], the presence of suitable Lewis bases is mandatory. The role of these bases has been supposed to vary from (i) a simple poisoning effect on achiral sites present on the surface to (ii) a more direct effect (participation in the formation of the stereoselective site). The second hypothesis is favored by the observation that a clear correlation exists between the interaction enthalpy of $TiCl_4$ with the Lewis base and the catalytic efficiency [47]. Other authors [48] hypothesize that the Lewis bases interact with the $MgCl_2$ surface creating a chiral pocket around the metal center as reported in Figure 1.8 for the DPB base.

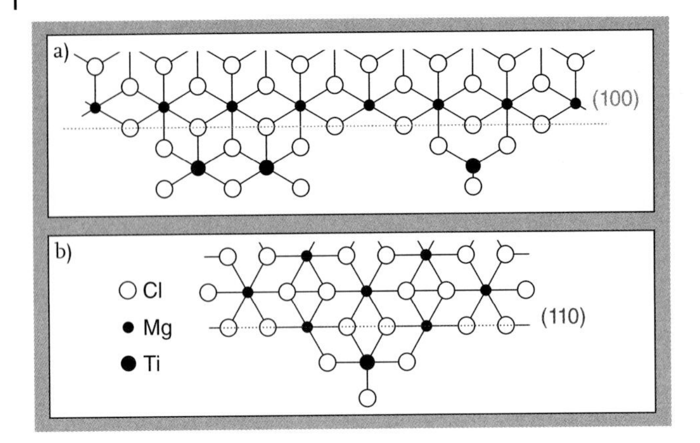

Figure 1.6 Models of $MgCl_2$ (100) and (110) surfaces and epitactic $TiCl_4$ surface adducts.

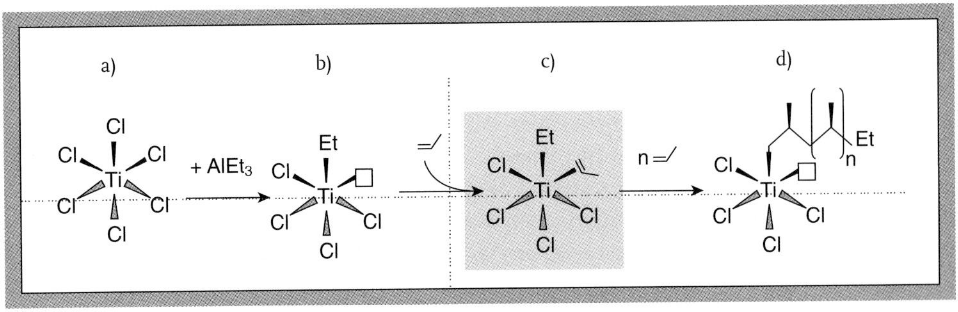

Figure 1.7 Schematic description of the formation of the really active site in case of $TiCl_4$/$MgCl_2$ Ziegler–Natta catalyst: precursor (a), coordinatively unsaturated reduced site (b), activated complex (or intermediate) with propene (c), and growing polymeric chain (d).

From the above considerations, it is clear that if on one side the single-site character of the centers is well established, on the other side the detailed structure of the active stereoselective centers is still unknown. This situation is quite common for heterogeneous catalysts, whose high efficiency and practical utility is in many cases more the result of empirical try and error approaches than of molecular engineering. It is so clear that to obtain an improved design of the catalyst, more sophisticated surface science investigations are needed. Along this line, a few contributions have appeared in the literature in the last decade [49–51], which brought new information on the structure of the sites on model Ziegler–Natta catalysts. However, the solution of the problem is still very far from completion.

1.2.2.4 The Cr/SiO_2 Phillips Catalyst for Ethylene Polymerization

The Cr/SiO_2 Phillips catalyst is by far the simplest single-site heterogeneous catalyst for ethylene polymerization synthesized so far [52, 53]. The synthesis receipt

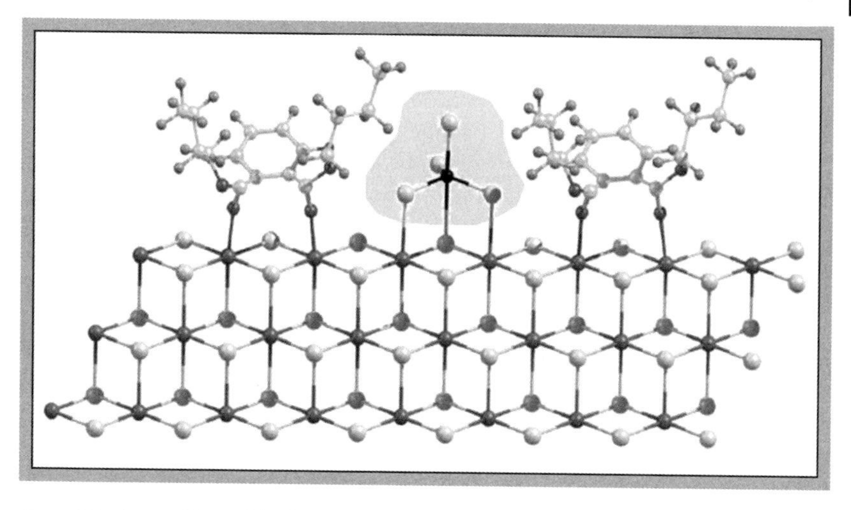

Figure 1.8 A possible model of $TiCl_4$ and DPB Lewis base co-adsorption on a $MgCl_2$ surface. Shaded area evidences the chiral pocket created by the Lewis base around $TiCl_4$.

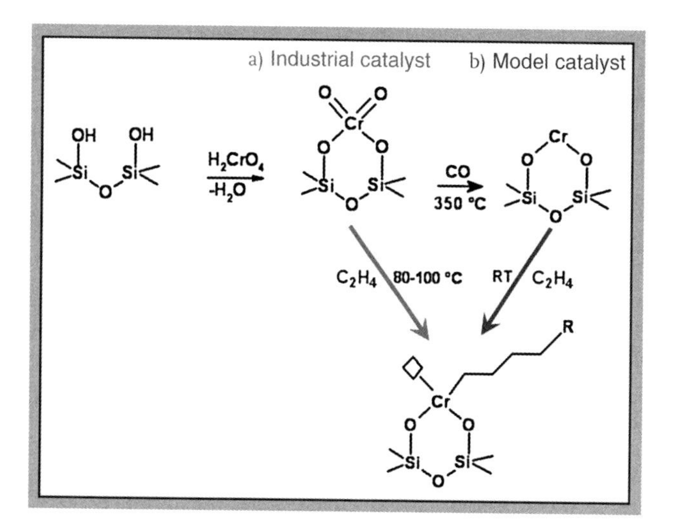

Figure 1.9 Schematic representation of the preparation and activation of Cr/SiO_2 Phillips catalyst for ethylene polymerization following the industrial (a) or model (b) route.

is very simple, as represented in Figure 1.9a. The surface hydroxyl groups of a silica particle react initially with a molecule of chromic acids forming a surface chromate. This step, which is conducted at high temperature under oxidizing conditions, does not only lead to chromate formation but is also accompanied by extensive surface dehydroxylation of the silica surface. The activation procedure can be performed in two different ways (routes a and b). Industrially (route (a))

[52], the surface chromates are treated with ethylene at about 80–100 °C; after an induction period, reduced Cr^{II} sites are formed, which are able to directly polymerize ethylene with a high efficiency. Following route (b) [53], the surface chromates are first reduced by CO (with concomitant formation of CO_2, i.e., released without interfere with the system). The so formed Cr^{II} sites are able to polymerize ethylene already at room temperature, without any induction period. In terms of productivity, the two activation ways are equivalent.

By analogy with the polymerization catalysts previously discussed, in the scheme displayed in Figure 1.9 it is arbitrarily assumed that the catalytic center contains a coordination vacancy (where the olefin molecule can insert) and a linear living hydrocarbon chain. The catalyst efficiency is comparable to that of metallocenes and Ziegler–Natta analogues. There is, however, an important difference: the catalyst cannot be used for stereospecific polymerizations. On the basis of the results illustrated for the previous catalysts, this fact can be easily understood when the structural simplicity of the ligand sphere is considered! This does not mean that the Phillips catalyst does not show selectivity properties. In fact, the molecular weight of the polymeric chain can be modulated by changing the hydroxyl groups population present on the silica surface (i.e., its surface strain), as explained by Groppo *et al.* [53, 54]. In the context of this short review, it is sufficient to recall that due to the amorphous character of the silica support, the coordination sphere of the anchored chromium ions is not completely known since, beside the two SiO^- groups linking Cr^{II} to the framework, a variable number of weaker ligands (oxygens of adjacent siloxane bridges) are present. For this reason, in this case it is better to speak of groups of single sites [55].

The Phillips catalyst is important not only because it is actually widely employed, but also because it is the only olefin polymerization catalyst where the formation of the active center does not involve the use of MAO or Al-alkyl activators. The sole reactant (ethylene) is doing the job. The questions concerning the structure of the active center and the polymerization mechanism have been the subject of many investigations [53], also with surface science methods [56–60]. The conclusion is that the initiation mechanism operative on this system, is different from that operative on homogeneous metallocenes and heterogeneous Ziegler–Natta catalysts and involves cyclic intermediates [55], as schematically shown in Figure 1.10. On this point, we shall return in the following, when *in situ* techniques under operando conditions will be illustrated. For the time being, we recall that the cyclic mechanism has been proposed also for Cr-based homogeneous dimerization and trimerization catalysts [61, 62].

1.2.3
Multiple-Site Heterogeneous Center

A multiple-site catalyst is a catalyst where the bonds breaking and forming occurs via intervention of a multitude of atoms. The hydrogenation reactions catalyzed by metal particles represent the simplest example of this type of catalyst. Figure 1.11a dis-

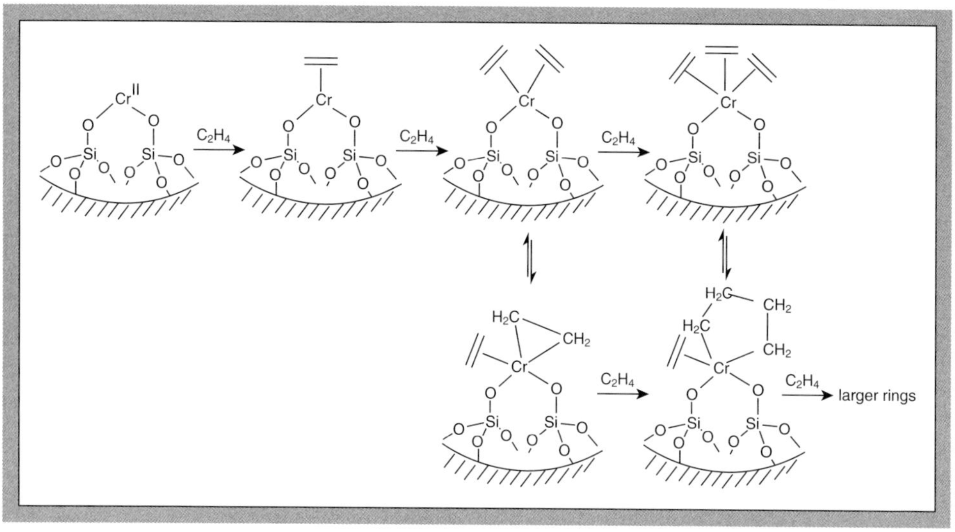

Figure 1.10 Schematic representation of the metallacycle mechanism hypothesized to explain the initiation mechanism for ethylene polimerization reaction over Cr/SiO_2 catalyst.

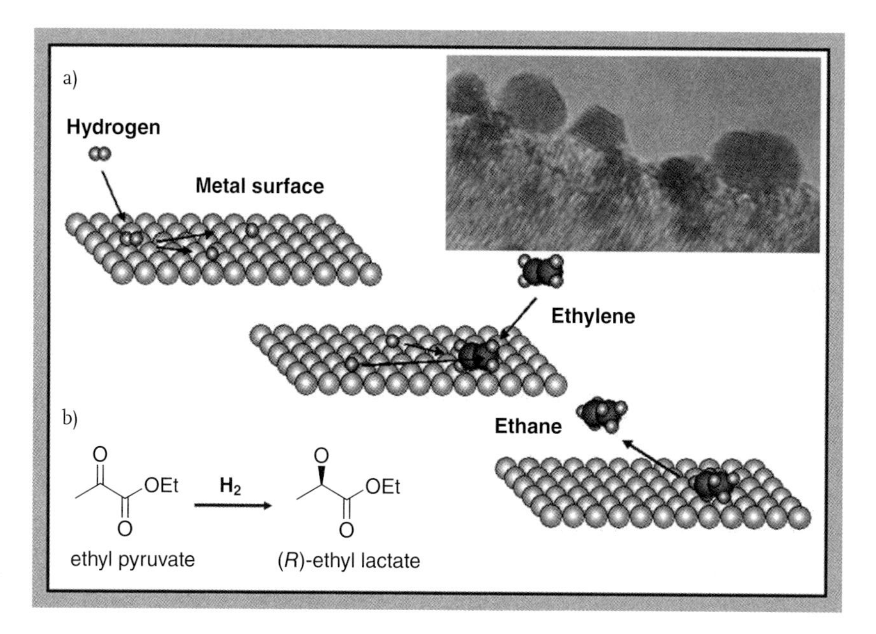

Figure 1.11 (a) Schematic representation of the ethylene hydrogenation mechanism over a metal surface. The inset shows a TEM image of supported metal nanoparticles. (b) Hydrogenation of ethyl pyruvate to ethyl lactate.

plays the hydrogenation of ethylene on a metal surface. The relevant point to be stressed here is the fact that hydrogen dissociation occurs on the surface atoms and that, due to the small surface energy barrier, the hydrogen atoms can easily migrate on the surface. This means that surface atoms not immediately adjacent to the site where ethylene is adsorbed can contribute to the reaction. In other words, a multitude of surface atoms participate to the reaction and the reaction rate is consequently very high.

Not only olefins but also other molecules containing double bonds are hydrogenated on metal surfaces such as Pd, Pt, Ru, Ir, etc. For instance, ketones can be reduced to alcohols. This is the case of ethyl pyruvate, which can be reduced to lactate in the presence of Pt particles (Figure 1.11b). The product is, however, fully racemic: this means that the metal surface is not showing enantioselectivity properties [63]. We shall see in the following that appropriate surface modification (surface engineering) can induce a satisfactory enatioselectivity, which is fully comparable with that shown by homogeneous hydrogenation catalysts.

Metal surfaces (in particular Ru) can also act as efficient catalysts for benzene hydrogenation with the formation of cycloexane (reaction (1.3) shown below) [64]. On bare metals, the formation of cyclohexene (reaction (1.4) shown below), an important feedstock for producing nylon and fine chemicals, is substantially negligible, although some information on the presence of some particle shape effect on the selectivity has been reported [65]. The absence of selectivity is mainly due to the following facts: (i) it is thermodynamically difficult to obtain cyclohexene with high selectivity since the standard free energy for cyclohexene formation by benzene hydrogenation is $-5.5\,\text{kcal mol}^{-1}$, while that for cycloexane formation is $-23.4\,\text{kcal mol}^{-1}$; (ii) the multiple-site character of the metal surface favors the multiple attack of the aromatic ring.

$$C_6H_6 + H_2 \rightarrow C_6H_{12} \quad (1.3)$$

$$C_6H_6 + H_2 \rightarrow C_6H_{10} \quad (1.4)$$

In the following, different strategies followed in literature to make the metal catalysts chemoselective for cyclohexene formation will be illustrated.

1.2.3.1 Surface Engineering and Selectivity in Heterogeneous Enantioselective Hydrogenation Catalysts

As discussed in the previous section, bare metal surfaces do not show appreciable enantioselectivity in hydrogenation reactions. In view of the increasing applicative interest of the chiral catalysis field, various strategies have been pursued to design solid enantioselective catalysts able to compete with the more established homogeneous counterparts.

Among the various proposed strategies, the most important and appealing for its synthetic potential is the strategy based on the *surface modification with chiral modifiers*. Important contributions on the subject have been published recently [66, 67]. The concept of chiral modification has been applied mainly to Pt group metals, and the used modifiers are the naturally occurring cinchona alkaloids and tartaric acid. The best example of this behavior is the hydrogenation of α-ketoesters

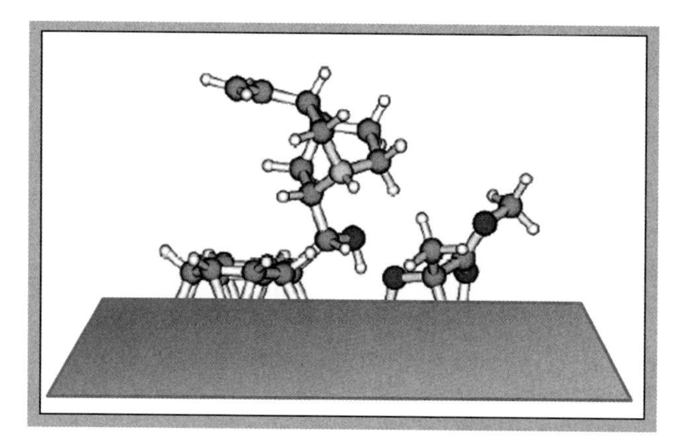

Figure 1.12 Schematic representation of a cinchonidine and methyl pyruvate adsorbed on a Pd metal particle, as modifiers to induce selectivity.

by platinum catalysts, where promotion with small amounts of cinchona alkaloids can lead to enantioselectivities exceeding 95%. The most popular mechanism by which enantioselectivity is attained is based on the idea that the cinchona modifier forms a weak hydrogen-bonded complex with the reactant and that places the reactant within its chiral pocket, forcing the carbonyl group to adopt a specific orientation. Hydrogenation of the carbonyl moiety can then lead to the preferential formation of one of the two possible enantiomers of the alcohol [68]. A representation of such interaction concerning adsorbed cinchonidine and methyl pyruvate is shown in Figure 1.12 [66, 69]. Other explanations on the effect of modifiers have been advanced. Among all, the one based on the *template model chirality* concept must be cited. Following this idea, arrays of modifiers are formed on the surface, which are responsible for the surface differentiation. The two explanations are not mutually exclusive. For the time being, it is important to stress that surface engineering at molecular level is the key point to obtain selectivity properties. Unfortunately, the level of definition of the real structure of the catalytic sites present in such heterogeneous systems is far from being completely understood and requires further efforts in surface characterization.

Another area where surface engineering is relevant and can be used to illustrate the empirical use of modifiers to increase the catalyst selectivity is the *selective hydrogenation of benzene to cyclohexene*, which is a reaction of considerable industrial utility. This reaction cannot be performed on the classical hydrogenation catalysts. Indeed it is an ascertained fact that in order to improve the selectivity of cyclohexene, the addition of surface modifiers is indispensable. These modifiers can be organic [70], inorganic [71, 72], or both [70]. So far, Ru-based catalysts modified with inorganic compounds like $ZnSO_4$ and $CdSO_4$ (or both) are the most efficient. About the function of the adsorbed modifiers in terms of local structure generated upon adsorption on the metal surface, no undisputed information is present in the literature. Even the valence state of the metal ions on the surface of Ru is debated [70, 73–75]. It is important to remark that if Zn and Cd are present as zero-valent elements, the resulting situation should be similar to that encountered with alloys, whose catalytic activity and selectivity are known to be different with respect to that of the bare metals [68, 76]. In conclusion, the case of benzene hydrogenation to cyclohexene well illustrates a situation frequently encountered for heterogeneous catalysts that are often characterized by an undefined knowledge of the structure of the active sites, which are generated on the surface upon promotion with modifiers more as the result of empirical efforts than of molecular design.

1.2.3.2 Is It Heterogeneous or Homogeneous? The Interplay between Homogeneous and Heterogeneous Catalysts

The determination of the heterogeneous or homogeneous character of the active species in catalysts containing Pd, Ru, Rh, Ir complexes (and others), involved in an enormous number of reactions (like selective arene hydrogenations [77, 78], Heck- and Suzuky-type reactions [79–82]) is not so straightforward. Phan *et al.* have discussed this problem for homogeneous Pd-based catalysts [83]. As pointed out by Weddle *et al.* [78] and Finke *et al.* [84], the problem of the real nature of many homogeneous catalysts is still an open question. It is a matter of fact that in many reactions catalyzed by monometallic complexes, the presence of an induction period suggests that the employed complex is simply a precursor species and that the real catalyst is formed successively, for instance via the intervention of hydrogen (in hydrogenation reactions). For Pd-based catalysts for Heck–Suzuki reactions, some authors think that the monometallic catalysts are in equilibrium with palladium particles in stabilized colloids [81, 82], which behave as Pd reservoirs, as represented in Figure 1.13 (derived from Ref. [81]). A similar scheme has been advanced also by Reetz [79] for Heck reactions in ionic liquids.

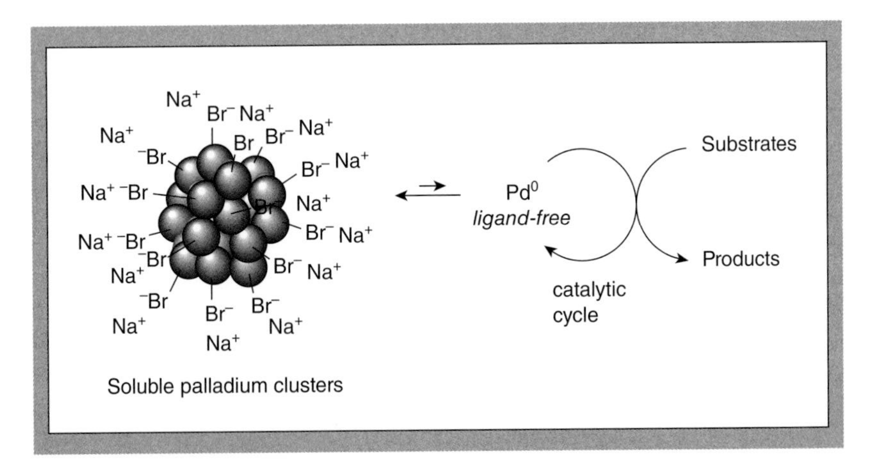

Figure 1.13 Palladium nanoclusters behaving as storage of Pd[0] in C–C bond forming reactions.

Following this hypothesis, monometallic species and clusters coexist but the active species are still the mononuclear homogeneous-type ones. On the contrary, following Reetz *et al.* [79], the catalytic events occur at defect sites (corners, steps, and kinks) of the particles. For this reason, these catalysts do not belong to the traditional area of homogeneous catalysis. The Reetz hypothesis can also be explained by hypothesizing that the Pd atoms are not free in the solution as reported in the previous scheme, but are adsorbed on the Pd surface. It is a matter of fact that an adsorbed atom, or group of atoms, on a flat surface simulates a defect site!

The discussion on this difficult point is still alive, and more sophisticated techniques and investigations are needed. A particularly important review article where the structure of hydrogenation catalysts containing not only noble metals but also Ni, Co, and Fe is discussed has been recently published [84]. As for the scope of this review, it is sufficient to underline that also for these apparently simple homogeneous catalysts, the working center appears only after an induction period (i.e., after a chemical transformation of a precursor species) and its structure looks like a very complex machine having nanotechnological character.

1.3 The Characterization Methods in Heterogeneous Catalysis (Including Operando Methods)

In the previous sections, we have seen that for many homogeneous and heterogeneous systems, the real catalyst is obtained by a well-known precursor after further reaction with promoters and/or after an induction period. The structure of the real catalyst is then more the result of reasonable hypotheses concerning the nature of the reactions occurring during the promotion/induction phase, than of direct determination by means of structural methods. In other much more complex and undefined cases, particularly involving heterogeneous systems, the real catalyst is the result of complex pretreatments and empirical surface engineering involving the use of cocatalysts and promoters. From these considerations, it is evident that the application of accurate characterization methods able to elucidate the structure of active centers and the catalytic mechanism is mandatory. However, the problem is made complex by the fact that in many cases the active centers represent a small fraction of the species present in solution (homogeneous catalysts) or on the surface (heterogeneous catalysts).

Many physical methods are used to characterize the catalytic systems. Among them, vibrational (IR and Raman), UV-Vis NIR, XANES and EXAFS, magnetic resonance (ESR and NMR), and electron (EELS, XPS, etc.) spectroscopies must be cited. The choice of the physical method which can be used to investigate the structure of the species present in a specific system depends upon the type of catalyst and reaction under investigation. The only rule which can be advanced having general validity is the following: *the characterization method must be able to distinguish between spectator and active species*. About the mechanism of the reactions occurring at the catalytic centers, the choice of suitable spectroscopic methods under operando conditions sensitive enough to give information on the structure of elusive reaction intermediates is also mandatory. These concepts will be illustrated by three examples taken from our experience on heterogeneous catalysts.

1.3.1
TS-1-(H_2O, H_2O_2) Interaction: *in situ* XANES Experiments

As illustrated in Section 1.2.2.1, TS-1 is definitely a well-defined single-site catalyst, which operates in aqueous solution. For this reason, some characterization techniques (like FTIR) cannot be used to investigate the chemical events occurring at the catalytic site by interaction with substrates. Useful spectroscopic techniques are UV-Vis, XANES, EXAFS, and Raman spectroscopies, which are not impeded (such as FTIR) by the presence of water in the channels. In the following, only the results obtained with XANES will be briefly illustrated. As for the complementary results obtained with the other techniques, the reader is referred to the literature [14–20].

Figure 1.14 illustrates the effect on the XANES spectrum of two relevant substrates present in the TS-1 channels during the oxidation reactions: H_2O and H_2O_2. It can be noticed that in the presence of water the intense and narrow XANES peak at 4971 eV (characteristic of Ti[4+] in tetrahedral coordination) is eroded. This is due to expansion of the Ti[4+] coordination sphere from tetrahedral to octahe-

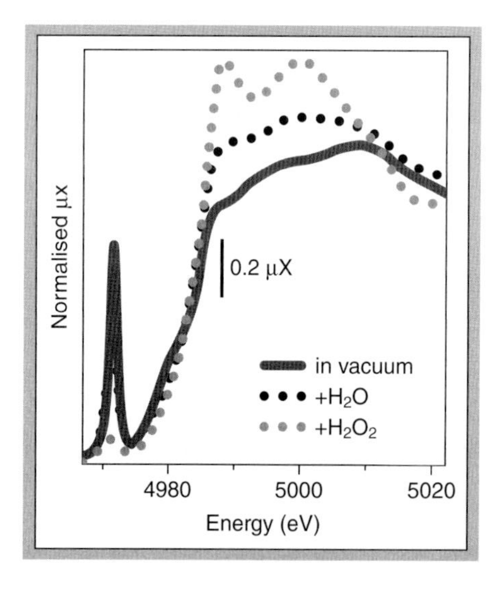

Figure 1.14 XANES spectra of TS-1 in vacuum (full gray line), immediately after contact with H_2O_2/H_2O solution (dotted gray line), and after subsequent H_2O dosage (dotted black curve).

dral, which is associated with hydrolysis of a ≡Si–O–Ti≡ bridge and coordination of H_2O to the Ti[4+] center. The process is reversible, that is, upon removal of water the original signal is restored. In the presence of H_2O_2 the erosion of the pre-edge peak is practically complete because of the formation of the final yellow colored peroxo structure reported in Figure 1.4c. Because of the presence of the solvent, in this complex the coordination sphere is complete. As the H_2O ligand is reversibly bound to the Ti[4+] center, it can be easily substituted by suitable substrates like olefins, which are then oxidized by H_2O_2. Similar conclusions have been reached by analysis of the results obtained with the other spectroscopic methods cited above. The results illustrated here can be classified as *in situ*, and well describe the chemical events occurring at the Ti[4+] center.

1.3.2
H-ZSM5-Propene Interaction: An Example of Operando Experiment by Fast Scanning FTIR Spectroscopy

As briefly described previously in Section 1.2.2.2, the interaction of olefins with the Brønsted sites present in H-ZSM5 channels leads to the formation of protonated cationic-like species. The formation of these species should occur via a short-lived hydrogen-bonded intermediate. When the interaction of H-ZSM5 with propene molecule is studied by conventional FTIR, only the spectra of the protonated product can be observed. A plausible explanation is that the reaction is so fast that the observation of the hydrogen-bonded precursor is not possible with a conventional FTIR technique. For this reason, we have performed FTIR experiment under fast scanning conditions [85]. The whole sequence of spectra reported in Figure 1.15 is lasting for 90 s. They clearly show that (i) upon interaction with propene, the Brønsted sites are immediately consumed (negative absorption band at $3610\,cm^{-1}$) with the formation of a broad positive absorption band at $3070\,cm^{-1}$ due to the hydrogen-bonded species; (ii) the IR signature of the hydrogen-bonded propene is also clearly observable at $1620\,cm^{-1}$ [ν(C=C)]; (iii) the IR absorption bands of the hydrogen-bonded precursors quickly decline and disappear completely after about 2 s; and (iv) the IR signature of the protonated species simultaneously appear. When a similar experiment is conducted in the presence of ethylene, the evolution of the hydrogen-bonded precursor is slow, thus allowing its observation with conventional IR spectroscopy [85].

Another intriguing example is represented by the propene olygomerization inside the H-MORD channels. In this case, the reaction is so fast at room temperature that even a fast scanning experiment is not sufficient to clearly observe the hydrogen-bonded precursor, which have been detected not only acting on the scanning time conditions, but also on the temperature (lower than ambient temperature) [23].

In conclusion, the results illustrated above show that spectroscopic measurements performed under real operando conditions can be of extreme utility for the elucidation of the reaction mechanisms occurring at the active centers. This utility reaches a maximum for well-defined single-site catalysts.

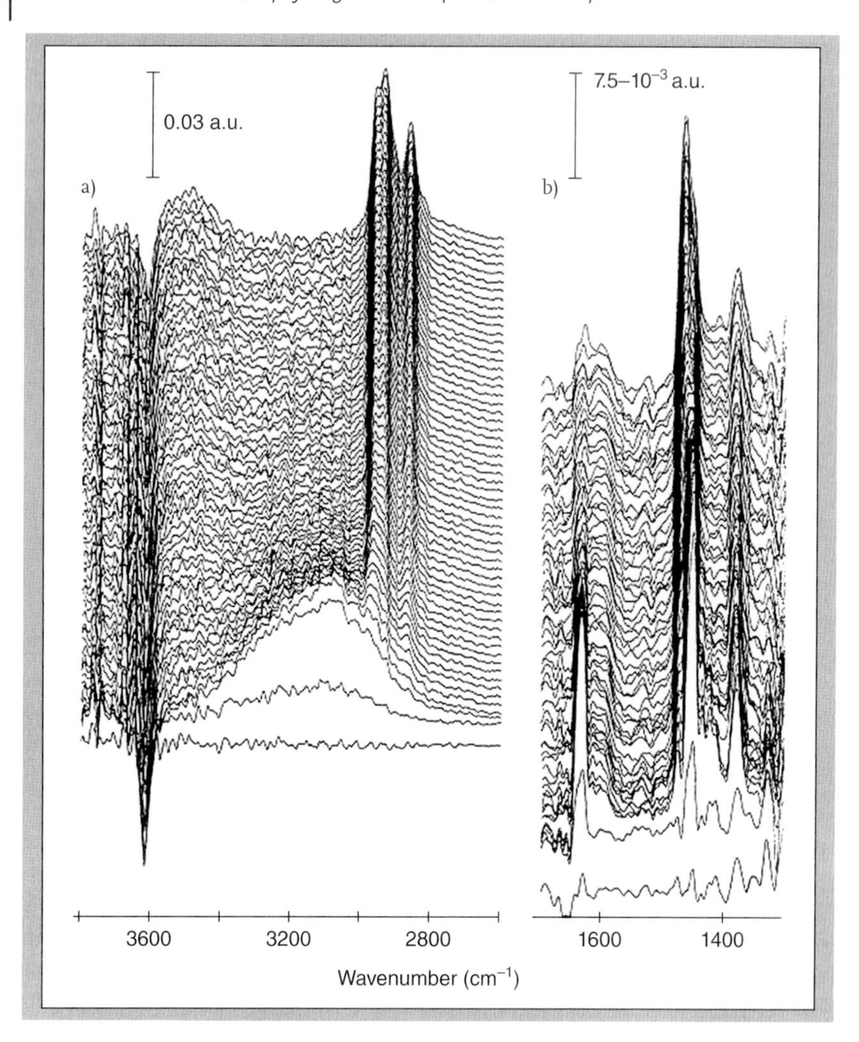

Figure 1.15 FTIR spectra, recorded in fast acquisition conditions, showing the initial stages of propene oligomerization on H-ZSM-5. The whole sequence of spectra is lasting 90 s. Part (a) reports the $\nu(CH_3)$ and $\nu(CH_2)$ region, while part (b) reports $\nu(C{=}C)$ and $\delta(CH_3)$ and $\delta(CH_3)$ region.

1.3.3
Phillips Catalyst: The Search of Precursors and Intermediates in Ethylene Polymerization Reaction by Temperature- and Time-Dependent FTIR Experiments

The major problem concerning the Phillips catalyst for ethylene polymerization (see Section 1.2.2.4) is represented by its peculiar and mysterious initiation mechanism. It is a matter of fact that the Cossee mechanism usually accepted for metallocenes and Ziegler–Natta catalysts cannot be extended to the Phillips one due to

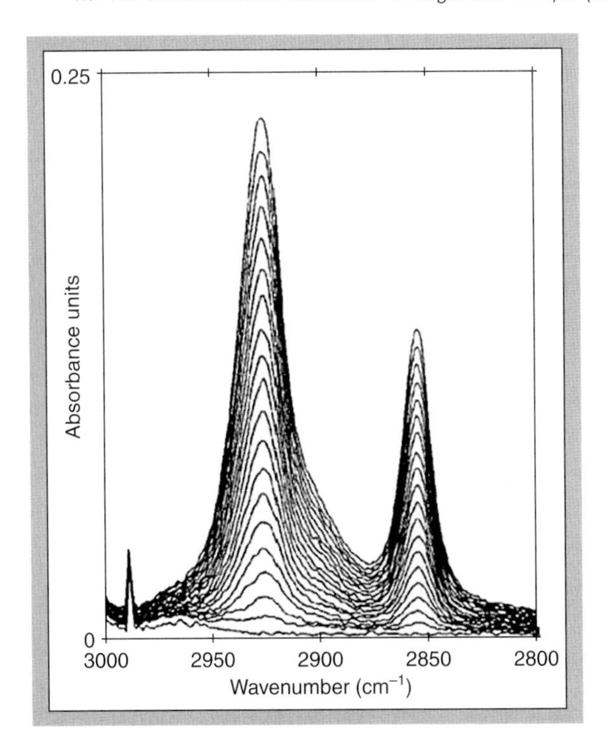

Figure 1.16 Fast scanning FTIR spectra, in the ν(CH_2) region, collected during ethylene polymerization over Cr/SiO_2 Phillips catalyst. The entire sequence of spectra is collected in only 15 s.

the absence of cocatalysts like MAO or Al trialkyls, which introduce an alkyl group into the coordination sphere of the metal center (see Figure 1.7). Previous attempts to observe the vibrational modes of terminal groups belonging to the growing polymeric chains by fast scanning IR spectroscopy failed. A typical result is illustrated in Figure 1.16, reporting a sequence of FTIR spectra of the growing polymeric chains in the 0–15 s interval. The spectra show only the IR absorption bands due to ν(CH_2) groups, without clear sign of methyl or other terminal groups. This result can have multiple explanations. Among all, two possible justifications can be advanced: (i) the reaction is so fast that the polymeric chains are always so long that the terminal groups cannot be observed; (ii) there are no $-CH_3$ or $=CH_2$ terminal groups because the mechanism is not of the Cossee or allylic type.

As completely documented in Refs. [53, 55], many hypotheses have been advanced to explain the chain initiation mechanism, and an exhaustive picture of the resulting situation is depicted in Figure 1.17. Without entering into too many details, the situation can be summarized as follows. (i) Ethylene coordinates to $Cr^{(II)}$ centers with the formation of mono-, di-, and tri- adducts. These species are the real initial precursors. The concentration of these species in the reaction conditions could be small because of the subsequent propagation steps. Consequently,

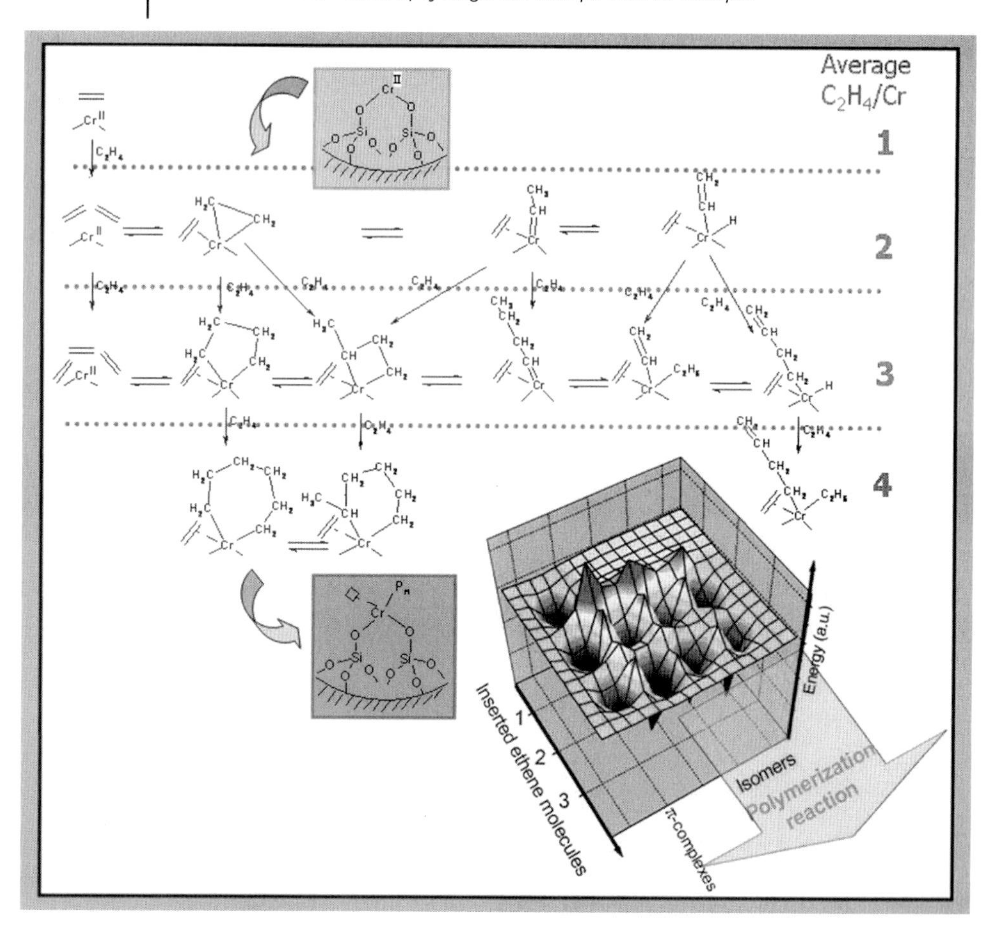

Figure 1.17 The "precursor pool" proposed in literature for the Cr(II)/SiO_2 catalyst. In the vertical direction, the evolution of the initial species upon addition of one ethylene molecule is represented. In horizontal direction, all the possible isomeric structures characterized by an average C_2H_4/Cr ratio equal to 1, 2, and 3 are reported. The inset displays a qualitative representation of the energetics of the initiation mechanisms reported in the scheme. The intermediate species are represented by potential wells of different depth, mutually separated by specific activation energy barriers of different heights. Energy values for both wells and barriers are arbitrary in depth and height, and are coded in a color scale from dark blue to red when going from negative to positive values.

they could escape detection by IR during experiments in operando conditions. (ii) The mono-ethylene adducts evolve into cyclic, ethylidene, or ethynil hydride species. These species could be in mutual equilibrium. In the presence of fast insertion and propagation reactions, their concentration under reaction conditions could also be very small and hence not detectable under the operando conditions illustrated in Figure 1.16. (iii) Ethylene insertion occurs with the formation of metallacycles, methyl-, or vinyl-terminated chains. Also these species could be in mutual equilibrium. These species could also be directly formed starting from di- adducts. About their concentration under reaction conditions, the considerations advanced above are still valid.

Propagation can then occur by insertion into the metallacycles, or via Cossee- or allylic-type mechanisms. The metallacyclic species could be in mutual equilibrium and be directly formed from the tri-coordinated precursor. If the propagation reaction is very fast, these species are the sole which can be observed under the operando conditions illustrated in Figure 1.16. From this picture, it is evident that (i) we are dealing with a complex network of potential energy barriers (see inset in Figure 1.17) and (ii) it is difficult to make any reasonable hypothesis about the reaction intermediates that are sufficiently abundant and long lived to be observable by FTIR under operando conditions.

Considering the sequence of reactions illustrated in Figure 1.17, an obvious idea to make some precursor species observable is to decrease the propagation rate, by acting on the temperature conditions. We have so designed a particular FTIR operando experiment, named "temperature resolved FTIR spectroscopy" [55]. During this experiment, the sample temperature was gradually increased from 100 to 300 K, while the equilibrium pressure of ethylene was kept constant. The results are illustrated in Figure 1.18a. To further decrease the propagation

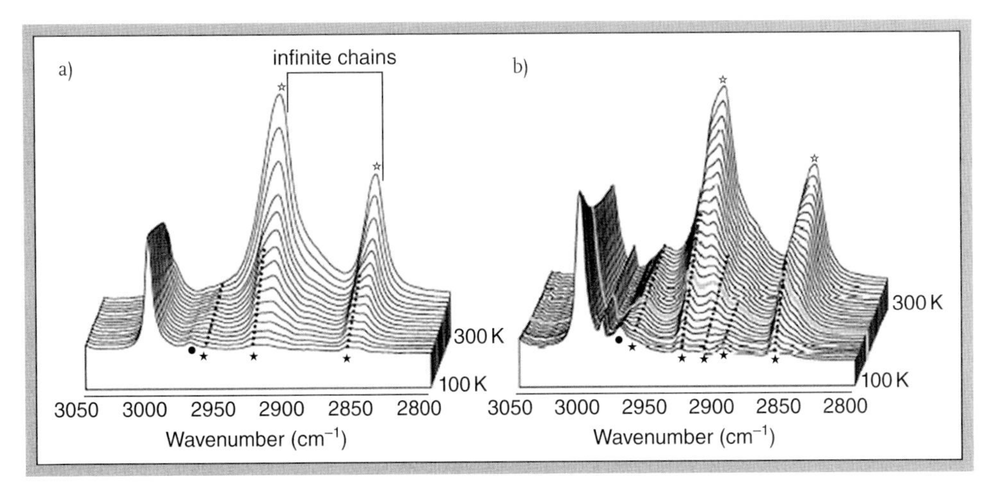

Figure 1.18 Temperature-resolved ethylene polymerization on $Cr(II)/SiO_2$, in the 100–300 K range, in absence (a) and in the presence of a CO poison (b). Only the C–H stretching region is shown. The first curve is dominated by the ethylene π-complexes at the highest ethylene pressure (peak at 3004 cm^{-1}) and by the almost total absence of polymerization products, which progressively appear in the successive spectra (empty stars). For comparison, the frequency position of infinite polymeric chains is also shown (black lines in (a)). The weak component at 2975 cm^{-1} (full circle) is due to the residual C_2H_4 molecules still in interaction with the silanol groups. The "anomalous bands" present in the first stages of the polymerization are evidenced by a full star and their evolution by dotted lines.

rate, the sequence of spectra illustrated in Figure 1.18b has been recorded in the presence of a small amount of CO acting as a reaction poison. When compared with the sequence illustrated in Figure 1.16, the new sequences illustrated in Figure 1.18a and b show new important features. In particular, (i) the Cr(II)(C_2H_4)$_{1,2}$ adducts can be clearly observed (IR absorption band around 3000 cm^{-1}) at low temperature and then gradually decline (as expected when the polymerization reaction starts). These species are real intermediates. (ii) New IR absorption bands which can be ascribed to terminal groups are now visible (full stars). As these "anomalous" bands cannot be assigned to methyl or vinyl terminal groups, they are assigned to CH_2 groups of small and strained cyclic species directly bonded to the Cr centers. From these results, the metallacycle mechanisms receive strong support.

From these results, it can be concluded that specifically designed operando methodologies are extremely precious to reveal the finest details about reaction intermediates and to make a choice between different reaction mechanisms. The more general conclusion deriving from the results illustrated in the previous paragraphs can be condensed as follows: (i) the physical methodology used to explore a given catalyst must be properly chosen on the basis of the nature of the catalyst to be investigated; (ii) great attention must be paid to use specifically designed operando conditions able to maximize the concentration of reaction intermediates.

1.4 Conclusions

Although the examples illustrated in the previous paragraphs have been arbitrarily selected among the numerous cases forming the homogeneous and heterogeneous catalysis realm, in our opinion they are sufficiently differentiated to be considered as representative of the structural complexity at the molecular level in the working catalytic centers. From this short review, the following conclusions can be derived:

(i) The real active center is very often the result of pretreatment and activation reactions performed on precursor structures. This explains the presence of induction periods in both homogeneous and heterogeneous reactions.

(ii) In order to build up a catalytic site highly selective for a specific reaction, the structure must be designed and engineered at a molecular level. The engineering procedure can be the result of (i) a rational approach based on the detailed knowledge of the catalyst precursor and of the subsequent reactions leading to effective formation of the active center and (ii) an empirical approach with trial-and-error character substantially based on chemical intuition. The "rational" approach is more frequent for homogeneous catalysts, while the empirical one is more frequently verified for the heterogeneous counterparts.

(iii) The role of ancillary structures surrounding the metal center is to direct the reaction along a well-defined and specific path. The more demanding in terms of selectivity is the catalyzed reaction and the more complex is the structure of the ligands sphere around the metal center. The resulting structure can be considered as a nanomachine for molecular assembling. Indeed in most cases, the ensemble constituted by the metal center, the sphere of ancillary ligands (both covalently and electrostatically bonded), and the coordinated substrates reaches the nanometer dimension.

(iv) The characterization of the structure of active centers is a difficult task and appropriate highly sensitive physical methods must be chosen. In this context, operando methods capable to distinguish between spectator species and real intermediates can be of great help.

References

1 Knowles, W.S. (2003) *Adv. Synth. Catal.*, **345**, 3.

2 Kiefer, J., Obert, K., Himmler, S., Schulz, P.S., Wasserscheid, P., and Leipertz, A. (2008) *Chemphyschem*, **9**, 2207.

3 Cui, X.H. and Burgess, K. (2005) *Chem. Rev.*, **105**, 3272.

4 Pena, D., Minnaard, A.J., Boogers, J.A.F., de Vries, A.H.M., de Vries, J.G., and Feringa, B.L. (2003) *Org. Biomol. Chem.*, **1**, 1087.

5 Macchioni, A. (2005) *Chem. Rev.*, **105**, 2039.

6 Zecchina, A., Groppo, E., and Bordiga, S. (2007) *Chem. Eur. J.*, **13**, 2440.

7 Buriak, J.M., Klein, J.C., Herrington, D.G., and Osborn, J.A. (2000) *Chem. Eur. J.*, **6**, 139.

8 Thomas, J.M. and Thomas, W.J. (1996) *Principles and Practice of Heterogeneous Catalysis*, Wiley-VCH Verlag GmbH, Weinheim.

9 Kaminsky, W. (2001) *Adv. Catal.*, **46**, 89.

10 Kaminsky, W. and Laban, A. (2001) *Appl. Catal. A*, **222**, 47.

11 Resconi, L., Cavallo, L., Fait, A., and Piemontesi, F. (2000) *Chem. Rev.*, **100**, 1253.

12 Correa, A. and Cavallo, L. (2006) *J. Am. Chem. Soc.*, **128**, 10952.

13 Busico, V., Castelli, V.V.A., Aprea, P., Cipullo, R., Segre, A., Talarico, G., and Vacatello, M. (2003) *J. Am. Chem. Soc.*, **125**, 5451.

14 Ricchiardi, G., Damin, A., Bordiga, S., Lamberti, C., Spano, G., Rivetti, F., and Zecchina, A. (2001) *J. Am. Chem. Soc.*, **123**, 11409.

15 Lamberti, C., Bordiga, S., Zecchina, A., Artioli, G., Marra, G., and Spano, G. (2001) *J. Am. Chem. Soc.*, **123**, 2204.

16 Lamberti, C., Bordiga, S., Arduino, D., Zecchina, A., Geobaldo, F., Spano, G., Genoni, F., Petrini, G., Carati, A., Villain, F., and Vlaic, G. (1998) *J. Phys. Chem. B*, **102**, 6382.

17 Bordiga, S., Coluccia, S., Lamberti, C., Marchese, L., Zecchina, A., Boscherini, F., Buffa, F., Genoni, F., Leofanti, G., Petrini, G., and Vlaic, G. (1994) *J. Phys. Chem.*, **98**, 4125.

18 Geobaldo, F., Bordiga, S., Zecchina, A., Giamello, E., Leofanti, G., and Petrini, G. (1992) *Catal. Lett.*, **16**, 109.

19 Prestipino, C., Bonino, F., Usseglio, S., Damin, A., Tasso, A., Clerici, M.G., Bordiga, S., D'Acapito, F., Zecchina, A., and Lamberti, C. (2004) *Chemphyschem*, **5**, 1799.

20 Bonino, F., Damin, A., Ricchiardi, G., Ricci, M., Spano, G., D'Aloisio, R., Zecchina, A., Lamberti, C., Prestipino, C., and Bordiga, S. (2004) *J. Phys. Chem. B*, **108**, 3573.

21 Collman, J.P., Boulatov, R., Sunderland, C.J., and Fu, L. (2004) *Chem. Rev.*, **104**, 561.
22 Merkx, M., Kopp, D.A., Sazinsky, M.H., Blazyk, J.L., Muller, J., and Lippard, S.J. (2001) *Angew. Chem. Int. Ed.*, **40**, 2782.
23 Geobaldo, F., Spoto, G., Bordiga, S., Lamberti, C., and Zecchina, A. (1997) *J. Chem. Soc. Faraday Trans.*, **93**, 1243.
24 Bordiga, S., Lamberti, C., Geobaldo, F., Zecchina, A., Palomino, G.T., and Arean, C.O. (1995) *Langmuir*, **11**, 527.
25 Zecchina, A., Bordiga, S., Spoto, G., Scarano, D., Petrini, G., Leofanti, G., Padovan, M., and Arean, C.O. (1992) *J. Chem. Soc. Faraday Trans.*, **88**, 2959.
26 Kvisle, S., Fuglerud, T., Kolboe, S., Olsbye, U., Lillerud, K.P., and Vora, B.V. (2008) *Handbook of Heterogeneous Catalysis*, vol. 6 (eds H. Ertl, H. Knözinger, F. Schüth, and J. Weitkamp), Wiley-VCH Verlag GmbH, Weinheim, Ch 13, p. 14.
27 Stocker, M. (1999) *Microporous Mesoporous Mater.*, **29**, 3.
28 Olsbye, U., Bjorgen, M., Svelle, S., Lillerud, K.P., and Kolboe, S. (2005) *Catal. Today*, **106**, 108.
29 Dahl, I.M. and Kolboe, S. (1994) *J. Catal.*, **149**, 458.
30 Lesthaeghe, D., Van Speybroeck, V., Marin, G.B., and Waroquier, M. (2006) *Angew. Chem. Int. Ed.*, **45**, 1714.
31 Song, W.G., Marcus, D.M., Fu, H., Ehresmann, J.O., and Haw, J.F. (2002) *J. Am. Chem. Soc.*, **124**, 3844.
32 Arstad, B. and Kolboe, S. (2001) *J. Am. Chem. Soc.*, **123**, 8137.
33 Song, W.G., Haw, J.F., Nicholas, J.B., and Heneghan, C.S. (2000) *J. Am. Chem. Soc.*, **122**, 10726.
34 Bjorgen, M., Bonino, F., Kolboe, S., Lillerud, K.P., Zecchina, A., and Bordiga, S. (2003) *J. Am. Chem. Soc.*, **125**, 15863.
35 Bjorgen, M., Olsbye, U., Petersen, D., and Kolboe, S. (2004) *J. Catal.*, **221**, 1.
36 Arstad, B., Nicholas, J.B., and Haw, J.F. (2004) *J. Am. Chem. Soc.*, **126**, 2991.
37 Svelle, S., Joensen, F., Nerlov, J., Olsbye, U., Lillerud, K.P., Kolboe, S., and Bjorgen, M. (2006) *J. Am. Chem. Soc.*, **128**, 14770.
38 Svelle, S., Olsbye, U., Joensen, F., and Bjorgen, M. (2007) *J. Phys. Chem. C*, **111**, 17981.
39 Svelle, S., Ronning, P.O., Olsbye, U., and Kolboe, S. (2005) *J. Catal.*, **234**, 385.
40 Bjorgen, M., Svelle, S., Joensen, F., Nerlov, J., Kolboe, S., Bonino, F., Palumbo, L., Bordiga, S., and Olsbye, U. (2007) *J. Catal.*, **249**, 195.
41 Ziegler, K., Holzkamp, E., Breil, H., and Martin, H. (1955) *Angew. Chem.*, **57**, 541.
42 Natta, G. (1956) *Angew.Chem.*, **68**, 393.
43 Soga, K. and Shiono, T. (1997) *Progr. Polym. Sci.*, **22**, 1503.
44 Busico, V., Causa, M., Cipullo, R., Credendino, R., Cutillo, F., Friederichs, N., Lamanna, R., Segre, A., and Castellit, V.V. (2008) *J. Phys. Chem. C*, **112**, 1081.
45 Busico, V., Corradini, P., De Martino, L., Proto, A., and Albizzati, E. (1986) *Makromol. Chem.*, **187**, 1115.
46 Busico, V., Cipullo, R., Monaco, G., Talarico, G., Vacatello, M., Chadwick, J.C., Segre, A.L., and Sudmeijer, O. (1999) *Macromol.*, **32**, 4173.
47 Cavallo, L., Ducere, J.M., Fedele, R., Melchior, A., Mimmi, M.C., Morini, G., Piemontesi, F., and Tolazzi, M. (2008) *J. Therm. Anal. Cal.*, **91**, 101.
48 Stukalov, D.V., Zakharov, V.A., Potapov, A.G., and Bukatov, G.D. (2009) *J. Catal.*, **266**, 39.
49 Freund, H.J., Baumer, M., Libuda, J., Risse, T., Rupprechter, G., and Shaikhutdinov, S. (2003) *J. Catal.*, **216**, 223.
50 Kim, S.H. and Somorjai, G.A. (2006) *Proc. Natl. Acad. Sci. USA*, **103**, 15289.
51 Andoni, A., Chadwick, J.C., Niemantsverdriet, J.W., and Thune, P.C. (2009) *Catal. Lett.*, **130**, 278.
52 McDaniel, M.P. (1985) *Adv. Catal.*, **33**, 47.
53 Groppo, E., Lamberti, C., Bordiga, S., Spoto, G., and Zecchina, A. (2005) *Chem. Rev.*, **105**, 115.
54 Groppo, E., Lamberti, C., Spoto, G., Bordiga, S., Magnacca, G., and Zecchina, A. (2005) *J. Catal.*, **236**, 233.
55 Groppo, E., Lamberti, C., Bordiga, S., Spoto, G., and Zecchina, A. (2006) *J. Catal.*, **240**, 172.

56 Agostini, G., Groppo, E., Bordiga, S., Zecchina, A., Prestipino, C., D'Acapito, F., van Kimmenade, E., Thune, P.C., Niemantsverdriet, J.W., and Lamberti, C. (2007) *J. Phys. Chem. C*, **111**, 16437.

57 Thune, P.C., Linke, R., van Gennip, W.J.H., de Jong, A.M., and Niemantsverdriet, J.W. (2001) *J. Phys. Chem. B*, **105**, 3073.

58 Thune, P.C., Loos, J., de Jong, A.M., Lemstra, P.J., and Niemantsverdriet, J.W. (2000) *Top. Catal.*, **13**, 67.

59 Thune, P.C., Loos, J., Lemstra, P.J., and Niemantsverdriet, J.W. (1999) *J. Catal.*, **183**, 1.

60 Thune, P.C., Loos, J., Lemstra, P.J., and Niemantsverdriet, J.W. (2000) *Abstr. Pap. Am. Chem. Soc.*, **219**, U390.

61 Ruddick, V.J., Dyer, P.W., Bell, G., Gibson, V.C., and Badyal, J.P.S. (1996) *J. Phys. Chem.*, **100**, 11062.

62 Dixon, J.T., Green, M.J., Hess, F.M., and Morgan, D.H. (2004) *J. Organom. Chem.*, **689**, 3641.

63 Heitbaum, M., Glorius, F., and Escher, I. (2006) *Angew. Chem. Int. Ed.*, **45**, 4732.

64 Hartog, F. and Zwietering, P. (1963) *J. Catal.*, **2**, 79.

65 Bratlie, K.M., Lee, H., Komvopoulos, K., Yang, P.D., and Somorjai, G.A. (2007) *Nano Lett.*, **7**, 3097.

66 Mallat, T., Orglmeister, E., and Baiker, A. (2007) *Chem. Rev.*, **107**, 4863.

67 Bonalumi, N., Vargas, A., Ferri, D., and Baiker, A. (2007) *Chem. Eur. J.*, **13**, 9236.

68 Zaera, F. (2009) *Acc. Chem. Res.*, **42**, 1152.

69 Vargas, A., Burgi, T., and Baiker, A. (2004) *J. Catal.*, **226**, 69.

70 Fan, G.Y., Li, R.X., Li, X.J., and Chen, H. (2008) *Catal. Commun.*, **9**, 1394.

71 Kluson, P. and Cerveny, L. (1995) *Appl. Catal. A*, **128**, 13.

72 Struijk, J., Moene, R., Vanderkamp, T., and Scholten, J.J.F. (1992) *Appl. Catal. A*, **89**, 77.

73 Yuan, P.Q., Wang, B.Q., Ma, Y.M., He, H.M., Cheng, Z.M., and Yuan, W.K. (2009) *J. Mol. Catal. A*, **309**, 124.

74 Yuan, P.Q., Wang, B.Q., Ma, Y.M., He, H.M., Cheng, Z.M., and Yuan, W.K. (2009) *J. Mol. Catal. A*, **301**, 140.

75 Liu, J.-L., Zhu, Y., Liu, J., Pei, Y., Li, Z.H., Li, H., Li, H.-X., Qiao, M.-H., and Fan, K.-N. (2009) *J. Catal.*, **268**, 100.

76 Serna, P., Concepcion, P., and Corma, A. (2009) *J. Catal.*, **265**, 19.

77 Lin, Y. and Finke, R.G. (1994) *Inorg. Chem.*, **33**, 4891.

78 Weddle, K.S., Aiken, J.D., and Finke, R.G. (1998) *J. Am. Chem. Soc.*, **120**, 5653.

79 Reetz, M.T. and Westermann, E. (2000) *Angew. Chem. Int. Ed.*, **39**, 165.

80 Na, Y., Park, S., Han, S.B., Han, H., Ko, S., and Chang, S. (2004) *J. Am. Chem. Soc.*, **126**, 250.

81 Alimardanov, A., de Vondervoort, L.S.V., de Vries, A.H.M., and de Vries, J.G. (2004) *Adv. Synth. Catal.*, **346**, 1812.

82 Cassol, C.C., Umpierre, A.P., Machado, G., Wolke, S.I., and Dupont, J. (2005) *J. Am. Chem. Soc.*, **127**, 3298.

83 Phan, N.T.S., Van Der Sluys, M., and Jones, C.W. (2006) *Adv. Synth. Catal.*, **348**, 609.

84 Alley, W.M., Hamdemir, I.K., Johnson, K.A., and Finke, R.R.G. (2010) *J. Mol. Catal. A*, **315**, 1.

85 Spoto, G., Bordiga, S., Ricchiardi, G., Scarano, D., Zecchina, A., and Borello, E. (1994) *J. Chem. Soc. Faraday Trans.*, **90**, 2827.

2
Supported Nanoparticles and Selective Catalysis: A Surface Science Approach

Wangqing Zhang, Da Wang, and Rui Yan

2.1
General Introduction

Metal particles show unique physical and chemical properties as their dimensions are reduced to nanoscale. Metal nanoparticles, especially noble metal nanoparticles, have a characteristic high surface-to-volume ratio, and consequently large fraction of surface atoms that are exposed to reactant molecules, which makes them to be a promising catalyst in chemical synthesis [1–5]. However, the direct application of the metal nanoparticles in catalysis is often difficult due to their ultra-small size and, therefore, high tendency toward agglomeration because of van der Waals forces. Consequently, the metal nanoparticles are deposited or *in situ* synthesized on a suitable support such as polymer [6, 7], metal oxide [8, 9], carbon materials [10], and mesoporous silica [11, 12]. The catalytic performance such as activity and selectivity of the supported metal nanoparticles is dependent on the size, morphology, and surface of the metal nanoparticles. For example, 80% of the atoms are exposed to the interface when the size of the metal nanoparticles is 1.5 nm. However, the value sharply decreases when the size of the metal nanoparticles increases above 10 nm. It is proposed that tuning the activity and selectivity is feasible by changing the size of metal nanoparticles because the properties relevant for catalytic behavior, like the electronic spectrum, the symmetry, or the spin multiplicities, can thus be tuned for obtaining the desired activation of the reactant molecules relevant in a catalytic process [13]. Besides, it is speculated that the support material plays a great role in the catalytic performance of the supported metal nanoparticles. Firstly, it prevents agglomeration of the metal nanoparticles. Secondly, the support may also exert important influence on the size- and shape-controlled synthesis of the metal nanoparticles. Finally, the nature of the support itself, such as the size, the morphology, the chemical structure of the support, and the interaction between the support and the metal nanoparticles, may also affect the catalysis of metal nanoparticles.

In this chapter, the size- and shape-controlled synthesis of supported metal nanoparticles is reviewed, and the survey of the size and surface structure effect on the selective catalysis of supported metal nanoparticles is briefly introduced.

Selective Nanocatalysts and Nanoscience, First Edition. Edited by Adriano Zecchina, Silvia Bordiga, Elena Groppo.

2.2
Synthesis of Supported Metal Nanoparticles: Size and Shape Control

2.2.1
General Synthesis of Metal Nanoparticles

Various methods to synthesize metal nanoparticles are reported in literatures [14, 15]. Generally, these methods can be classified into gas-phase and liquid-phase based methods. In the gas-phase method, bulk material is evaporated to obtain a supersaturated gas phase, which then produces nuclei and further grows to become metal nanoparticles. In the liquid-phase method, also known as the wet method, precursors react to form a supersaturated solution of zero-valent metal atoms, which nucleate and further grow to metal nanoparticles.

Compared with the gas-phase method, the wet synthesis method is much more attractive at least for two reasons: (i) they are more energy efficient and (ii) they can be used to produce metal nanoparticles using the standard apparatus available in a laboratory. So far, two strategies have been followed for the wet synthesis. In the first strategy of micelle template synthesis, metal nanoparticles are produced in the surfactant micelles. The size of metal nanoparticles is controlled by the rates of nucleation and growth, and stabilization is provided by the micelles. Nucleation is correlated to the size of micelles, and growth is determined by the rate of material exchange between micelles. Both these features can be altered through chemical nature and concentration of the surfactant. In the second strategy, precipitation is carried out in the presence of stabilizers that adsorb onto the surface of metal nanoparticles and prevent coagulation of the metal nanoparticles. Most of these methods were first demonstrated for their ability to the synthesis of Au nanoparticles and then extended to the synthesis of other types of nanoparticles including semiconductors such as CdS nanoparticles. Three widely used precipitation-based techniques for the synthesis of Au nanoparticles are listed as follows: (i) citrate method of Turkevich *et al.* [16–18], (ii) citrate-tannic acid method of Muhlpfordt [19], and (iii) Brust–Schiffrin method of Brust *et al.* [20].

In the simplest Turkevich method for the preparation of Au nanoparticles, tetrachloroauric acid is reduced with trisodium citrate. Citrate acts as both the reducing agent and the stabilizer. It is expected that the relative amounts of the gold precursor to the reducing agent, the concentrations of the precursor, the addition rate of the precursor, and the state of mixing in the reactor can influence the particle size. Generally, nanoparticles form by first nucleation and further growth as shown in Figure 2.1 [21]. However, a quantitative model to predict the size and size distribution of the synthesized metal nanoparticles is not available.

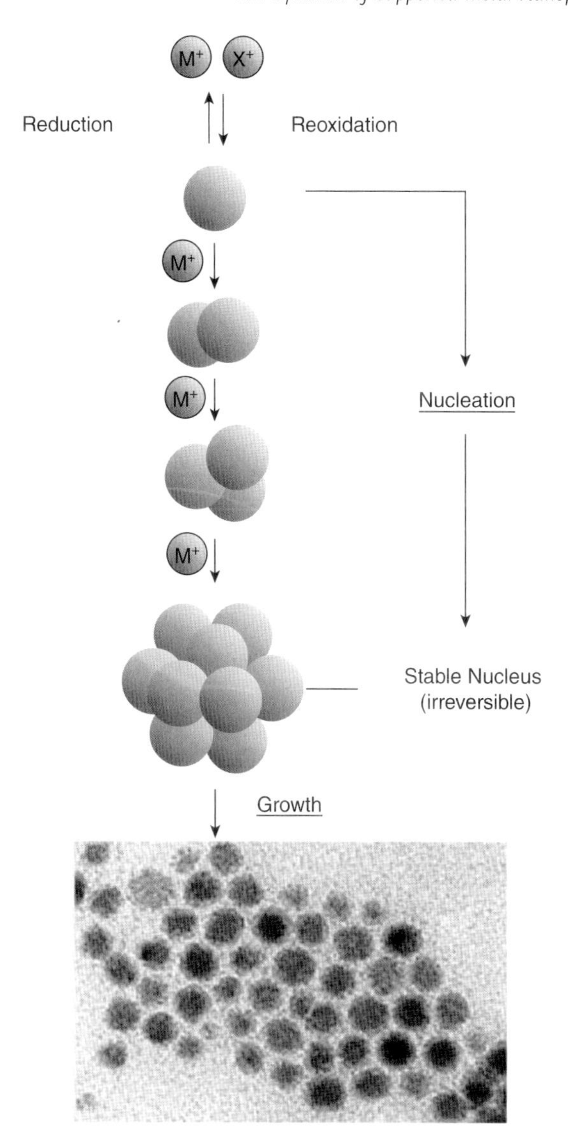

Figure 2.1 Wet chemical synthesis of metal nanoparticles. Adopted from Ref. [21].

2.2.2 Synthetic Methodologies for Supported Metal Nanoparticles

2.2.2.1 Metal Nanoparticles Stabilized by Polymeric Materials

2.2.2.1.1 Metal Nanoparticles Stabilized by Soluble Polymers

Of all the soluble polymers, poly(*N*-vinyl-2-pyrrolidone) (PVP), a water-soluble amine-based polymer, may be the most popular one to stabilize metal nanoparticles.

The PVP macromolecules are thought to be adsorbed on the surface of the metal nanoparticles, and it is deemed that this polymeric stabilizer establishes many weak bonds with the surface of the nanoparticles rather than forming less strong bonds at the specific sites [22, 23]. In the presence of PVP, various metal nanoparticles such as Au [24], Ag [25], Pd [26–28], Pt [28, 29], and Rh [30], and bimetallic Pt/Au nanoparticles [31], and Ag nanowires [32] can be produced by reducing the corresponding precursors of metal salts with alcohol and polyol having an α-hydrogen.

Besides being a stabilizer, PVP is also suggested to be a reducing agent for the preparation of some Au and Ag hydrosols [33–36]. It is proposed that the formation of metal nanoparticles without addition of reducing agent is ascribed to the macroradicals formed during the partial degradation of the polymer. However, studies by Xia *et al.* [37–39] completely contradict these reports and the authors claim that PVP plays no role as a reducing agent at all but acts simply as a protecting agent.

It should be noted that the alcohol having an α-hydrogen has double functions as both reductive agent and the solvent. The reduction of the metal salt is usually performed at reflux temperature resulting in fast metal particle formation. Now, it is generally thought that the strength of the metal–metal bond, the molar ratio of metal salt to the PVP stabilizer, the structure and quantity of the applied alcohol or polyol, the extent of conversion or the reaction time, and the applied temperature and pH are essential to the size- and shape-controlled synthesis of metal nanoparticles [29, 40–47]. Besides, surfactants may also be added to the polyol process to control the particle morphology and size [48, 49]. Furthermore, Xia *et al.* [50, 51] found that Pd or Ag nanoparticles with the uniform sizes and shapes could be obtained if oxidative etching was carried out during the polyol synthesis. For example, 8-nm colloidal octahedral Pd nanoparticles were generated when the Pd nanoparticles produced by reduction with polyol were simultaneously exposed to Cl^-/O_2. Oxidative etching gradually converted the nanoparticles from the presynthesized 4–8 nm twinned cubo-octahedra nanoparticles into 8-nm cubo-octahedra nanoparticles.

Besides PVP, some other polymers such as poly(acrylic acid) [52], poly(aniline-2-carboxylic acid) [53], poly(*N*-isopropylacrylamide) (PNIPAM) [54], etc., have also been used to stabilize metal nanoparticles. Usually, these polymers are coordination polymer containing a chelating group. Readers can refer to the reviews on the coordination polymer for more information [55, 56].

Highly dispersed metal nanoparticles such as Au, Ag, and Pd are also synthesized in the presence of thiol-terminated polymers such as poly(methyl methacrylate), poly(*n*-butyl acrylate), poly(ethylene glycol), and PNIPAM [57–62]. It is found that the size of synthesized polymer-grafted metal nanoparticles can be tuned by the molar ratio of the thiol-terminated polymer to the metal precursor [61]. The polymer-grafted metal nanoparticles have a well-defined core–corona structure as shown in Figure 2.2, where the polymer forms the corona and the metal nanoparticle forms the core. The interaction between the thiol group and the metal nanoparticles is much stronger than those between PVP and metal nanoparticles, and thus the metal nanoparticles grafted with thiol-terminated

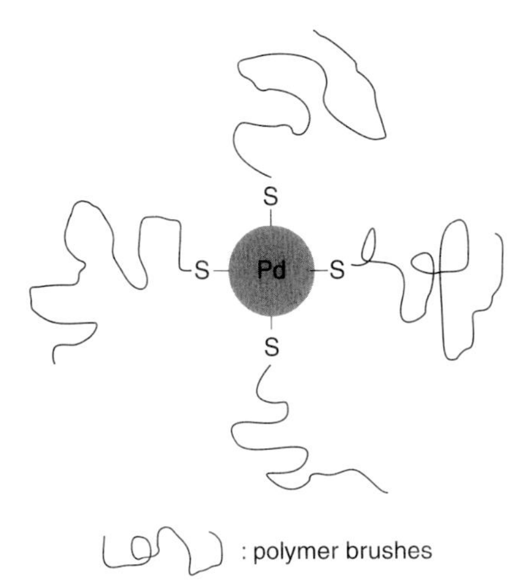

Figure 2.2 Schematic core–corona structure of the thiol-terminated polymer grafted metal nanoparticles.

polymer brushes are expected to be more stable. The polymer-grafted metal nanoparticles have an advantage to be dispersed both in aqueous phase and organic phase depending on the hydrophility or hydrophobility of the polymer brushes. For example, PNIPAM-grafted Au nanoparticles can be highly dispersed both in water and in $CHCl_3$ [61]. When two thiol-terminated polymers are used, two kinds of polymer brushes can be grafted to the synthesized metal nanoparticles [63]. These hybrid polymer-grafted metal nanoparticles have special characters such as selective solubility in solvents.

2.2.2.1.2 **Metal Nanoparticles Stabilized by Self-Assembled Block Copolymer Micelles**

In the block-selective solvent, amphiphilic block copolymers can self-assemble into core–corona micelles, wherein the soluble block forms the corona and the insoluble block forms the core [64–69]. The self-assembled micelles can be considered as nanoreactors or templates, within which nucleation and growth of metal nanoparticles upon reduction are restricted to the mesoscale level, and the size and morphology of the resultant metal nanoparticles depend on the size and morphology of the template micelles [70–73]. Generally, the synthesis of metal nanoparticles within the micellar nanoreactor involves at least two steps, that is, loading of the metal precursor and chemical reduction followed by a nucleation and growth process.

Loading of the metal precursors into the micellar core is usually achieved by simply stirring the metal precursor in the aqueous dispersion of self-assembled micelles. The metal precursor is either bound directly to the polymeric ligand in the micellar core or indirectly as the counterion [74, 75]. For example, $HAuCl_4$ protonates the poly(4-vinylpyridine) (P4VP) block in the core of the micelles self-assembled by polystyrene-*b*-poly(4-vinylpyridine) (PS-*b*-P4VP); thereby Au is loaded into the P4VP core by binding $AuCl_4^-$ as the counterion. For $PdCl_2$ or $Pd(AcO)_2$, Pd can be easily loaded into the core of PS-*b*-P4VP micelles due to the coordination between the core-forming segment of P4VP with Pd [70, 74].

Reduction of metal precursors loaded into the micellar core leads to the formation of primary metal atoms. These primary metal atoms subsequently aggregate to form larger nanoparticles by nucleation and growth processes. The morphology and size of the resultant metal nanoparticles depend on the size and morphology of the template micelles and the amount of the metal precursors loaded within the micellar core. It is deemed that the number of metal nanoparticles formed within one microcompartment depends on the degree of supersaturation c/c_0 with the primary particles, the interfacial tension γ of the particle/polymer interface, and the diffusivity D of the particles [75]. That is, a small γ and large c/c_0 will lead to the formation of a large number of small nanoparticles within a microdomain of the micellar core. The degree of supersaturation c/c_0 depends on the rate of the chemical reduction. Fast chemical reduction leads to a high supersaturation c/c_0. The interfacial tension γ depends on the stabilization of the polymer/inorganic interface. By using appropriate functional blocks to stabilize the polymer/inorganic interface, the value of γ can be made small and, therefore, leads to good solubility of the primary particles.

In the organic solvent of tetrahydrofuran (THF), Antonietti *et al.* synthesized Au nanoparticles using the PS-*b*-P4VP micelles as template [74]. As shown in Figure 2.3a, the amphiphilic block copolymer of PS-*b*-P4VP self-assembled into 30-nm micelles in THF, which was the block-selective solvent for the polystyrene (PS) block. The coordination block of P4VP formed the core and the PS block formed the corona of the micelles. After loading the Au precursor by adding $HAuCl_4$ into the micellar dispersion and then stirring for about 24 h, the strong reducing reagent of $LiAlH_4$ was added. As shown in Figure 2.3b, a number of Au nanoparticles with size at about 3 nm were synthesized within one microcompartment of the micelles. These raspberry-like Au nanoparticles were generally quite stable. Precipitation, redispersion, and heating at temperature below the glass transition temperature of the block copolymer did not affect the size distribution

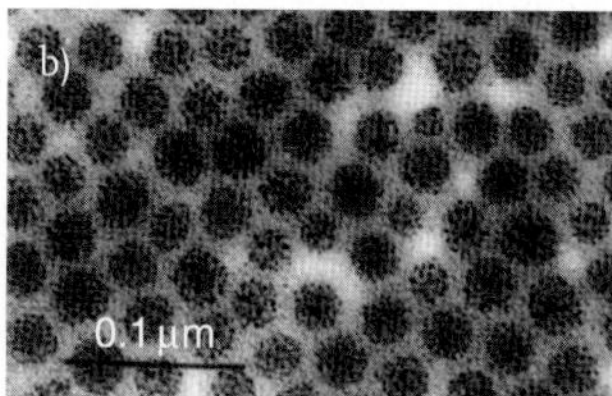

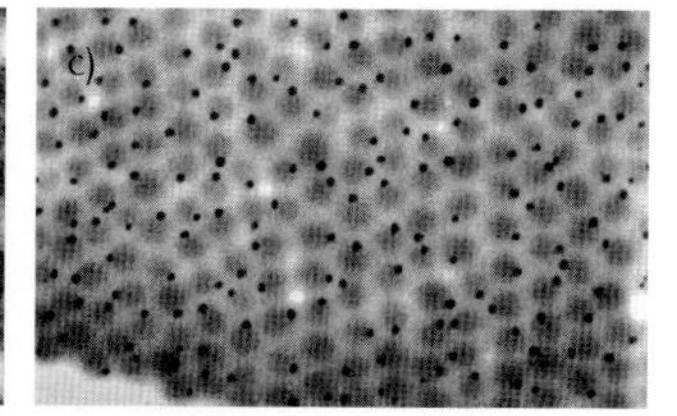

Figure 2.3 TEM images of the PS-*b*-P4VP micelles containing the solubilized precursor ($HAuCl_4$) (a), small Au colloids after fast reduction with $LiAlH_4$ (b), and one single Au colloid after slow reduction with Et_3SiH loaded within single micelles (c). Adopted with permission from Ref. [74]. Copyright 1998 John Wiley & Sons.

of the nanoparticles. However, when a weak reducing reagent of Et_3SiH was used, single Au nanoparticle with size at 8 nm was produced in one template micelle as shown in Figure 2.3c. Besides Au nanoparticles, Pd [70] and Pd-Au [76, 77], and Pd-Pt [77] bimetallic nanoparticles were also synthesized within the PS-*b*-P4VP micelles and were demonstrated to be an effective catalyst for Heck coupling reaction and hydrogenation [70, 77].

In aqueous solution, various metal nanoparticles such as Pd [78–84], Au [85–87], and Pt [88] were also synthesized using block or random copolymer micelles as template. For examples, Taton *et al.* [89] prepared Au nanoparticles using the cross-linked block copolymer micelles as template in aqueous solution. We have also reported the synthesis of Au nanoparticles mediated with a thermo- and pH-responsive coordination triblock copolymer of poly(ethylene glycol)-*b*-poly(4-vinyl pyridine)-*b*-poly(*N*-isopropylacrylamide) (PEG-*b*-P4VP-*b*-PNIPAM) [90]. It was found that the size and morphology of the resultant Au nanoparticles could be easily tuned by changing pH or temperature of the aqueous solution of the triblock copolymer. At pH 6.5 and at 25 °C, the triblock copolymer self-assembled into 30-nm core–corona micelles with the hydrophobic P4VP block as core and the hydrophilic PNIPAM and poly(ethylene glycol) blocks as corona. After loading the Au precursor into the P4VP core of the core–corona micelles and then reducing with $NaBH_4$ aqueous solution, discrete Au nanoparticles with size at 10 nm as shown in Figure 2.4a were synthesized. At pH 2.5 and at 50 °C, the triblock copolymer formed large compound micelles or micelle clusters, and clusters of Au nanoparticles as shown in Figure 2.4b were synthesized using these large compound micelles or micelle clusters as template.

As similar as the metal nanoparticles stabilized by soluble polymers, the metal nanoparticles stabilized by amphiphilic block copolymer micelles can be highly dispersed in solution, which is useful in catalysis. Compared with the metal

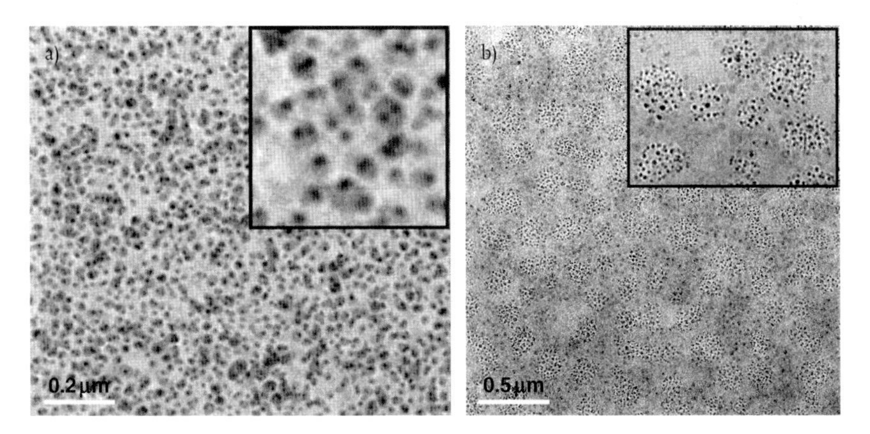

Figure 2.4 Discrete Au nanoparticles (a) and clusters of Au nanoparticles (b) synthesized using the PEG-*b*-P4VP-*b*-PNIPAM micelles as template in aqueous solution. Adopted with permission from Ref. [90]. Copyright 2006 American Chemical Society.

nanoparticles stabilized by soluble polymers, the catalyst of metal nanoparticles stabilized by micelles is much more stable and therefore more easily reusable than those stabilized by single soluble polymers. For example, the catalyst of Au nanoparticles stabilized by the thermoresponsive block copolymer micelles of poly(*N*-isopropylacrylamide)-*b*-poly(4-vinylpyridine) can be first separated from the reaction system just by heating above the lower critical solution temperature (LCST) of the block copolymer micelles and then recovered by redispersion of the micelles/Au nanoparticles in the solvent of water at temperature below LCST [91].

Up to now, the application of metal nanoparticles stabilized by self-assembled amphiphilic block copolymer micelles is limited, possibly due to the relatively difficult synthesis and therefore the expensive amphiphilic block copolymer. Recently, Lee *et al.* [92] have found that the block copolymer micelles can work not only as a support for Pd nanoparticles but also as a supramolecular nanoreactor for Suzuki coupling reaction as shown in Figure 2.5. It was declared that Suzuki coupling reaction within the self-assembled micelle nanoreactor runs efficiently

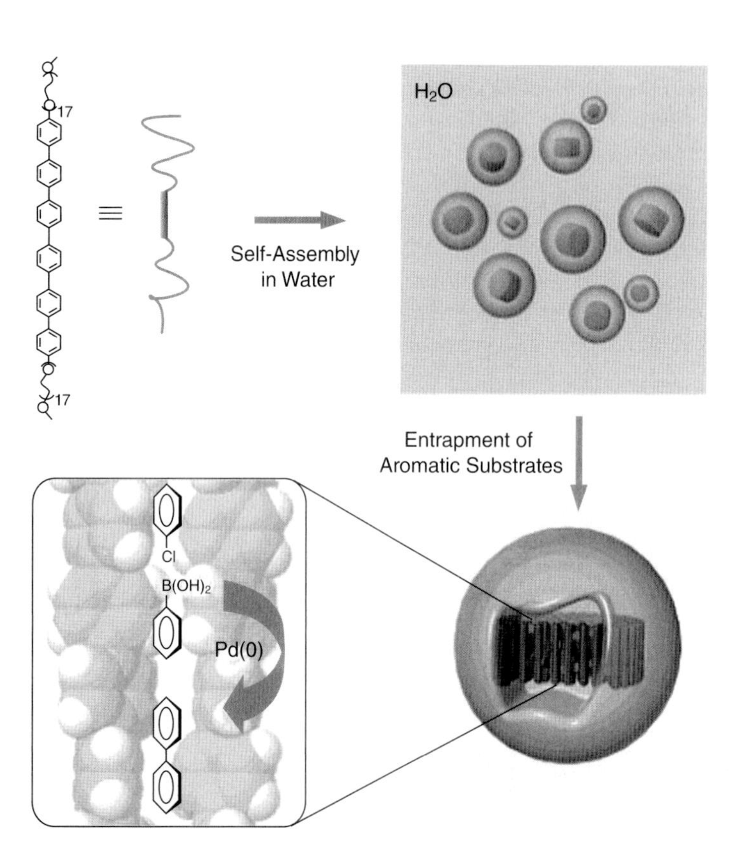

Figure 2.5 Schematic representation of supramolecular nanoreactor of the micelle/Pd nanoparticles within which Suzuki reaction is performed. Adopted with permission from Ref. [92]. Copyright 2004 American Chemical Society.

at room temperature. This possibly suggests that further research on the application of the micelles/metal nanocatalyst system is worthy.

2.2.2.1.3 **Metal Nanoparticles Supported on Cross-Linked Functional Polymers**

Cross-linked functional polymers (CFPs) are isotropic materials which are normally originated by the homo- or copolymerization of vinyl monomers, the most important of which is styrene, in the presence of a cross-linking agent such as divinylbenzene (DVB). CFPs are normally manufactured as submillimetric beads or powders [93], and the gel-type and macroreticular resins as shown in Figure 2.6 are the most popular [94]. The proposal of CFPs as supports for metal nanoparticles to be employed in catalytic applications dates back to 1969 [95].

Generally, the introduction of the active metal nanoparticles into the CFPs can be fulfilled in three ways:

1) Introduction of a suitable metal precursor into the CFPs support and generation of the metal nanoparticles therein through reduction or decomposition of the metal precursor (RIMP);
2) Synthesis of the CFPs support in the presence of a suitable metal precursor and generation of the metal nanoparticles therein through reduction or decomposition of the precursor (incorporation of metal during the polymerization and reduction, IMPR);
3) Introduction of preformed metal nanoparticles into the CFPs support (IPMN).

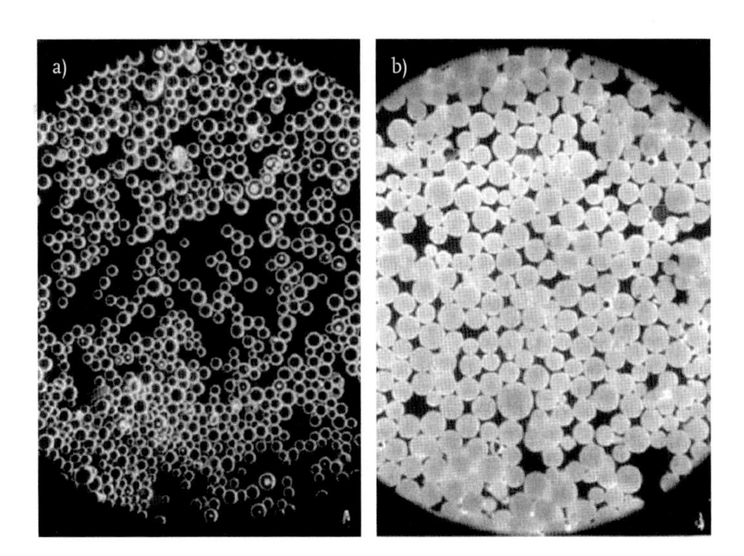

Figure 2.6 Representative optical micrographs of the macroreticular resin of poly-HEMA cross-linked with EDMA produced by suspension copolymerization in the presence of a porogen of toluene in the dry (a) and swollen (b) state. Adopted with permission from Ref. [94]. Copyright 1996 Elsevier.

Reduction of Immobilized Metal Precursors (RIMP) The RIMP is the most widely employed method to synthesize metal nanoparticles supported on CFPs. Generally, these CFPs possess functional groups such as $-SO_3H$ or $-COOH$ suited for ion exchange or metal coordination group such as amino, cyano, pyridyl, thiol, sulfide, benzimidazolyl, or pyrrolidyl. These functional groups can be introduced during the polymerization upon using a properly functionalized monomer or after modification of the presynthesized CFPs. As the metal amount in the catalyst required for the catalytic application is generally low (a few percent in weight), the amount of functional groups for metal "docking" can be low as well.

The treatment of CFPs with a solution of a proper metal precursor allows the introduction of the latter in an easy way, with ion exchange or ligand addition/replacement processes providing the driving force for the metal uptake. Under these conditions, excess amount of solvent and relatively low metal precursor amount can be employed. The solvent should be a good swelling agent for the CFPs support. Under full swelling conditions, the whole or a large proportion of the polymer framework is accessible to the metal centers. Metal uptake is often quantitative, and the desired final metal weight percentage in the catalyst (0.1–10%, w/w, for catalytic application) can be attained with no or little metal waste. Generally, higher metal loadings and homogeneous metal distribution throughout the support beads can be attained in a good solvent than in a poor one.

As early as in 1974, Hanson *et al.* [96] reported the size-controlled synthesis of Pd, Pt, Ag, and Ni nanoparticles supported on Amberlyst 15 by means of the RIMP method. The size-controlled synthesis of metal nanoparticles was achieved by reducing the immobilized metal precursor employed by different reduction protocols. The metalated resin was either treated with flowing H_2 in the dry state or with a basic aqueous solution of hydrazine or ethylformate. In the preparation of the catalysts by H_2 reduction, 2-nm metal nanoparticles were synthesized and the size distribution was pretty uniform. In the synthesis of the catalysts prepared with aqueous solutions of the reducing agent, in addition of the 2-nm nanoclusters, a small number of large metal nanoparticles with size ranged from 10 to 30 nm were also observed. It was deemed that the small nanoclusters were formed within the polymer framework and the large nanoparticles were formed in the permanent pores (whose diameter was 20–60 nm on average). The formation of metal nanoparticles in the pores was possible only in the swollen state of the CFPs in solution employing the reducing agent.

Sidorov *et al.* [97, 98] found that highly cross-linked resin (HPS) of PS or PS-DVB which had a very rigid and permanently nanoporous framework (a few nanometers) could also be used as support to synthesize metal nanoparticles, even they did not possess functional groups apt to ion exchange or ligand addition/substitution reactions. It was found that the size of the synthesized nanoparticles did not exceed the diameter of the cavities where they were generated. Owing to the absence of specific functional groups, adsorption of the metal precursors onto HPS was carried out with an incipient wetness impregnation procedure. For examples, when an HPS with 2.0-nm nanopores was impregnated with a THF or methanol solution of H_2PtCl_6, dried and treated with H_2 for 3 h at room temperature, 1.3 nm (THF) and 1.4 nm (methanol) Pt nanoclusters were obtained, with an overall metal loading of 7.5–8.3% (w/w) [97]. Using the same support of the HDS resin, 2.0-nm Co nanoparticles were synthesized by first impregnation with isopropanol or dimethylformamide (DMF) solution of $Co_2(CO)_8$ and subsequent drying and thermal decomposition of the precursors [98].

The nanocavities within the HPS resin are deemed as nanoreactors, within which metal atoms grow and nucleate to form metal nanoparticles. The size and even the morphology of the metal nanoparticles are determined by the template of nanocavities. In view of the absence of functional groups able to strongly interact with the metal centers or nanoparticles, a simple steric effect was proposed for the immobilization of the catalyst in the HPS resin. For this peculiar way of size control synthesis of metal nanoparticles, Ziolo *et al.* introduced the term "template-controlled synthesis" (TCS) [99, 100]. Up to now, following this TCS strategy, noble metal nanoparticles such as Au, Pd, Pt nanoparticles, and even the magnetic nanoparticles of ferric oxide have been synthesized with the RIMP method, and the diameter of the synthesized nanoparticles exhibits a remarkable agreement with the size of the cavities of the swollen polymeric supports [101–106].

For the metal nanoparticles immobilized on CFPs, only the accessible ones are catalytically active. This implies that the CFPs-supported metal nanocatalyst is more suitable to be used in the presence of a swelling agent under solid–liquid conditions. In the swollen state, ascribed to the temporary nanopores and mesopores of the CFPs even in the macroreticular materials appear as shown in Figure 2.7, the accessibility of the catalytically active site of the immobilized metal nanoparticles is ensured [107]. These temporary nanopores and mesopores have actually diameters in the order of the nanometers or a few tens of nanometers (<50 nm), that is, one or two orders of magnitude larger than in zeolites. While in the absence of a swelling agent, the polymer framework is collapsed over the metal nanoparticles. Under these conditions, molecules from another phase cannot diffuse into CFPs and therefore cannot interact with the metal nanoparticles.

Incorporation of Metal Precursors during Polymerization and Reduction Different from the immobilization of metal catalyst following the RIMP method, in the IMPR strategy the metal precursor is introduced within the support matrix during the polymerization of monomers with which the metal precursor is usually coordinated. Compared with the RIMP method, the IMPR method is much less popular possibly since the metal precursor will disturb the polymerization. Figure 2.8 shows the typical example of incorporation of metal precursors to synthesize a commercial Pd EnCat™ catalyst employing the IMPR method [108], in which EnCat represents those of metal nanoparticles encapsulated or entrapped within polymeric support. As shown in Figure 2.8, incorporation of the Pd precursor is achieved by microencapsulation within the polymerurea matrix generated by interfacial polymerization of isocyanate oligomers in the presence of $Pd(OAc)_2$. Microencapsulated 2- and 5-nm Pd nanoparticles of the Pd EnCat™ catalyst can be obtained thereafter by reduction with either H_2 or formic acid, respectively.

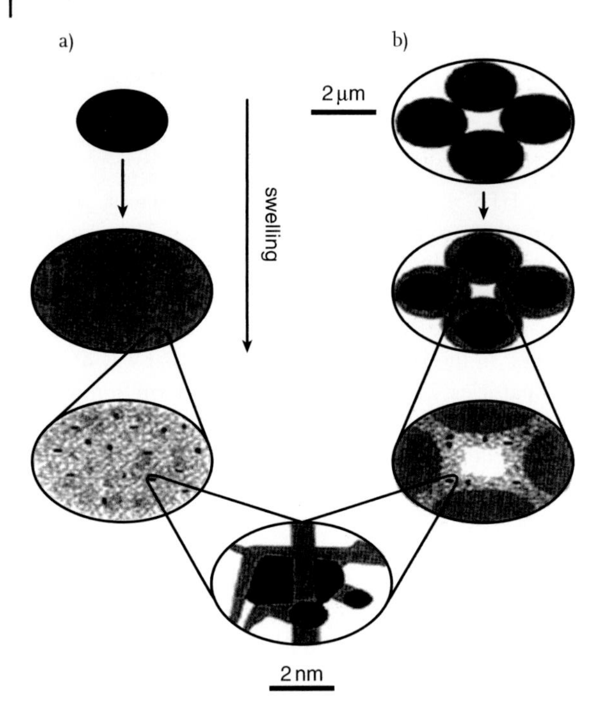

Figure 2.7 Schematic representation of the metal nanoparticles supported on gel-type (a) and macroreticular (b) resins of CFPs during different swelling state, wherein the metal nanoparticles are represented as black spots and the upper represents the dried CFPs and the lower represents the swollen CFPs. Swelling increases from the top down. Adopted with permission from Ref. [107]. Copyright 2005 John Wiley & Sons.

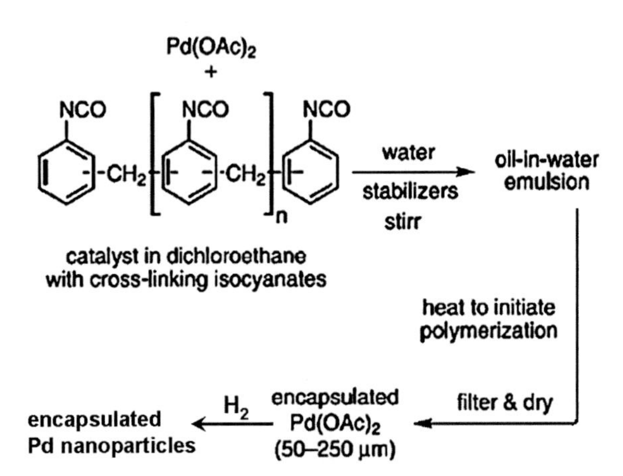

Figure 2.8 Synthesis of the Pd EnCat™ catalyst following the RIMP method initially by microencapsulation of $Pd(OAc)_2$ during the polymerization followed reduction with H_2. Adopted from Ref. [108].

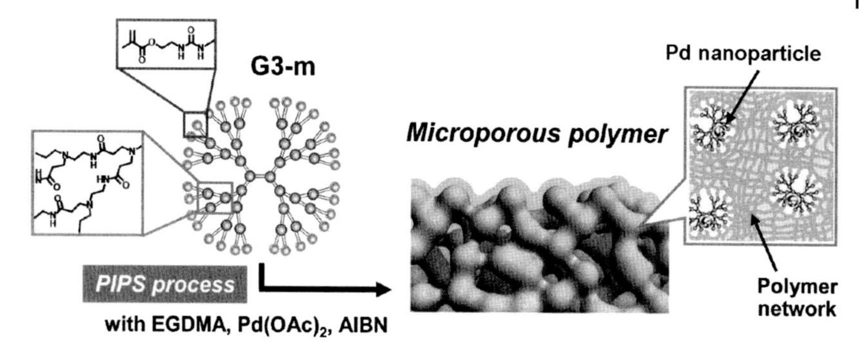

Figure 2.9 Schematic incorporation of Pd nanoparticles encapsulated within a microporous polymeric network. Adopted with permission from Ref. [109]. Copyright 2010 American Chemical Society.

Ogasawara *et al.* proposed another IMPR strategy based on polymerization-induced phase separation (PIPS) techniques to fabricate palladium nanoparticles encapsulated within microporous polymeric network [109]. As shown in Figure 2.9, the Pd precursor of $Pd(OAc)_2$ was initially premixed with a monomer having a poly(amidoamine)-based dendrimer ligand (G3-M), and subsequently the Pd coordinated monomer was thermally polymerized with an excess amount of ethylene glycol dimethacrylate (EGDMA) under PIPS conditions. In this system, the formation of encapsulated Pd nanoparticles occurred concurrently with the microporous polymer synthesis in a one-pot process even with no additional reducing reagent.

Immobilization of Preformed Metal Nanoclusters (IPMN) The IPMN method has the advantage that the size and shape control can be achieved during the preformation of the metal nanoparticles, provided that no sintering or coagulation occurs during the introduction. Up to now, several methods are available for the synthesis of size-controlled naked or soluble metal nanoparticles. For examples, the soluble Pd or Pt nanoparticles stabilized by linear polymer of PVP [22–31], polyvinylalcohol (PVA) [110], or citrate [16–19] are reported. Even more, naked metal nanoparticles stabilized by solvent molecules are synthesized [111–115]. These successes make IPMN a promising method. Generally, immobilization of preformed metal nanoparticles is performed by adsorption of the preformed metal nanoparticles on CFPs. For example, Yu *et al.* [116, 117] initially treated the polymeric supports of PS-DVB with the Pt/PVP solution at room temperature and the preformed Pt nanoparticles were adsorbed on the CFPs. Then, after filtration, the collected solid was Soxhlet extracted to remove the protected polymer of PVP, and finally dried under vacuum at room temperature. Allegedly, the Pt catalyst immobilized on CFPs generally contained 0.5% (w/w) and no variation of the Pt nanoparticles size was observed.

Toshima *et al.* [118] reported another method of immobilizing preformed metal nanoparticles on CFPs, which is highlighted in Figure 2.10. This strategy involves

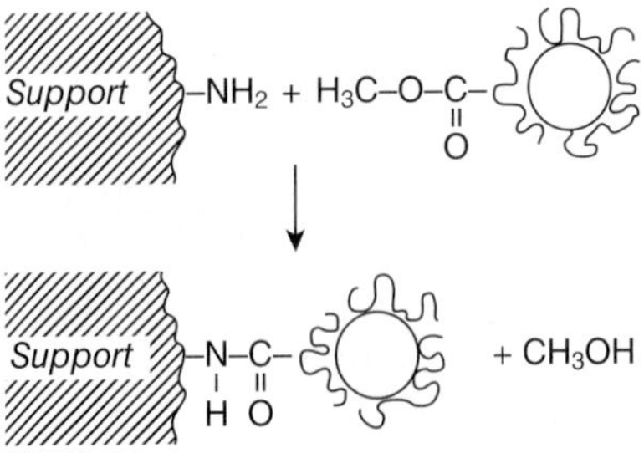

Figure 2.10 Covalent attachment of the preformed polymer-stabilized Rh nanoparticles to an amino-functionalized gel-type CFP. Adopted with permission from Ref. [118]. Copyright 1991 American Chemical Society.

a CFP with amino group and the preformed rhodium nanoparticles protected with the copolymer of methyl acrylate and *N*-vinyl-2-pyrrolidone (PMA-*co*-PVP). The Rh nanoparticles were covalently immobilized onto the CFPs by forming the amide bond between the primary amino group contained in the CFP support and the methyl acrylate residue in the protective polymer on the surface of the stabilized Rh nanoparticles.

Metal Nanoparticles Supported on Other Polymer Materials with Special Structure or Morphology Recently, some polymer materials with special structure or morphology are synthesized and used as support for metal nanoparticles. These polymer materials including colloidal nano- and microspheres, core–shell nano- and microspheres, and polymer–inorganic composites are briefly introduced as below.

Zheng *et al.* [119] synthesized core–shell microspheres of poly(styrene-*co*-2-[acetoacetoxy]ethyl methacrylate-*co*-methyl acrylic acid) (PS-*co*-PAEMA-*co*-PMAA), which contained a pH-responsive shell of PMAA segment and a coordinative core of PAEMA and PS segments (Figure 2.11a). Pd nanoparticles as shown in Figure 2.11b were immobilized in the core-layer of the core–shell microspheres firstly by introducing the Pd precursor through coordination between the β-diketo ligand in the PAEMA segment and the Pd ions, and then reducing the immobilized Pd precursor with $NaBH_4$ aqueous solution. The PS-*co*-PAEMA-*co*-PMAA core–shell microspheres had three advantages to be used as support for metal nanoparticles. First, the glass transition temperature (T_g) of the core-forming segments of PAEMA and PS is ~3 °C and ~103 °C, respectively. The combined use of the two polymer segments with a relatively lower T_g and a higher T_g can accelerate diffusion of the reactants and therefore increases the catalytic accessibility of the immobilized Pd nanoparticles, and can keep the mechanical and thermal stability of the catalyst. Second, the shell-forming segment of PMAA in the core–shell microspheres is hydrophilic and therefore the PS-*co*-PAEMA-*co*-PMAA core–shell microspheres can form stable colloidal dispersion in water and basic aqueous solution. The immobilized Pd nanocatalyst is quasi-homogeneous and therefore efficient for Suzuki reaction performed in basic aqueous solution. Third, also due

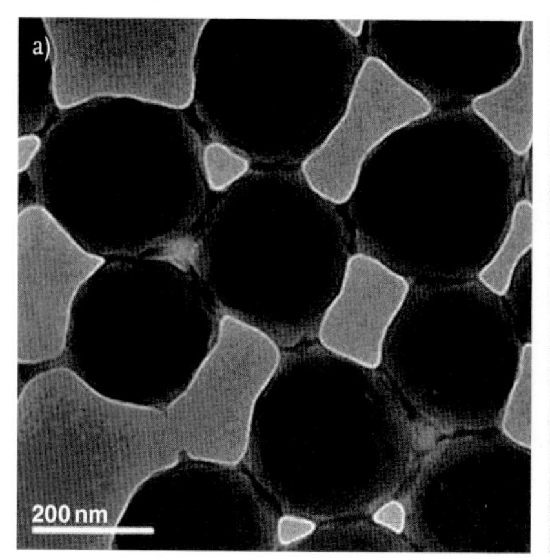

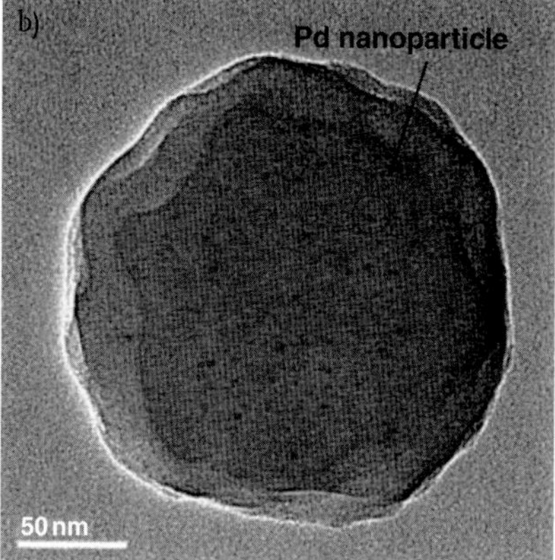

Figure 2.11 TEM images of PS-*co*-PAEMA-*co*-PMAA core–shell microspheres (a) and Pd nanoparticles immobilized in the core layer of the PS-*co*-PAEMA-*co*-PMAA core–shell microspheres (b). Adopted with permission from Ref. [119]. Copyright 2007 Elsevier.

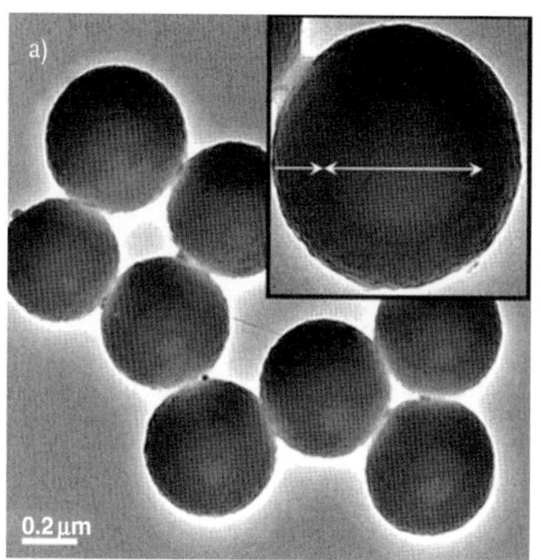

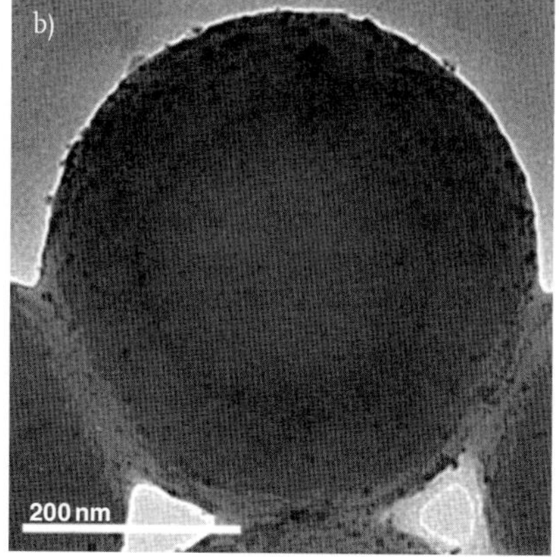

Figure 2.12 TEM images of the PS-*co*-P4VP core–shell microspheres (a) and Pd nanoparticles immobilized in the shell layer of the PS-*co*-P4VP core–shell microspheres (b). Adopted with permission from Ref. [120]. Copyright 2008 American Chemical Society.

to the pH-responsive shell-forming segment of PMAA, the quasi-homogeneous Pd nanocatalyst can be separated from the reaction mixture and reused by acidifying the reaction mixture.

To increase the accessibility of the immobilized metal nanoparticles, Wen *et al.* [120] synthesized core–shell microspheres of poly(styrene-*co*-4-vinylpyridine) (PS-*co*-P4VP) as shown in Figure 2.12a and used as scaffold for Pd, Au, and Ag nanoparticles. The PS-*co*-P4VP core–shell microspheres contain a polystyrene core and a coordination P4VP shell. Pd, Au, and Ag nanoparticles can be easily immobilized

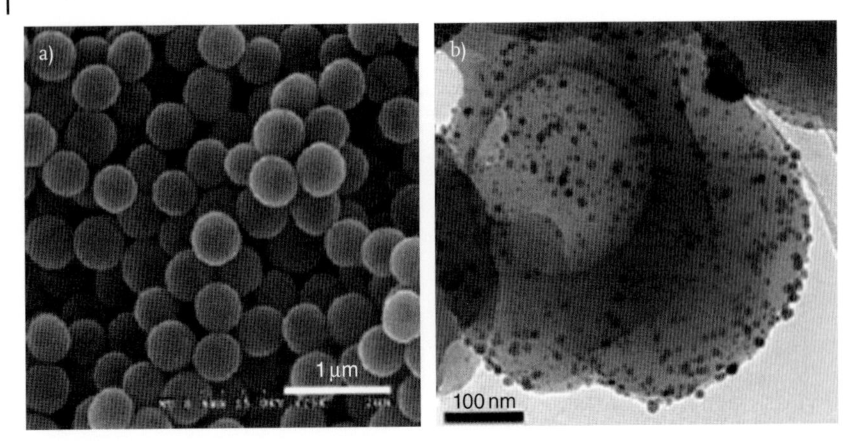

Figure 2.13 SEM images of hollow polymeric microspheres (a) and Pd nanoparticles immobilized on the hollow microspheres (b). Adopted with permission from Ref. [121]. Copyright 2008 American Chemical Society.

on the out-shell layer of the core–shell microspheres first by introducing the metal precursors through coordination with the shell-forming P4VP segment followed by reducing with $NaBH_4$ aqueous solution. As shown in Figure 2.12b, the 3-nm Pd nanoparticles are uniformly dispersed in the shell layer of the core–shell microspheres and are demonstrated to be an efficient and easily recycled catalyst for Suzuki reaction performed in water.

Miao *et al.* [121] used hollow polymeric microspheres (Figure 2.13a) as scaffold to prepare immobilized noble metal nanoparticles with a supercritical route. In this method, the metal precursors were first adsorbed on the hollow polymeric microspheres in a supercritical CO_2/ethanol solution, followed by H_2 reduction to generate metal nanoparticles. It was indicated that the Pd nanoparticles with a size of about 5 nm were uniformly attached to the surface of the hollow polymeric microspheres (Figure 2.13b), and exhibited a high activity and stability in hydrogenation and Heck reactions.

To increase the mechanical and thermal stability of the supported metal nanocatalyst, Corain *et al.* [122, 123] used polymer–silica composites as support for metal nanoparticles, wherein the silica was generated by hydrolysis of tetraethoxylsilane (TEOS) inside either Figure 2.13a the polymeric support or Figure 2.13b the Pd nanoparticles/CFPs catalyst after swelling the organic materials with a hydroalcoholic solution of TEOS. The catalysts obtained along route Figure 2.13b exhibited a much better mechanical stability and easier reusability, although it was less efficient than the parent polymer-supported materials in the hydrogenation of cyclohexene.

2.2.2.2 Metal Nanoparticles Supported on Carbon Nanotubes

Carbon materials may be one of the most widely used materials for heterogeneous catalysis applications due to their low cost, high thermal and chemical stability, the possibility of forming them into various shapes and sizes, and tunable textural properties such as surface area, porosity, and surface chemistry. The synthesis of metal nanoparticles deposited on general carbon materials such as active carbon can be referred elsewhere. Herein, the application of carbon nanotubes (CNTs, including multiwalled carbon nanotubes (MWCNTs), and single-walled carbon nanotubes (SWCNTs)) as support of metal nanoparticles is briefly introduced, although the active carbon may be the most popular one. CNTs have intrinsic properties such as high surface area, unique physical properties and morphology, and high electrical conductivity, which make them extremely attractive as supports for metal nanoparticles [124, 125]. Up to date, many ingenious methods of depositing metal nanoparticles onto CNTs are reported and these methods can be categorized to electrochemical, chemical, and physical methods.

2.2.2.2.1 **Electrochemical Methods**

Electrochemistry is a powerful technique for the deposition of metals and/or the surface modification of CNTs, and thus allows the chemist to easily control the nucleation and growth of metal nanoparticles [126–133]. Recently, Guo *et al.* [132, 133] proposed an elegant synthesis of Pt or Pd nanoparticles deposited on CNTs as shown in Figure 2.14. They carried out a gentler electrochemical pretreatment to oxidize SWCNTs, thus introducing the required oxygen-containing functionalities such as quinonyl, carboxyl, or hydroxyl groups to the SWCNTs without damaging them. This method elided the harsh treatments with concentrated mixtures of strong oxidizing acids such as sulfuric and nitric acid and/or ozonolysis, which might lead to severe damage to the CNTs. On these functioned SWCNTs, 5-nm Pd or Pt nanoparticles were prepared by electrochemical deposition.

Guo *et al.* [132] further proposed a strategy of surface modification of CNTs with a monolayer of 4-aminobenzene molecules covalently attached to CNTs via the direct electrochemical reduction of the corresponding diazonium salt of nitrobenzene as shown in Figure 2.15. Then the grafting molecules of 4-aminobenzene were reduced to the corresponding amines and the palladium salt of $PdCl_6^{2-}$ was then adsorbed onto the aminobenzene monolayer via electrostatic interaction. Lastly, 2.5-nm Pd nanoparticles deposited on the surface of the modified CNTs were produced via further potentiostatic reduction.

2.2.2.2.2 **Chemical and Physical Methods**

Various chemical and physical methods are suggested to synthesize metal nanoparticles supported on CNTs. Readers can refer to a recent review by Compton *et al.* [124]. Herein, only the method to deposit metal nanoparticles on polymer-wrapped SWCNTs proposed by Vinodgopal *et al.* is briefly introduced [134]. This method is based on the finding that SWCNTs can be solubilized in water with the help of linear polymers to wrap around the SWCNTs. Figure 2.16a schematically shows the structure of the polymer-wrapped SWCNTs and the Pt nanoparticles supported on the polymer-wrapped SWCNTs, wherein the soluble polymer of

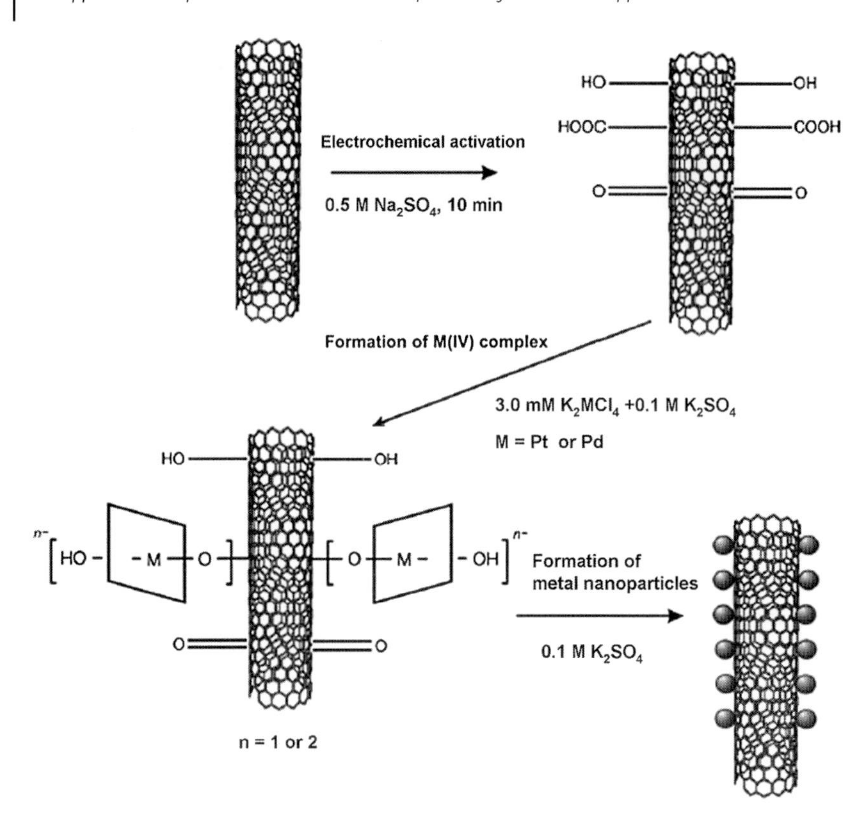

Figure 2.14 Schematic synthesis of Pt or Pd nanoparticles deposited onto SWCNTs through the electrochemical deposition method. Adopted with permission from Ref. [133]. Copyright 2005 Elsevier.

poly(sodium 4-styrenesulfonate) (PSS) is employed. The PSS-wrapped SWCNTs were synthesized firstly by sonicating the SWCNTs dispersed in DMF in the presence of PSS and then by washing with water. The PSS-wrapped SWCNTs were believed to form loose bundles in water due to the electric double-layer effect of the ionic polymer of PSS. Deposition of Pt nanoparticles on the PSS-wrapped SWCNTs was achieved by simply mixing the PSS-wrapped SWCNTs with Pt colloids. Figure 2.16b and c shows the TEM images of the Pt nanoparticles deposited on the PSS-wrapped SWCNTs. Clearly, 2–3-nm Pt nanoparticles are uniformly dispersed on the nanotube bundles. Compared with the general supported Pt nanocatalysts of Pt/C and Pt/MWCNT, the catalyst of Pt nanoparticles deposited on the PSS-wrapped SWCNTs had a relatively higher electrochemically active surface area (ECSA), and therefore more accessible and efficient. Besides, possibly due to the good solubility of the wrapping polymer of PSS, the resultant supported Pt nanocatalyst was easily dispersed in water, which provided potential of quasi-homogeneous catalysis in water.

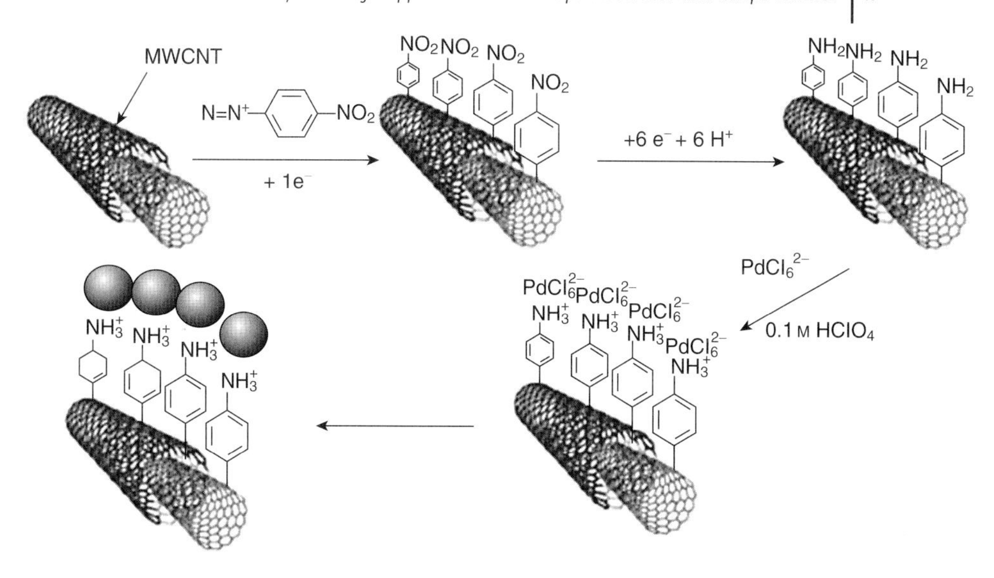

Figure 2.15 Schematic synthesis of Pt nanoparticles deposited on CNTs modified with aminobenzene. Adopted with permission from Ref. [132]. Copyright 2004 Elsevier.

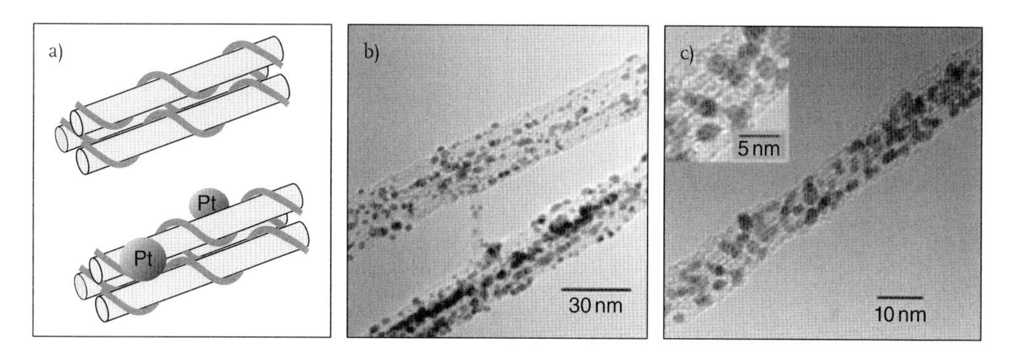

Figure 2.16 Schematic structure (a) and TEM images of Pt nanoparticles deposited on the PSS-wrapped SWCNTs (b, c). Adopted with permission from Ref. [134]. Copyright 2006 American Chemical Society.

2.2.2.3 Metal Nanoparticles Supported on Metal Oxides

The impregnation [135], deposition–precipitation [136, 137], coprecipitation [138], and vapor deposition methods [139, 140] are usually applied to synthesize metal nanoparticles supported on metal oxides such as Al_2O_3, SiO_2, TiO_2, MgO, CeO_2, etc. Besides these conventional methods, some other methods are also proposed to deposit metal nanoparticles on metal oxides [141–144]. Herein, two typical methods are briefly introduced.

Stucky *et al.* [145] proposed a general method to synthesize noble metal nanoparticles supported on metal oxides. This method involves adsorption of the dodecanethiol-capped noble metal nanoparticles on the metal oxides in an aprotic solvent through weak dipole charge, dipole–dipole, and dipole-induced dipole interactions. The presynthesized noble metal nanoparticles were capped by the long-chain alkyl thiols (e.g., dodecanethiol) and therefore were hydrophobic and soluble in an aprotic solvent such as chloroform, dichloromethane, toluene, and hexanes. When oxide powders were added to the dispersion of the dodecanethiol-capped gold nanoparticles in an aprotic solvent, the adsorption of the dodecanethiol-capped noble metal nanoparticles on the oxides occurred. It was declared that the presynthesized Au nanoparticles with different sizes were found to be highly dispersed on the oxides after removing the capping agent by thermal treatment at 300 °C.

This method is deemed to have several advantages. Firstly, a wide range of oxide supports can be used since the synthesis of supported metal nanocatalysts is performed in an aprotic solvent. For example, the general synthesis of Au nanocatalyst with the deposition–precipitation method requires the adjustment of pH value within the range of 6–10 and is not applicable to acidic and hydrophobic supports such as SiO_2, WO_3, Al_2O_3, and activated carbon. Secondly, the interaction between the metal oxide support and the noble metal nanoparticles is weak, and therefore the adsorption and desorption of metal nanoparticles kinetically occurs on the oxide surface in the presence of an aprotic solvent. This kinetic process leads to a homogeneous dispersion of metal nanoparticles on the oxide support. Thirdly, this method allows immobilization of different amounts of the same-sized metal nanoparticles by simply changing the amount of the oxide materials added into the dispersion of metal nanoparticles, which is technically difficult by conventional preparation methods. However, the as-prepared metal-oxide composite does not possess catalytic properties due to the capping agent of organic thiols. Only after removing the capping ligands by heating at 300 °C in air, the metal-oxide composite becomes catalytically efficient.

The other method of flame synthesis including vapor fed flame reactors [146], flame spray pyrolysis (FSP) [147], and liquid flame spray (LFS) [48] is a relatively new method for the one-step production of supported metal nanocatalysts. The working principle of flame spray synthesis is that an oxidizing gas containing an atomized organic solution of organometallic precursor compound is led into a flame zone where the droplets are combusted at a high temperature (T_{max} ~3000 °C) and the precursors convert into nanometer-sized metal or metal-oxide particles, depending on the metal and the operating conditions. Figure 2.17 shows the typical experimental setup for the synthesis of supported metal nanocatalysts by FSP [148]. One of the advantages is that the supported metal nanocatalysts, even the bimetallic nanocatalysts, can be prepared within one step. Up to date, several groups such as Baiker [147, 149, 150], Quaade [148], Moser [151], Johannessen [152], and Jensen *et al.* [153] are focused on the flame synthesis of supported metal nanocatalysts.

Of all the flame synthesis methods, it is said that the FSP offers possibility of using nonvolatile precursors and is applied for synthesis of the Pt/Al_2O_3 nanocatalysts [147] at a considerably higher production rate of 15 g h^{-1}, Au/TiO_2, and

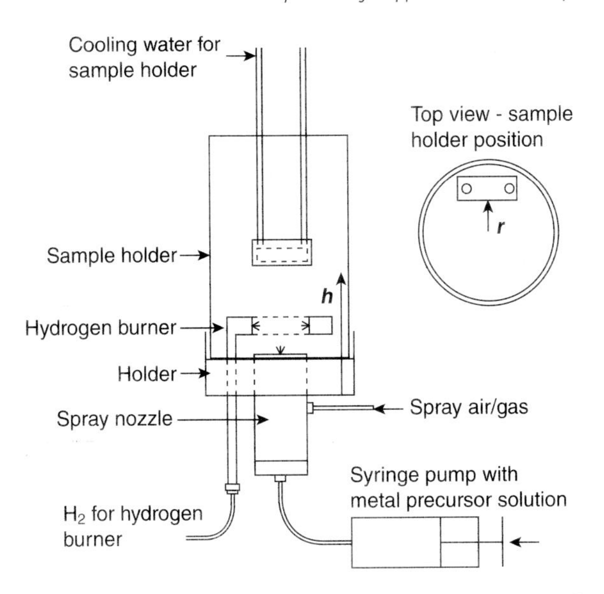

Figure 2.17 The experimental setup for flame spray synthesis of supported metal nanocatalyst. Adopted with permission from Ref. [148]. Copyright 2004 Elsevier.

Au/SiO_2 nanocatalysts [154]. Baiker *et al.* [150] reported the flame spray synthesis of the Pd/Al_2O_3 nanocatalyst, wherein the precursor solutions were prepared by dissolving appropriate amounts of aluminum *sec*-butoxide (Al[s-BuO]$_3$) and palladium acetylacetonate (Pd[acac]$_2$) in xylene/acetonitrile mixture (70/30 vol.%). FSP of the palladium/alumina precursor solutions resulted in spherical alumina particles in the size range of 10–30 nm with well-dispersed 1–5 nm Pd nanoparticles attached to the surface. For synthesis of the Pd/Al_2O_3 nanocatalyst, it was found that changing the flame conditions by applying different dispersion or sheath gases (O_2, air, N_2) did not have a major effect on the morphology of the alumina support and the Pd nanoparticles. Whereas the specific surface area and therefore the performance of the Pd/Al_2O_3 nanocatalyst could be tuned by changing the liquid flow rate of the palladium/alumina precursor solution.

2.2.2.4 **Metal Nanoparticles in Mesoporous Silica**

Because of their high surface areas, regular pore channels, and suitable pore diameters (2–50 nm), the ordered mesoporous materials such as FSM-16 [155, 156], MCM-41 [157], and SBA-15 [158] have been used as suitable scaffolds for the dispersion of metal nanoparticles since their discovery in the early 1990s. Up to now, various methods including impregnation [159], interaction with functional groups of the mesoporous solids (e.g., silanol groups) [160–162], chemical vapor deposition inside the pores [163], template ion exchange with transition metal cations [164], infusion of prefabricated metal nanoparticles [165], and templating over metal-containing templates [166, 167] have been proposed to immobilize metal nanoparticles on the support of mesoporous silica. Herein, the infusion of prefabricated metal nanoparticles into mesopores of mesoporous silica and the immobilization of metal nanoparticles through the interaction with functional groups of the mesoporous silica are briefly introduced.

Infusion of prefabricated metal nanoparticles into the mesopores of mesoporous silica may be one of the most popular methods to synthesize the mesoporous silica supported metal nanocatalysts. As similar as those of the immobilization of prefabricated metal nanoparticles, the infusion method has the advantages of precise control of the metal particle size, crystallinity, shape, and surface properties if no particle agglomeration occurs during the infusion. However, there are two primary challenges in the infusion of the prefabricated metal nanoparticles into mesopores of mesoporous silica. First, the dispersion of the metal nanoparticles must be transported through the pores by favorable capillary wetting, and the metal nanoparticles must not block the small mesopore entrances of mesoporous silica. Once the metal nanoparticles penetrate the pores, the second challenge is to achieve sufficient adsorption of the metal nanoparticles from the solution phase to the substrate surface. Generally, the diffusion of metal nanoparticles through the solvent into the mesopores of mesoporous silica is undesirably slow, which suggests that effort to accelerate the metal nanoparticles diffusion is one of the key factors.

Rioux *et al.* [41] reported the sonication-aid infusion method as shown in Figure 2.18 to prepare the Pt metal nanoparticles immobilized on SBA-15. The Pt nanoparticles in the size range of 1.7–7.1 nm, which was smaller than the mesopore size of SBA-15 (9.0 nm), were produced by alcohol reduction methods in the presence of the stabilizing polymer of PVP. After incorporation of the preprepared Pt nanoparticles into the mesoporous SBA-15 using low-power sonication, the resultant was initially calcined to remove the stabilizing polymer and then reduced by H_2. It was found that Pt nanoparticles were well dispersed in the entire channels of SBA-15. Furthermore, it was found that the size of Pt nanoparticles immobilized on SBA-15 was almost the same as those in solution in the presence of PVP even when the resultant nanocatalyst was calcined at 623–723 K. It was demonstrated that sonication was necessary for the incorporation of Pt nanoparticles into the mesopores of SBA-15. Without sonication, Pt nanoparticles were not able to infuse into the mesopores and primarily attached on the external surface of SBA-15, and therefore became large aggregates after high-temperature treatment.

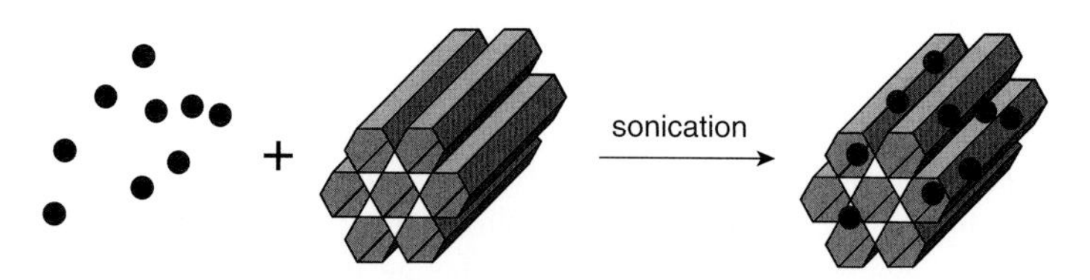

Figure 2.18 Synthetic scheme for sonication-aid infusion of prefabricated metal nanoparticles into the mesopores of SBA-15.

Different from the sonication-aid method, Gupta *et al.* [168] reported another infusion method to immobilize the prefabricated metal nanoparticles into mesoporous silicas. They found that supercritical CO_2 as antisolvent helped overcome both transport and thermodynamic challenges in infusion of metal nanoparticles into mesoporous silica. With the help of the supercritical CO_2, the dispersion of Au or Ir nanoparticles in supercritical CO_2/toluene mixtures was much accelerated, and relatively high loading of metal nanoparticles in mesoporous silica was achieved. While without the aid of supercritical CO_2, the infusion of Au nanoparticles into S-MCM41, L-MCM41, and SBA-15 was negligible.

Now, we tend to the introduction of modification of mesoporous silica with ligands. Generally, surface modification of mesoporous silica was performed through the reaction of silanols with alkoxysilyl compounds. Figure 2.19 shows a typical method on the surface modification of MCM-41 with *N*-(2-aminoethyl)-3-(aminopropyl)trimethoxysilane and the immobilization of the metal precursor [169]. The modification is due to the condensation of the (methoxysilyl)-propylamine with the accessible silanols on MCM-41. The condensation reaction

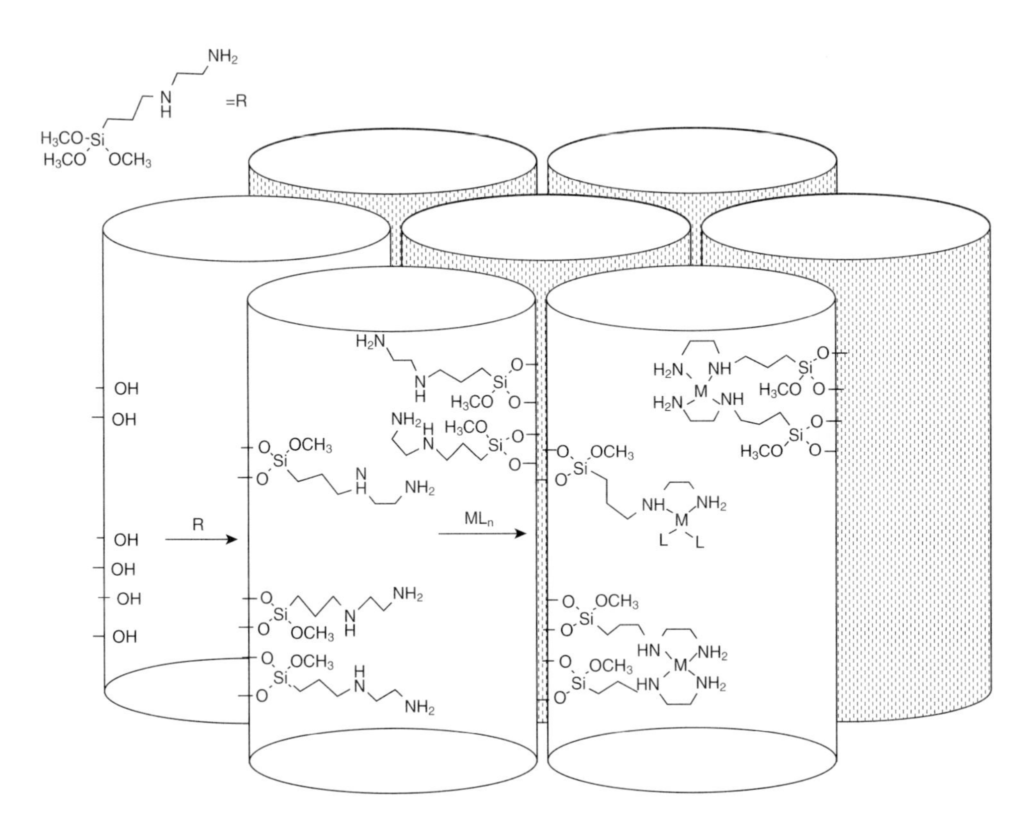

Figure 2.19 Schematic surface modification of MCM-41 with the amine ligand and immobilization of metal precursor in MCM-41. Adopted with permission from Ref. [169]. Copyright 1997 American Chemical Society.

modifies both external and inner channel silanols. However, internal functionalization of the channel walls should outnumber the external sites due to the high surface area of MCM-41. After the immobilization of the metal precursors, metal nanoparticles or nanowires could be synthesized by reducing the metal precursor. Besides amine, other ligands such as amino acids and alkanethiols had also been grafted on mesoporous silica [170].

Different from the surface modification of mesoporous silica with ligands, Xie *et al.* [171] reported the specific modification of the internal channel of SBA-15. To prevent formation of big Ag or Au nanoparticles on the external surface of SBA-15, the external silanols in the mesopores of SBA-15 were firstly modified with hexamethyldisilazane and then the internal surface of channels was grafted with 3-aminopropyl groups through the condensation of 3-aminopropyltriethoxysilane with the accessible silanols located in the internal surface of channels. Subsequently, the amine groups were converted into imines or hemiaminals by reacting them with CH_3CHO or HCHO, and then SBA-15 with internal ligands of imines or hemiaminals was prepared. It was found that Ag and Au nanoparticles or nanorods located within the channels of SBA-15 were synthesized by *in situ* reduction of the coordinated gold or silver precursors as shown in Figure 2.20. Figure 2.21 shows the typical TEM images and the size distribution of the Ag nanoparticles and nanorods located within the channels of the modified SBA-15. It was found that the loading and the aspect ratios of the nanostructures were simply tuned by changing the types of metal complexes, by varying the concentration of the solution, and/or by changing the reaction times [171].

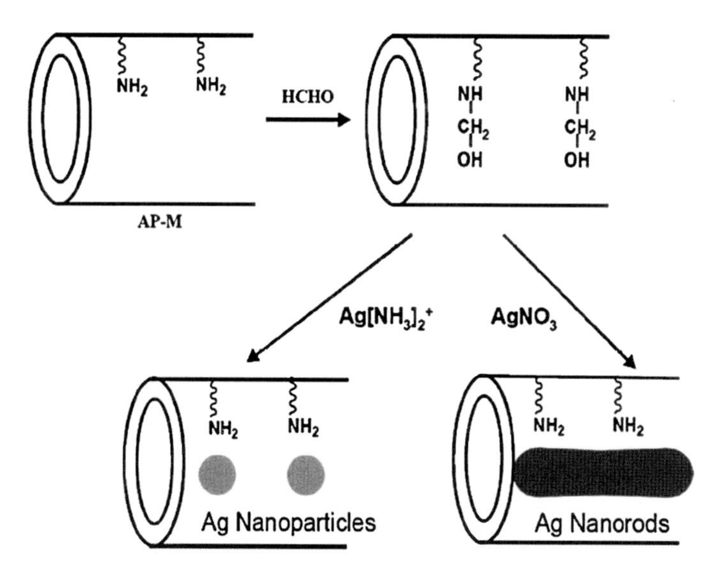

Figure 2.20 Schematic modification of the internal wall of SBA-15 and synthesis of Ag nanoparticles and nanorods within the channels of the internal wall modified SBA-15. Adopted with permission from Ref. [171]. Copyright 2008 American Chemical Society.

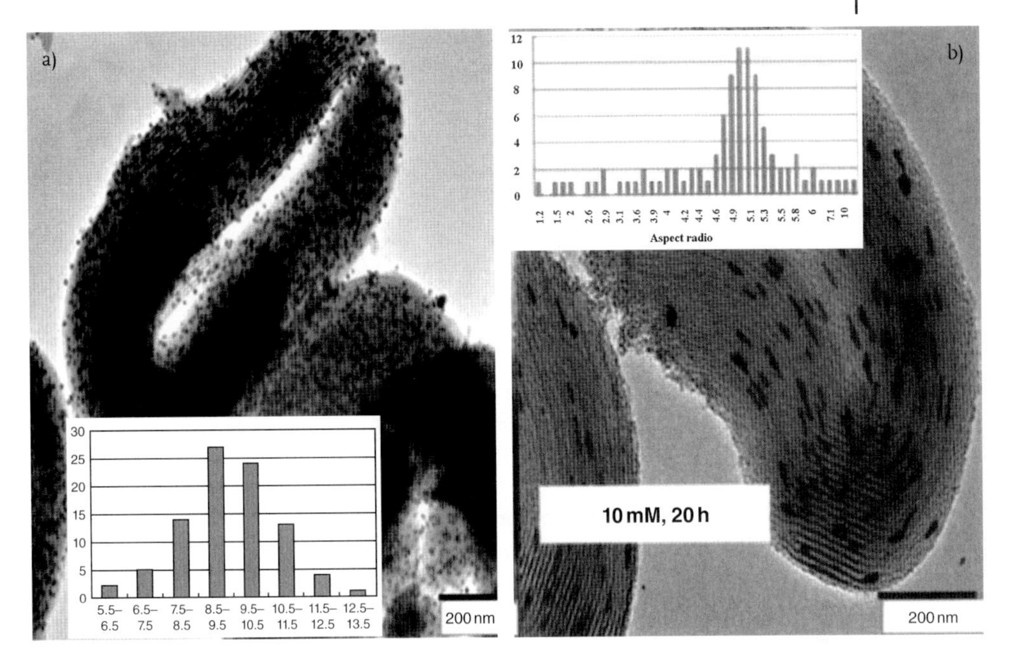

Figure 2.21 TEM images of the Ag nanoparticles (a) and Ag nanorods (b) in the internal wall modified SBA-15. Insets: Size distribution of the Ag nanoparticles. Adopted with permission from Ref. [171]. Copyright 2008 American Chemical Society.

2.3 Selective Catalysis of Supported Metal Nanoparticles

In the above-mentioned section, synthesis of size- and shape-controlled metal nanoparticles supported on various supports including polymeric and inorganic materials is introduced. For the metal nanocatalysts supported on polymeric materials, it has been demonstrated that some of these metal nanocatalysts have advantages including easy synthesis, good dispersion in the reaction mixture, and high catalytic efficiency. Whereas the polymeric supports are not as thermally stable as those of the inorganic ones at high temperature such as 150 °C, and some polymeric supports swell or deswell in solvent, which will greatly affect the catalysis of metal nanoparticles, and these make the metal nanocatalysts supported on polymeric materials not to be suitable model to explore the selective catalysis of metal nanoparticles. Therefore, in this section we focus on the selective catalysis of the metal nanoparticles supported on inorganic support.

For catalysis, one major focus is to improve the "activity" of the catalytic processes, which is generally expressed as turnover frequency (TOF). The other focus or challenge is to understand the molecular features of the catalytic structure that control and dictate the reaction selectivity. Benefiting from the emergence of some powerful characterization facilities, the surface structure, the electrical and thermal characterization of the supported metal nanoparticles can be detected by applying a wide range of surface-sensitive imaging (e.g., scanning tunneling microscopy, STM; TEM) and spectroscopic (e.g., X-ray photoelectron spectroscopy, XPS; infrared reflection absorption spectroscopy, IRAS; sum frequency generation, SFG; temperature programmed desorption, TPD) techniques. With the help of these facilities, many researchers including Freund and Rupprechter explored the correlation between the size and surface structure of the supported metal nanoparticles and their performance in catalysis [172–175]. Generally, the catalytic performance including the activity and selectivity of metal nanoparticles is dependent on the size, shape, and/or surface structure of metal nanoparticles. For supported metal nanocatalysts, the support or the interaction between the support and the active metal nanoparticles also plays an important role in the catalytic performance. In the following, these factors are briefly introduced.

2.3.1 Shape or Surface Structure Effect on Selective Hydrogenation of Cinnamaldehyde and Benzene

The success in shape-controlled synthesis of metal nanoparticles provides the potential of studying the effect of metal nanoparticles' shape on the catalytic activity and selectivity, and it is found that the shape or surface structure of metal nanoparticles can greatly affect the catalytic activity and selectivity at some cases [176]. For examples, hydrogenation [177–179], CO oxidation [180–182], Suzuki reactions [183], and electron-transfer reaction [184–186] have revealed the importance of controlling the shape of metal nanoparticles. Herein, the effect of the shape or surface structure of supported Pt nanoparticles on selective hydrogenation of cinnamaldehyde (CALD) and methanol decomposition are briefly introduced as typical examples.

The selective hydrogenation of carbonyl group is generally preferred because unsaturated alcohols are the valuable intermediates for the production of perfumes and pharmaceuticals. While the complete hydrogenation of both C=C and C=O groups to yield the primary alcohol or the selective hydrogenation of the C=C bond to give the saturated aldehyde is often obtained when a usual nanocatalyst is used. For example, low selectivity toward cinnamyl alcohol (CALC) is generally obtained with general supported Pt catalysts, although it can be increased by using different approaches such as an increment of the platinum particle size and by the use of a reducible oxide such as TiO_2 as support [187–189]. Recently, Serrano-Ruiz *et al.* [190] synthesized carbon-supported polyoriented Pt nanoparticles (Pt/C, 3 nm, Figure 2.22a), (100) preferentially oriented Pt nanoparticles (Pt[100]/C, 10 nm, Figure 2.22b), and (111) preferentially oriented Pt nanoparticles (Pt[111]/C, 10 nm, Figure 2.22c). The presence of the cubic nanoparticles in Figure 2.22b is an indication of the higher presence of Pt(100) surface atoms. Whereas the existence of tetrahedral and hexagonal Pt nanoparticles in Figure 2.22c is again evidence of a higher presence of Pt(111) surface atoms. The hydrogenation of CALD leads to the formation of saturated hidrocinnamaldehyde (HALD) from

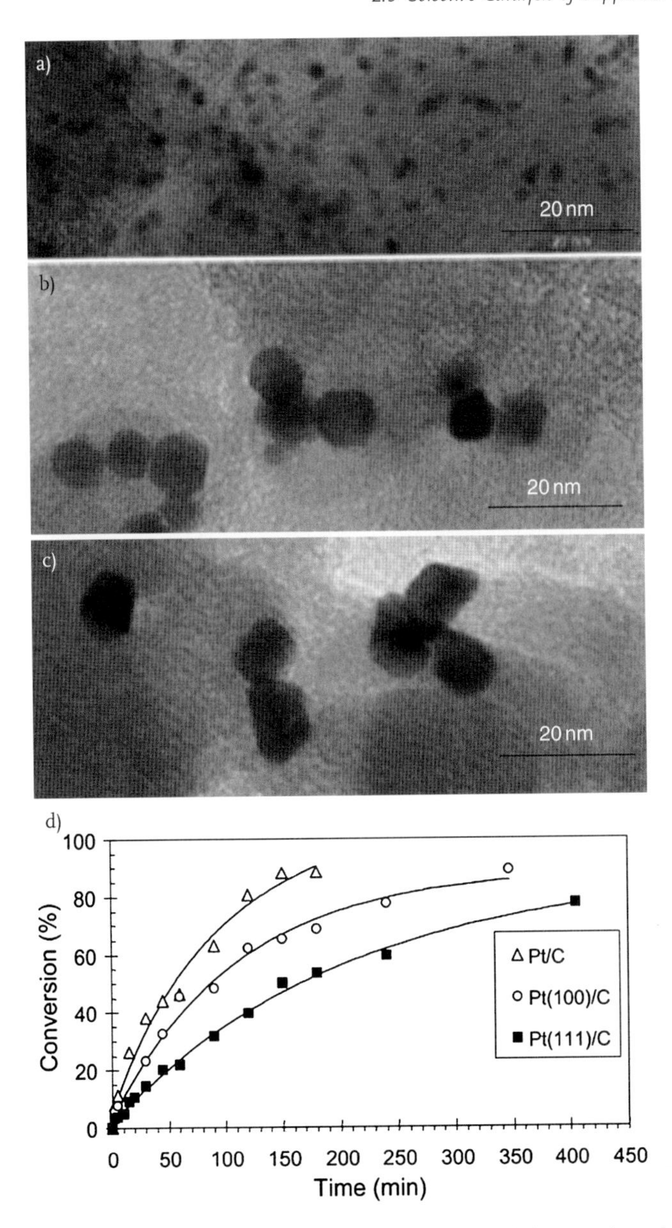

Figure 2.22 TEM images of the carbon-supported polyoriented Pt nanoparticles of Pt/C (a), and preferentially oriented Pt nanoparticles of Pt(100)/C (b) and Pt(111)/C (c), and the time-dependent conversion of CALD catalyzed the three supported Pt nanocatalysts. Adopted with permission from Ref. [190]. Copyright 2008 Elsevier.

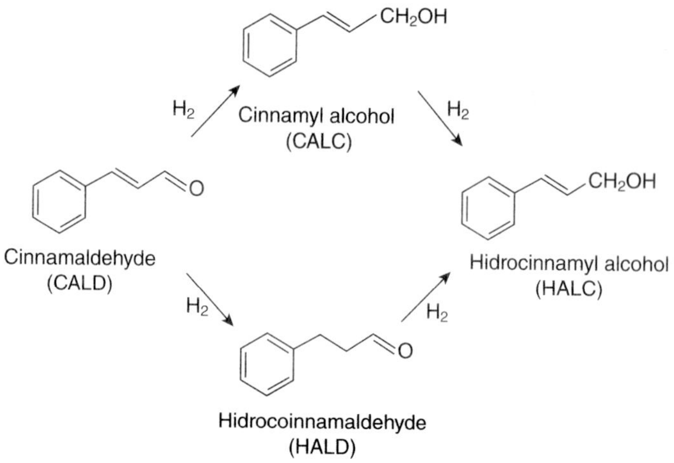

CH_2OH

H_2 Cinnamyl alcohol (CALC) H_2

CH_2OH

Cinnamaldehyde (CALD)

Hidrocinnamyl alcohol (HALC)

H_2 H_2

Hidrocoinnamaldehyde (HALD)

Scheme 2.1 Hydrogenation of cinnamaldehyde.

hydrogenation of the conjugated C=C bond and of CALC from hydrogenation of the C=O bond through two parallel reactions (Scheme 2.1). It was found that both the rate constant and the conversion of CALD reached for similar reaction times (Figure 2.22d) followed the trend Pt/C > Pt(100)/C > Pt(111)/C, and the normalized TOF value followed the trend Pt(100)/C > Pt(111)/C > Pt/C. As to the selectivity, it was found that the 3-nm polycrystalline Pt/C catalyst showed much lower selectivity to CALC than the 10-nm preferentially oriented catalysts, suggesting that metal particle size might have a strong influence on the selectivity to CALC. Repulsive interactions between the phenyl ring and the metal crystallites were proposed to explain the size effect [191]. That is, higher metal crystallites hindered the interaction via the C=C bond and enhanced the interaction via C=O bond and therefore gave higher selectivity to CALC. Furthermore, it was found that the 10-nm Pt(111)/C catalyst showed higher selectivity to CALC than the 10-nm Pt(100)/C catalyst (48% vs. 36%), indicating that the amount of Pt(111) surface atoms played a crucial role in increasing the selectivity to unsaturated alcohol.

The dependence of the selectivity of benzene hydrogenation on the shape of Pt nanoparticles was studied by Bratlie *et al.* [179]. Cubic and cubo-octahedra Pt nanoparticles with the almost same size but with different surface structure stabilized by tetradecyltrimethylammonium bromide (TTAB) were prepared. The cubic Pt nanoparticles consist of only Pt(100) surface, whereas the cubo-octahedra ones consist of both Pt(100) and Pt(111) surface. It was found that benzene hydrogenation turned out to produce two molecules of cyclohexane and cyclohexene on the Pt(111) surface but only one molecule of cyclohexene on the Pt(100) face. The selective hydrogenations of benzene and the TOF value of the cubic and cubo-octahedra Pt nanoparticles are summarized in Figure 2.23.

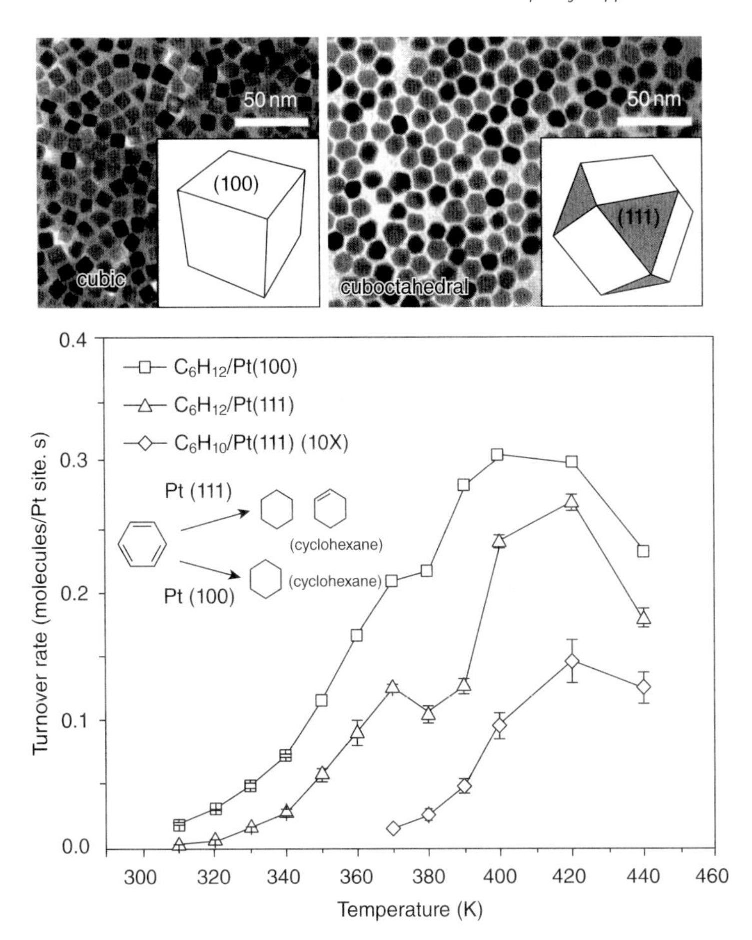

Figure 2.23 The TEM images of tetradecyltrimethylammonium bromide-stabilized cubic and cubo-octahedra Pt nanoparticles and turnover rates of cyclohexane (C_6H_{12}) and cyclohexene (C_6H_{10}) formation in the benzene hydrogenation on Pt(100) and Pt(111) surfaces under 10 Torr C_6H_6, 100 Torr H_2, and 650 Torr Ar. Adopted with permission from Ref. [179]. Copyright 2009 American Chemical Society.

2.3.2 Shape or Surface Structure Effect on the Selective Decomposition of Methanol

To explore the effect of the exact catalytic cite or the surface structure of the supported metal nanoparticles on the selective catalysis, model catalysts such as Pd nanoparticles deposited on the surface of Al_2O_3/NiAl(110) or MgO(110) film are usually used to simplify the analysis. Figure 2.24a shows an STM image of such a model catalyst of cubo-octahedra Pd nanoparticles supported on Al_2O_3/NiAl(110) (Pd/Al_2O_3/NiAl[110]) [192, 193]. The Pd nanoparticles are crystalline and expose a

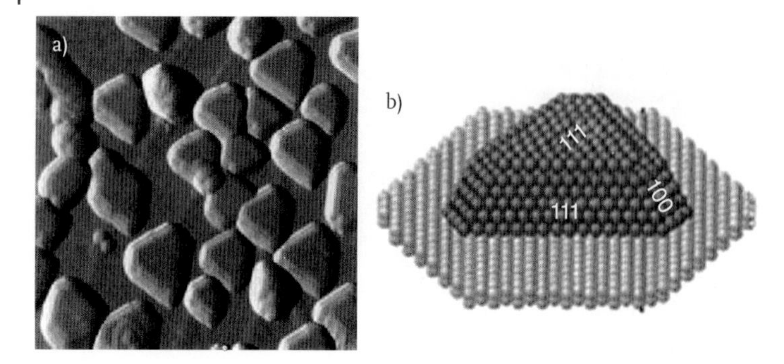

Figure 2.24 STM image (a) and the schematic representation (b) of the model nanocatalyst of Al_2O_3/NiAl(110). Adopted with permission from Ref. [193]. Copyright 2003 Elsevier.

(111) facet on its top. Also, on the side, (111) facets, typical for cubo-octahedra particles, can be discerned. The cluster on the oxide support is schematically represented in Figure 2.24b.

For the model reaction of methanol decomposition, two competing decomposition pathways exist, wherein the dehydrogenation to CO represents the dominant reaction channel and the slow carbon–oxygen bond breakage leads to the formation of adsorbed carbon and CH_x species (Figure 2.25) [172]. For the first reaction channel, CO adsorbed on the cubo-octahedra Pd nanoparticles, whose STM image is also shown in Figure 2.25, can be monitored via the *in situ* time-resolved reflection absorption IR spectroscopy (TR-RAIRS) ($\nu = 1900\,cm^{-1}$, $1840\,cm^{-1}$ due to C=O). As to the second pathway, the slow cleavage of the carbon–oxygen bond leading to the formation of adsorbed carbon and CH_x species ($\nu = 2945\,cm^{-1}$, $2830\,cm^{-1}$) can also be detected by TR-RAIRS.

Based on the TR-RAIRS spectra shown in Figure 2.25, it was concluded that the adsorption of CO at the cubo-octahedra Pd nanoparticle defect sites (i.e., steps and edges) was blocked by carbon, suggesting that the carbon–oxygen bond breakage was preferentially accumulated at the Pd particle defect sites. Further based on the surface fraction covered by carbon ($\theta_C(t) = \theta_{CO}(0) - \theta_{CO}(t)$) as a function of exposure time to methanol, it was concluded that the initial rate of carbon formation was high, but dropped rapidly with increasing carbon coverage. The kinetic molecular beam study involving isotope exchange experiments further showed that whereas the rate of carbon–oxygen bond breakage drastically decreased with increasing carbon coverage, the rate constant for CO exchange remained nearly unaffected by this process. Furthermore, the rate of the dehydrogenation and carbon–oxygen bond breakage, R_{CO}/R_C, increased from 30 on the pristine sample to approximately 1000 on the carbon-contaminated sample, as indicated in Figure 2.26. Therefore, it was concluded that the carbon–oxygen bond breakage preferentially occurred at the defect sites of the cubo-octahedra Pd nanoparticles such as edges and steps and the Pd(100) facets, and meanwhile the dehydrogenation to CO at the carbon-free Pd(111) facets.

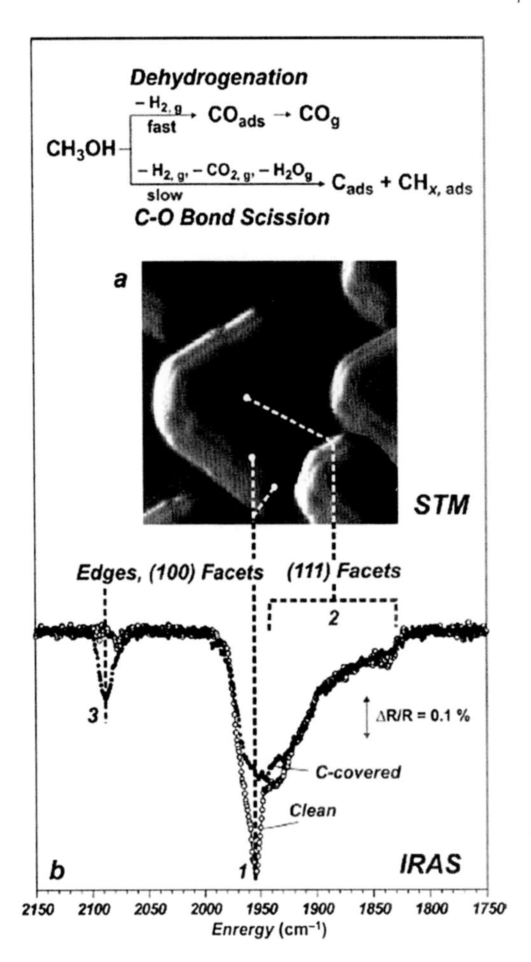

Figure 2.25 Two competing reaction pathways for methanol decomposition on the model catalyst of $Pd/Al_2O_3/NiAl(110)$: fast dehydrogenation and slow C–O bond scission. (a) STM image of the $Pd/Al_2O_3/NiAl(110)$ model catalyst (15×15 nm); (b) IR reflection absorption spectra for CO adsorbed on the $Pd/Al_2O_3/NiAl(110)$ catalyst (sample temperature 100 K, after CO exposure at 300 K). Open symbols for clean Pd nanoparticles immediately after preparation; solid symbols for partially C-covered Pd nanoparticles after prolonged exposure to methanol at 440 K. Adopted with permission from Ref. [172]. Copyright 2003 Elsevier.

2.3.3
Size Effect on the Selective Hydrogenation of 1,3-Butadiene and Pyrrole

Attempts to determine how the activity and/or the selectivity depend on the metal nanoparticles size have been undertaken for many decades. In 1960s, Boudart proposed a definition for structure sensitivity [194]. A heterogeneously catalyzed reaction was considered to be structure sensitive if the TOF value depended on

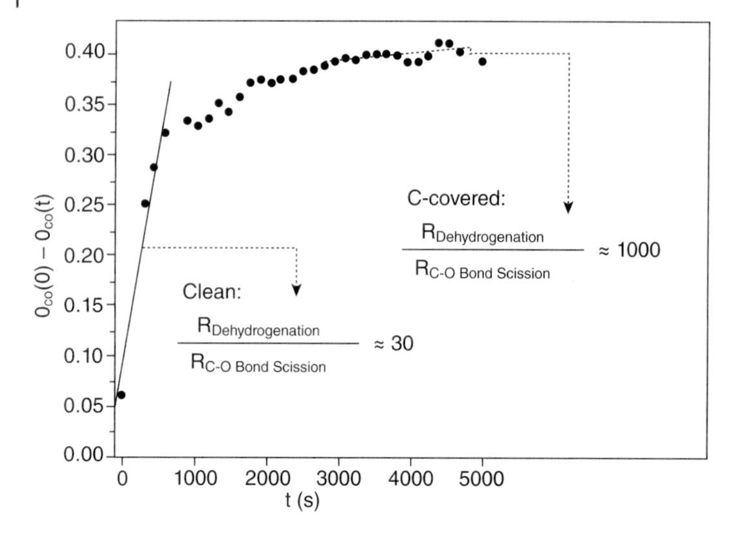

Figure 2.26 Carbon coverage on the cubo-octahedra Pd nanoparticles deposited on Al_2O_3/NiAl(110) as a function of exposure time to the methanol beam and the ratio of the rates of dehydrogenation and C–O bond scission. Adopted with permission from Ref. [172]. Copyright 2003 Elsevier.

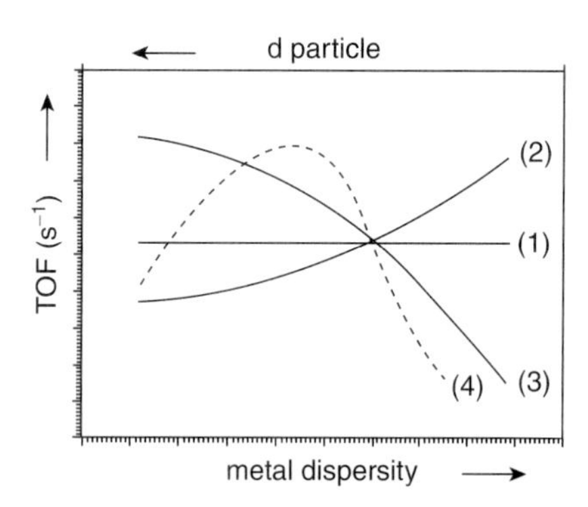

Figure 2.27 Structure-insensitive reaction (1), sympathetic (2), antipathetic structure sensitivity (3), and TOF reaching a maximum with varying particle size (4). Adopted with permission from Ref. [197]. Copyright 1989 Elsevier.

the particle size of the active component or a specific crystallographic orientation of the exposed catalyst surface. Boudart later expanded this model and proposed that structure sensitivity was related to the number of surface atoms of metal nanoparticles [195]. According to Bond [196], structure-sensitive reactions can be divided into four categories, thereby broadening the original classification built by Boudart [194], that is, TOFs can either be independent of particle size (structure-insensitive reactions), increase (antipathetic structure sensitivity), or decrease (sympathetic structure sensitivity) with growing particle size, or cross a maximum as shown in Figure 2.27 [197].

Till now, various structure-sensitive reactions such as hydrogenation of olefin [190, 198–200], CO oxidation [201, 202], alcohol oxidation [26, 203], and C–C coupling reaction [204], etc., have been reported. Herein, the size of supported Pd nanoparticles on the selective semihydrogenation of 1,3-butadiene is briefly introduced as a typical example. Due to its practical importance and theoretical significance, the selective semihydrogenation of 1,3-butadiene was extensively studied in the last decade, and palladium-based catalysts were frequently used for the selective hydrogenation of 1,3-butadiene [205, 206].

Semihydrogenation of 1,3-butadiene employing Pd catalyst can take place by either 1,2 or 1,4 addition to produce 1-butene or 2-butene, respectively. The proposed reaction intermediates and the reaction routes suggested by Wells *et al.* [207, 208] are shown in Scheme 2.2. The *trans:cis* ratio in the range of 8–12 indicates that the interconversion between the conformers of 1,3-butadiene and that of the surface adsorbed species 8 and 9 are limited. The *trans*-isomer, consequently, is formed from the dominant, more stable *S-trans* conformer through 1,4 addition with the involvement of π-allyl intermediate 11. The *S-cis* conformer, in turn, gives π-allyl intermediate 10 and ultimately yields *cis*-2-butene.

Boitiaux *et al.* [209] have proposed another interpretation as shown in Scheme 2.3. It assumes that the conformational interconversion of diene species is slow on all metals in comparison to hydrogenation and the adsorbed diene mimics the concentrations of *S-trans-* and *S-cis*-1,3-butadiene in the gas phase. Their

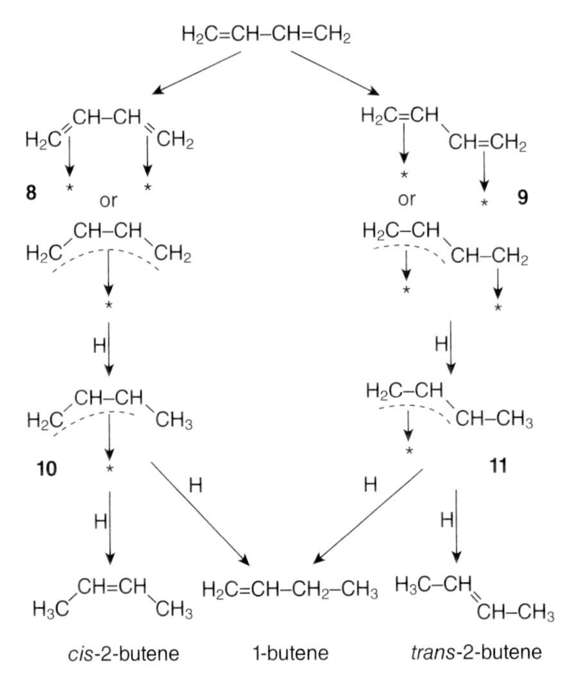

Scheme 2.2 One possible mechanism of hydrogenation of 1,3-butadiene proposed by Wells.

Scheme 2.3 One possible mechanism of hydrogenation of 1,3-butadiene proposed by Boitiaux.

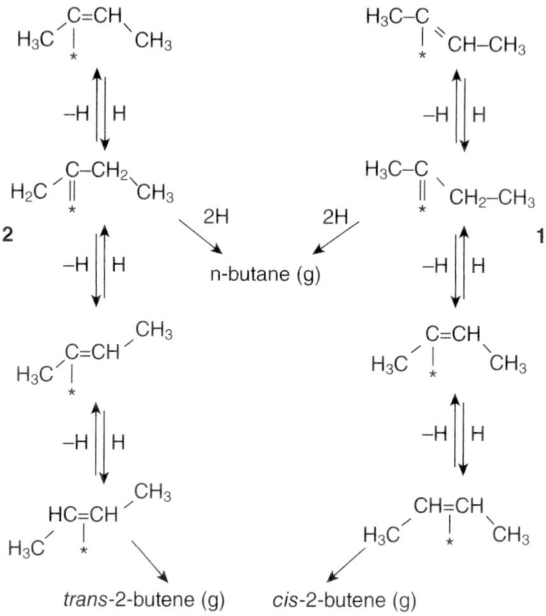

mechanism includes formation of carbine species (12 and 13) to explain the formation of *n*-butane as a direct product of 1,3-butadiene hydrogenation and the comparable *cis-* and *trans-*selectivity for 2-butenes over group VIII metals. According to this scheme, the metal, for example, Pd, does not promote formation of intermediate carbene, neither allows high conformational interconversion between *trans-* and *cis-*species nor hydrogenates directly to *n*-butane.

It is found that the hydrogenation of 1,3-butadiene is structure sensitive, that is, the activity of the supported Pd nanoparticles is correlated to their size and surface structure. In fact, the sites of threefold symmetry on Pd(111) favor multiple bonding, and apparently they exhibit very low value for hydrogen-sticking probability. Hydrogenation of 1,3-butadiene has been documented to be face sensitive and the activity increases in the order Pd(111) < Pd(100) < Pd(110) [210, 211]. Rupprechter *et al.* [174, 212] studied the Pd nanoparticles size effect on the selective hydrogenation of 1,3-butadiene employing the model catalyst of Pd/Al_2O_3/NiAl(110) [192, 193]. Pd nanoparticles were deposited on film of Al_2O_3/NiAl(110) by electron beam evaporation at 90 and 303 K. Through varying the amount of Pd, 2.1–4.4 and 4.2–7.7-nm Pd nanoparticles were deposited on the support at 90 K and 303 K, respectively. High-resolution STM images indicated that larger nanoparticles had a truncated cubo-octahedra shape with sometimes incomplete (111) facets (comprising 80% of the surface) and with (100) side facets. For small Pd particles (~2–3-nm mean size), they did not exhibit well-developed facets (the

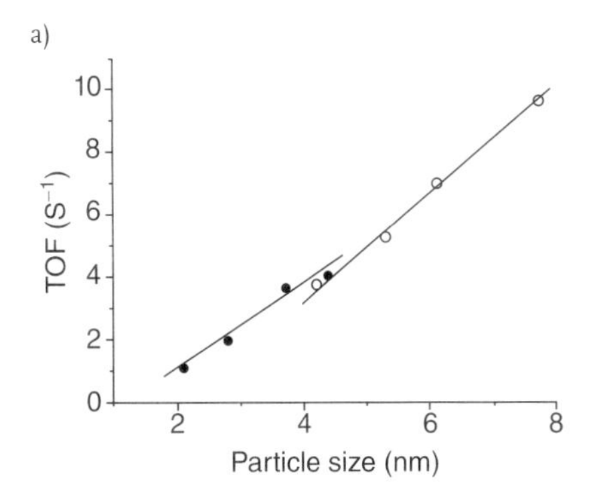

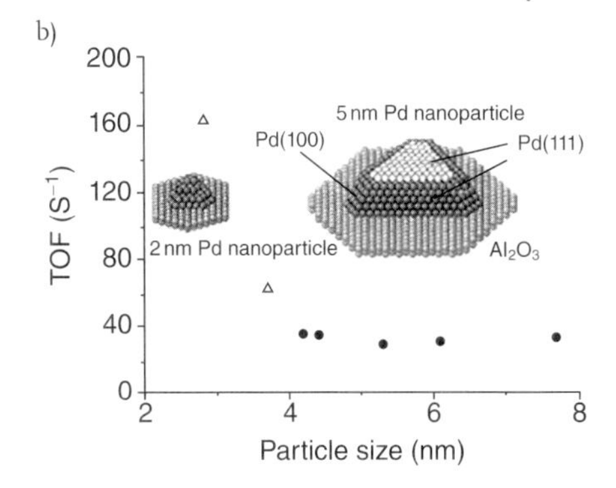

Figure 2.28 TOF for the two Pd/Al_2O_3/NiAl(110) catalysts as a function of mean particle size, (a) normalized by the total number of Pd surface atoms, and (b) normalized by using the number of Pd atoms on (111) facets, using a truncated cubo-octahedra as the structural model (shown as inset). Adopted with permission from Ref. [212]. Copyright 2006 Elsevier.

"facets" typically contain only 4–8 atoms). It was found that the TOF increased linearly with particle size when the total number of Pd surface atoms was used for normalization (Figure 2.28a). However, when the number of Pd surface atoms in the incomplete (111) facets was used for normalization, the TOF of butadiene hydrogenation was clearly particle size independent as shown in Figure 2.28b (just for Pd nanoparticles with size larger than 4 nm; for Pd nanoparticles with size smaller than 4 nm the abundance of the surface defects of low-coordinated sites might give rise to greater-than-expected activity).

As to the size effect on selectivity, an increase in size of the supported Pd nanoparticles from 2 to 8 nm induced a small increase in the selectivity toward 1-butene, suggesting that on larger Pd nanoparticles, 1,2 hydrogen addition was faster than 1,4 addition.

Kuhn *et al.* initially prepared Pt nanoparticles between 0.8 and 5 nm stabilized by polyamidoamine (PAMAM) dendrimers or PVP and then immobilized them on SBA-15 [213]. It was found that the catalytic selectivity during the multipath pyrrole hydrogenation was dependent on the size of the immobilized Pt nanoparticles as shown in Figure 2.29. For Pt nanoparticles smaller than 2 nm, the selectivity to pyrrolidine and *n*-butylamine was a strong function of the size of supported Pt nanoparticles as indicated by the easier formation of pyrrolidine over smaller Pt nanoparticles. As Pt nanoparticles size increased above 2 nm, behavior became independent of size with *n*-butylamine selectivity approaching 100%.

2.3.4
Support Effect on the Selective Catalysis

Besides the size, shape, or surface structure of supported metal nanoparticles as discussed above, the support itself also exerts obvious influence on the selective

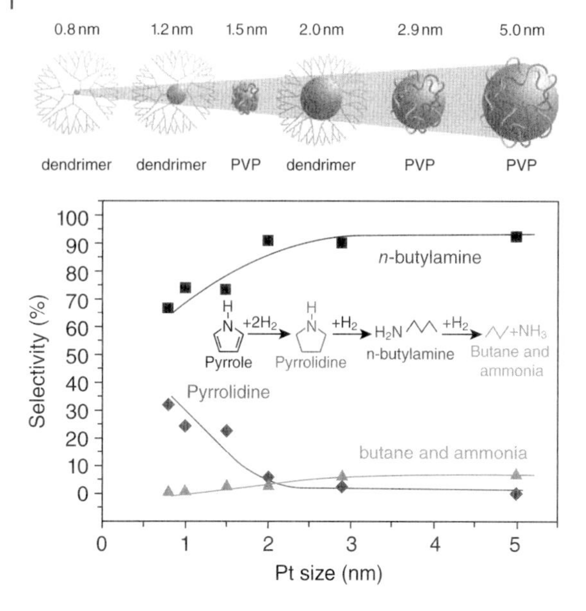

Figure 2.29 Schematic control on Pt nanoparticles size, and dependence of the selectivity of the pyrrole hydrogenation (4 Torr pyrrole, 400 Torr H_2, 413 K) on the size of the immobilized Pt nanoparticles. Adopted with permission from Ref. [213]. Copyright 2009 American Chemical Society.

catalysis, especially in the CO oxidation [180–182], in the hydrogenation of 1,3-butadiene [174, 205–212], in the hydrogenation of α,β-unsaturated aldehydes/ketones [187–191], and in the aerobic oxidation of alcohol employing the supported Au or Pd nanocatalysts [26, 203]. The strong metal–support interaction (SMSI) effect was first described in 1978 by Tauster, which was observed as a severely negative effect on CO and H_2 uptake on the catalyst after high-temperature calcinations [214]. Over the years after the first discovery of SMSI, it has been recognized that the support effect is not always negative. For instance, it appears to have a positive effect on the CO–H_2 reaction [215, 216]. A number of models have been proposed to explain this phenomenon, but there are two main proposals that the effect is due to an electronic perturbation of the metal function generally named charge transfer effect [214], and/or the effect is due to encapsulation by some form of the support which acts as a site blocker, a poison [217]. A number of researchers have shown that such encapsulation appears to happen both by indirect methods and direct imaging [218–220]. For example, Pesty *et al.* prepared a model catalyst of Pt nanoparticles deposited on single-crystal $TiO_2(110)$, and XPS analysis indicated an adsorbed layer of partially reduced titania species (denoted as TiO_x, with $1 < x < 2$) on the surface of the Pt nanoparticles [218]; therefore the catalysis of the supported Pt nanoparticles was modified. However, it should be pointed out that the catalysis of supported noble metal nanoparticles are not only dependent on the metal nanoparticles themselves such as size and shape of the nanoparticles, but the nature of the support materials, the preparation method, and even the activation procedure can also play a key role in the catalysis. Thus, the exact reason is relatively complex and the role of the support effect on selective catalysis is still under discussion.

For Au catalysis, since Haruta discovered a remarkable activity of supported Au nanoparticles in CO oxidation, it has received considerable interest [221]. Up to date, the catalysis origin of Au nanoparticles has been extensively debated in the literature, and a large number of explanations have been proposed. It is deemed that the catalytic activity of Au nanoparticles is ascribed to the nature of surface Au atoms, the interaction between the Au nanoparticles and the support, and/or the combination of the two effects. Unfortunately, there is still no clear picture with respect to the origin of Au catalysis, and often the results reported in literatures concerning Au catalysis are contradictory. Herein, the support effect on Au catalysis in selective hydrogenation of α,β-unsaturated ketone (UK) to α,β-unsaturated alcohol (UA) similar to those as shown in Scheme 2.1 is briefly introduced.

Before the introduction on support effect, the nature of Au nanoparticles on the selective hydrogenation of UK to UA is initially introduced. It has been argued that the intrinsic selectivity of gold can be due to the adsorption mode of the conjugated systems on this metal. It is proposed that among the different adsorption modes of the conjugated systems, Au nanoparticles are likely the catalyst on which the 1,2-C=O adsorption mode is preferred over the 1,4-C=C–C=O mode [200, 222–224]. The scarcity of the 1,4-C=C–C=O species on Au nanoparticles, which are responsible for the formation of saturated aldehydes or ketones, is in accordance with the higher intrinsic selectivity of Au nanocatalysts toward the formation of UA. However, it should be pointed out that the literature data on the effect of Au nanoparticles size on selective hydrogenation vary and are somewhat controversial. Claus *et al.* reported that the selectivity to crotyl alcohol in hydrogenation of acrolein employing the Au/TiO_2 and Au/ZrO_2 catalysts increased with increasing the size of Au nanoparticles [200, 224]. These authors concluded that the C=O group was preferentially activated on gold sites of high coordination (face atoms), whereas sites of low coordination, mostly present on small particles as corners and edges, strongly favored the activation of the C=C group.

To demonstrate the support effect on the selective hydrogenation of benzalacetone, Milone *et al.* deposited Au nanoparticles on the supports of goethite (FeO[OH]), maghemite (γFe_2O_3), hematite (αFe_2O_3), and iron oxyhydroxide [222]. It was found that the selectivity toward hydrogenation of the conjugated C=O bond did not depend on the size of Au nanoparticles but varied with the structural characteristics of the support ranking in the order of FeO(OH) > iron oxyhydroxide > γFe_2O_3 >> αFe_2O_3. That is, the selectivity toward the formation of UA increases with increasing reducibility of the support. It was suggested that an electron transfer from the reduced support to the metal created more electron-enriched Au nanoparticles on which back-bonding with the π^* C=O orbital was favored and therefore hydrogenation of the C=O group was faster than that of the C=C group.

2.4 Summary

A general survey on the preparative methodologies for the size- and shape-controlled synthesis of supported metal nanoparticles on polymeric materials, carbon nanotubes, metal oxides, as well as mesoporous silica is made and a special emphasis is focused on those methodologies employing polymers such as single polymer chains, copolymer micelles, the CFPs, the core–shell polymeric microspheres, and the hollow polymeric microspheres.

The shape, the surface structure, and the size of the supported metal nanoparticles and the support effect on the selective catalysis in hydrogenation of α,β-unsaturated aldehyde, 1,3-butadiene, benzene and pyrrole, and decomposition of methanol are introduced.

References

1 Lewis, L.N. (1993) *Chem. Rev.*, **93**, 2693.
2 Roucoux, A., Schulz, J., and Patin, H. (2002) *Chem. Rev.*, **102**, 3757.
3 Min, B.K. and Friend, C.M. (2007) *Chem. Rev.*, **107**, 2709.
4 Arcadi, A. (2008) *Chem. Rev.*, **108**, 3266.
5 Dioos, B.M.L., Ivo, F., Vankelecom, J., and Jacobs, P.A. (2006) *Adv. Synth. Catal.*, **348**, 1413.
6 Kralik, M. and Biffis, A. (2001) *J. Mol. Catal. A*, **177**, 113.
7 Akagi, T., Baba, M., and Akashi, M. (2007) *Polymer*, **48**, 6729.
8 Risse, T., Shaikhutdinov, S., Nilius, N., Sterrer, M., and Freund, H.-J. (2008) *Acc. Chem. Res.*, **41**, 949.
9 Chen, M.S. and Goodman, D.W. (2008) *Chem. Soc. Rev.*, **37**, 1860.
10 Lee, K., Zhang, J.J., Wang, H.J., and Wilkinson, D.P. (2006) *J. Appl. Electrochem.*, **36**, 507.
11 Stein, A. (2003) *Adv. Mater.*, **15**, 763.
12 Bronstein, L.M. (2003) *Top. Curr. Chem.*, **266**, 55.
13 Kaldor, A., Cox, D., and Zakin, M.R. (1988) *Adv. Chem. Phys.*, **70**, 211.
14 Cushing, B.L., Kolesnichenko, V.L., and Connor, C.J.O. (2004) *Chem. Rev.*, **104**, 3893.
15 Schmid, G. (2004) *Nanoparticles: From Theory to Applications*, Wiley-VCH Verlag GmbH, Weinheim.
16 Turkevich, J., Stevenson, P.C., and Hillier, J. (1951) *Discuss. Faraday Soc.*, **11**, 55.
17 Turkevich, J. and Kim, G. (1970) *Science*, **169**, 873.
18 Turkevich, J. (1985) *Gold Bull.*, **18**, 86.
19 Muhlpfordt, H. (1982) *Expirentia*, **38**, 1127.
20 Brust, M., Walker, M., Bethell, D., Schiffrin, D.J., and Whyman, R. (1994) *Chem. Commun.*, 801.
21 Maase, M. (1999) *Ph.D. Thesis*, Verlag Mainz, Aachen, ISBN: 3-89653-463-7.
22 Hirai, H., Nakao, Y., Toshima, N., and Adachi, K. (1976) *Chem. Lett.*, **5**, 905.
23 Toshima, N., Harada, M., Yonezawa, T., Kushihashi, K., and Asakura, K. (1991) *J. Phys. Chem.*, **95**, 7448.
24 Silvert, P.-Y. and Elhsissen, K.T. (1995) *Solid State Ionics*, **82**, 53.
25 Sales, E.A., Benhamida, B., Caizergues, V., Lagier, J.-P., Fievet, F., and Verduraz, F.B. (1998) *Appl. Catal. A Gen.*, **172**, 273.
26 Tsunoyama, H., Sakurai, H., Negishi, Y., and Tsukuda, T. (2005) *J. Am. Chem. Soc.*, **127**, 9374.
27 Sanguesa, C.D., Urbina, R.H., and Figlarz, M. (1993) *Solid State Ionics*, **63–65**, 25.
28 Elhsissen, T.K., Bonet, F., Grugeon, S., Lambert, S., and Urbina, R.H. (1999) *J. Mater. Res.*, **14**, 3707.
29 Bonet, F., Delmas, V., Grugeon, S., Urbina, R.H., Silvert, P.-Y., and

Elhsissen, K.T. (1999) *Nanostruct. Mater.*, **11**, 1277.
30 Viau, G., Brayner, R., Poul, L., Chakroune, N., Lacaze, E., Vincent, F.F., and Fievet, F. (2003) *Chem. Mater.*, **15**, 486.
31 Elhsissen, T.K., Bonet, F., Silvert, P.-Y., and Urbina, R.H. (1999) *J. Alloy. Comp.*, **292**, 96.
32 Piquemal, J.Y., Viau, G., Beaunier, P., Verduraz, F.B., and Fievet, F. (2003) *Mater. Res. Bull.*, **38**, 389.
33 Deivaraj, T.C., Lalla, N.L., and Lee, J.Y. (2005) *J. Colloid Interface Sci.*, **289**, 402.
34 Umar, A.A. and Oyama, M. (2006) *Cryst. Growth Des.*, **6**, 818.
35 Zhou, M., Chen, S., and Zhao, S. (2006) *J. Phys. Chem. B*, **110**, 4510.
36 Hoppe, C.E., Lazzari, M., Blanco, I.P., and Quintela, M.A.L. (2006) *Langmuir*, **22**, 7027.
37 Chen, J., Herricks, T., and Xia, Y. (2005) *Angew. Chem. Int. Ed.*, **44**, 2589.
38 Chen, J., Herricks, T., Geissler, M., and Xia, Y. (2004) *J. Am. Chem. Soc.*, **126**, 10854.
39 Lee, E., Chen, J., Yin, Y., Campbell, C., and Xia, Y. (2006) *Adv. Mater.*, **18**, 3271.
40 Teranishi, T. and Miyake, M. (1998) *Chem. Mater.*, **10**, 594.
41 Rioux, R.M., Song, H., Hoefelmeyer, J.D., Yang, P., and Somorjai, G.A. (2005) *J. Phys. Chem. B.*, **109**, 2192.
42 Frens, G. (1973) *Nat. Phys. Sci.*, **241**, 20.
43 Chen, S., Huang, K., and Stearns, J.A. (2000) *Chem. Mater.*, **12**, 540.
44 Zanella, R., Giorgio, S., Henry, C.R., and Louis, C. (2002) *J. Phys. Chem. B.*, **106**, 7634.
45 Zanella, R. and Louis, C. (2005) *Catal. Today*, **107**, 768.
46 Zanella, R., Giorgio, S., Shin, C.H., Henry, C.R., and Louis, C. (2004) *J. Catal.*, **222**, 357.
47 Zanella, R., Louis, C., Giorgio, S., and Touroude, R. (2004) *J. Catal.*, **223**, 328.
48 Sanguesa, C.D., Urbina, R.H., and Figlarz, M. (1999) *J. Solid State Chem.*, **100**, 272.
49 Silvert, P.-Y., Urbina, R.H., Duvauchelle, N., Vijayakrishnan, V., and Elhsissen, K.T. (1996) *J. Mater. Chem.*, **6**, 573.
50 Xiong, Y., Chen, J., Wiley, B., and Xia, Y. (2005) *J. Am. Chem. Soc.*, **127**, 7332.
51 Wiley, B., Herricks, T., Sun, Y., and Xia, Y. (2004) *Nano Lett.*, **4**, 1733.
52 Toshima, N., Shiraishi, Y., and Teranishi, T. (2001) *J. Mol. Catal. A*, **177**, 139.
53 Englebienne, P. and Van Hoonacker, A. (2005) *J. Colloid Interface Sci.*, **292**, 445.
54 Miyazaki, A. and Nakano, Y. (2000) *Langmuir*, **16**, 7109.
55 Kaliyappan, T. and Kannan, P. (2000) *Prog. Polym. Sci.*, **25**, 343.
56 Shan, J. and Tenhu, H. (2007) *Chem. Commun.*, 4580.
57 Wuelfing, W.P., Gross, S.M., Miles, D.T., and Murray, R.W. (1998) *J. Am. Chem. Soc.*, **120**, 12696.
58 Zhu, M., Wang, L., Exarhos, G.J., and Li, A.D.Q. (2004) *J. Am. Chem. Soc.*, **126**, 2656.
59 Corbierre, M.K., Cameron, N.S., and Lennox, R.B. (2004) *Langmuir*, **20**, 2867.
60 Shan, J., Nuopponen, M., Jiang, H., Kauppinen, E., and Tenhu, H. (2003) *Macromolecules*, **36**, 4526.
61 Wei, G., Wen, F., Zhang, X., Zhang, W., Jiang, X., Zheng, P., and Wei (2007) *J. Colloid Interface Sci.*, **316**, 53.
62 Wei, G., Zhang, W., Wen, F., Wang, Y., and Zhang, M. (2008) *J. Phys. Chem. C*, **112**, 10827.
63 Shan, J., Nuopponen, M., Jiang, H., Viitala, T., Kauppinen, E., Kontturi, K., and Tenhu, H. (2005) *Macromolecules*, **38**, 2918.
64 Zhang, L. and Eisenberg, A. (1995) *Science*, **268**, 1728.
65 Zhou, Z., Li, Z., Ren, Y., Hillmyer, M.A., and Lodge, T.P. (2003) *J. Am. Chem. Soc.*, **125**, 10182.
66 Borisov, O.V. and Zhulina, E.B. (2003) *Macromolecules*, **36**, 10029.
67 Tao, J., Stewart, S., Liu, G., and Yang, M. (1997) *Macromolecules*, **30**, 2738.
68 Zhang, W., Shi, L., An, Y., Gao, L., Wu, K., and Ma, R. (2004) *Macromolecules*, **37**, 2551.
69 Zhang, Q., Remsen, E.E., and Wooley, K.L. (2000) *J. Am. Chem. Soc.*, **122**, 3642.
70 Klingelhöfer, S., Heitz, W., Greiner, A., Oestreich, S., Förster, S., and Antonietti, M. (1997) *J. Am. Chem. Soc.*, **119**, 10116.
71 Noonan, K.J.T., Gillon, B.H., Cappello, V., and Gates, D.P. (2008) *J. Am. Chem. Soc.*, **130**, 12876.

72 Vriezema, D.M., Aragones, M.C., Elemans, J.A.A.W., Cornelissen, J.J.L.M., Rowan, A.E., and Nolte, R.J.M. (2005) *Chem. Rev.*, **105**, 1445.
73 Forster, S. and Antonietti, M. (1998) *Adv. Mater.*, **10**, 195.
74 Antonietti, M., Forster, S., Hartmann, J., and Oestreich, S. (1996) *Macromolecules*, **29**, 3800.
75 Antonietti, M., Wenz, E., Bronstein, L., and Seregina, M. (1995) *Adv. Mater.*, **174**, 795.
76 Seregina, M.V., Bronstein, L.M., Platonova, O.A., Chernyshov, D.M., Valetsky, P.M., Hartmann, J., Wenz, E., and Antonietti, M. (1997) *Chem. Mater.*, **9**, 923.
77 Bronstein, L.M., Chernyshov, D.M., Volkov, I.O., Ezernitskaya, M.G., Valetsky, P.M., Matveeva, V.G., and Sulmanz, E.M. (2000) *J. Catal.*, **196**, 302.
78 Sulman, E., Bodrova, Y., Matveeva, V., Semagina, N., Cerveny, L., Kurtc, V., Bronstein, L., Platonova, O., and Valetsky, P. (1999) *Appl. Catal. A Gen.*, **176**, 75.
79 Sulman, E., Matveeva, V., Usanov, A., Kosivtsov, Y., Demidenko, G., Bronstein, L., Chernyshov, D., and Valetsky, P. (1999) *J. Mol. Catal. A*, **146**, 265.
80 Okamoto, K., Akiyama, R., Yoshida, H., Yoshida, T., and Kobayashi, S. (2005) *J. Am. Chem. Soc.*, **127**, 2125.
81 Beletskaya, I.P., Kashin, A.N., Litvinov, A.E., Tyurin, V.S., Valetsky, P.M., and van Koten, G. (2006) *Organometallics*, **25**, 154.
82 Semagina, N., Joannet, E., Parra, S., Sulman, E., Renken, A., and Kiwi-Minsker, L. (2005) *Appl. Catal. A Gen.*, **280**, 141.
83 Underhill, R.S. and Liu, G. (2000) *Chem. Mater.*, **12**, 3633.
84 Semagina, N.V., Bykov, A.V., Sulman, E.M., Matveeva, V.G., Sidorov, S.N., Dubrovina, L.V., Valetsky, P.M., Kiselyova, O.I., Khokhlov, A.R., Stein, B., and Bronstein, L.M. (2004) *J. Mol. Catal. A*, **208**, 273.
85 Kuo, P.-L., Chen, C.-C., and Jao, M.-W. (2005) *J. Phys. Chem. B*, **109**, 9445.
86 Jaramillo, T.F., Baeck, S.-H., Cuenya, B.R., and McFarland, E.W. (2003) *J. Am. Chem. Soc.*, **125**, 7148.
87 Cuenya, B.R., Baeck, S.-H., Jaramillo, T.F., and McFarland, E.W. (2003) *J. Am. Chem. Soc.*, **125**, 12928.
88 Bronstein, L.M., Chernyshov, D.M., Timofeeva, G.I., Dubrovina, L.V., Valetsky, P.M., Obolonkova, E.S., and Khokhlov, A.R. (2000) *Langmuir*, **16**, 3626.
89 Kang, Y. and Taton, T.A. (2005) *Angew. Chem. Int. Ed.*, **44**, 409.
90 Zheng, P., Jiang, X., Zhang, X., Zhang, W., and Shi, L. (2006) *Langmuir*, **22**, 9393.
91 Wang, Y., Wei, G., Zhang, W., Jiang, X., Zheng, P., Shi, L., and Dong, A. (2007) *J. Mol. Catal. A*, **266**, 233.
92 Lee, M., Jang, C., and Ryu, J. (2004) *J. Am. Chem. Soc.*, **126**, 8082.
93 Guyot, A., Sherrington, D.C., and Hodge, P. (eds) (1988) *Synthesis and Separations Using Functional Polymers*, John Wiley & Sons, Inc., New York, p. 1.
94 Kesenci, K., Tuncel, A., and Piskin, E. (1996) *React. Funct. Polym.*, **31**, 137.
95 Wollner, J. and Neier, W. (1966) Bergbau und chemie (Homberg). German patent 1260454.
96 Hanson, D.L., Katzer, J.R., Gates, B.C., Schuit, G.C.A., and Harnsberger, H.F. (1974) *J. Catal.*, **32**, 204.
97 Sidorov, S.N., Volkov, I.V., Davankov, V.A., Tsyurupa, M.P., Valetsky, P.M., Bronstein, L.M., Karlinsey, R., Zwanziger, J.W., Matveeva, V.G., Sulman, E.M., Lakina, N.V., Wilder, E.A., and Spontak, R.J. (2001) *J. Am. Chem. Soc.*, **123**, 10502.
98 Sidorov, S.N., Bronstein, L.M., Davankov, V.A., Tsyurupa, M.P., Solodovnikov, S.P., and Valetsky, P.M. (1999) *Chem. Mater.*, **11**, 3210.
99 Ziolo, R.F., Giannelis, E.P., Weinstein, B.A., O'Horo, M.P., Ganguly, B.N., Mehrotra, V., Russell, M.W., and Huffman, D.R. (1992) *Science*, **257**, 219.
100 Winnik, F.M., Morneau, A., Ziolo, R.F., Stoever, H.D.H., and Li, W.-H. (1995) *Langmuir*, **11**, 3660.
101 Corain, B., Burato, C., Centomo, P., Lora, S., Meyer-Zaika, W., and Schmid, G. (2005) *J. Mol. Catal. A Chem*, **225**, 189.

102 Corain, B., Jerabek, K., Centomo, P., and Canton, P. (2004) *Angew. Chem. Int. Ed.*, **43**, 959.

103 Biffis, A., Orlandi, N., and Corain, B. (2003) *Adv. Mater.*, **15**, 1551.

104 Artuso, F., D'Archivio, A.A., Lora, S., Jerabek, K., Kralik, M., and Corain, B. (2003) *Chem. Eur. J.*, **9**, 5292.

105 Ding, S., Qian, W., Tan, Y., and Wang, Y. (2006) *Langmuir*, **22**, 7105.

106 Rodríguez-González, B., Salgueiriño-Maceira, V., García-Santamaría, F., and Liz-Marzán, L.M. (2002) *Nano Lett.*, **2**, 471.

107 Pozzar, F., Sassi, A., Pace, G., Lora, S., D'Archivio, A.A., Jerabek, K., Grassi, A., and Corain, B. (2005) *Chem. Eur. J.*, **11**, 7395.

108 Pears, D. and Smith, S.C. (2005) *Aldrichimica Acta*, **38**, 23.

109 Ogasawara, S. and Kato, S. (2010) *J. Am. Chem. Soc.*, **132**, 4608.

110 Porta, F., Prati, L., Rossi, M., and Scari, G. (2002) *J. Catal.*, **211**, 464.

111 Imizu, Y. and Klabunde, K.J. (1984) *Inorg. Chem.*, **23**, 3602.

112 Andrescu, D., Sau, T.K., and Goia, D.V. (2006) *J. Colloid Interface Sci.*, **298**, 742.

113 Fink, J., Kiely, C.J., Bethell, D., and Schiffrin, D.J. (1998) *Chem. Mater.*, **10**, 922.

114 Comotti, M., Della Pina, C., Matarrese, R., and Rossi, M. (2004) *Angew. Chem. Int. Ed.*, **43**, 5812.

115 Ozin, G.A. and Mitchell, S.A. (1983) *Angew. Chem. Int. Ed. Engl.*, **22**, 674.

116 Yu, W., Liu, H., and An, X. (1998) *J. Mol. Catal. A Chem.*, **129**, L9.

117 Yu, W., Liu, M., Liu, H., An, X., Liu, Z., and Ma, X. (1999) *J. Mol. Catal. A Chem.*, **142**, 201.

118 Ohtaki, M., Komiyama, M., Hirai, H., and Toshima, N. (1991) *Macromolecules*, **24**, 5567.

119 Zheng, P. and Zhang, W. (2007) *J. Catal.*, **250**, 324.

120 Wen, F., Zhang, W., Wei, G., Wang, Y., Zhang, J., Zhang, M., and Shi, L. (2008) *Chem. Mater.*, **20**, 2144.

121 Miao, S., Zhang, C., Liu, Z., Han, B., Xie, Y., Ding, S., and Yang, Z. (2008) *J. Phys. Chem. C*, **112**, 774.

122 Pozzar, F., Sassi, A., Pace, G., Lora, S., D'Archivio, A.A., Jerabek, K., Grassi, A., and Corain, B. (2005) *Chem. Eur. J.*, **11**, 7395.

123 Burato, C., Centomo, P., Pace, G., Favaro, M., Prati, L., and Corain, B. (2005) *J. Mol. Catal. A Chem.*, **238**, 26.

124 Wildgoose, G.G., Banks, C.E., and Compton, R.G. (2006) *Small*, **2**, 182.

125 Serp, P., Corrias, M., and Kalck, P. (2003) *Appl. Catal. A*, **253**, 337.

126 Quinn, B.M., Dekker, C., and Lemay, S.G. (2005) *J. Am. Chem. Soc.*, **127**, 6146.

127 Qu, J., Shen, Y., Qu, X., and Dong, S. (2004) *Chem. Commun.*, 34.

128 Day, T.M., Unwin, P.R., Wilson, N.R., and Macpherson, J.V. (2005) *J. Am. Chem. Soc.*, **127**, 10639.

129 Sivakumar, K., Lu, S., and Panchapakesan, B. (2004) *Mater. Res. Soc. Symp. Proc.*, **818**, 341.

130 He, Z., Chen, J., Liu, D., Zhou, H., and Kuang, Y. (2004) *Diamond Relat. Mater.*, **13**, 1764.

131 Guo, D.-J. and Li, H.-L. (2004) *J. Electroanal. Chem.*, **573**, 197.

132 Guo, D.-J. and Li, H.-L. (2004) *Electrochem. Commun.*, **6**, 999.

133 Guo, D.-J. and Li, H.-L. (2005) *J. Colloid Interface Sci.*, **286**, 274.

134 Kongkanand, A., Vinodgopal, K., Kuwabata, S., and Kamat, P.V. (2006) *J. Phys. Chem. B*, **110**, 16185.

135 Xu, Q., Kharas, K., and Datye, A.K. (2003) *Catal. Lett.*, **85**, 229.

136 Haruta, M., Tsubota, S., Kobayashi, T., Kageyama, H., Genet, M.J., and Delmon, B. (1993) *J. Catal.*, **144**, 175.

137 Tsubota, S., Haruta, M., Kobayashi, T., Ueda, A., and Nakahara, Y. (1991) *Stud. Surf. Sci. Catal.*, **63**, 695.

138 Haruta, M., Yamada, N., Kobayashi, T., and Iijima, S. (1989) *J. Catal.*, **115**, 301.

139 Okumura, M., Nakamura, S., Tsubota, S., Nakamura, T., Azuma, M., and Haratu, M. (1998) *Catal. Lett.*, **51**, 53.

140 Arrii, S., Morfin, F., Renouprez, A.J., and Rousset, J.L. (2004) *J. Am. Chem. Soc.*, **126**, 1199.

141 Kozlova, A.P., Sugiyama, S., Kozlov, A.I., Asakura, K., and Iwasawa, Y. (1998) *J. Catal.*, **176**, 426.

142 Okumura, M. and Haruta, M. (2000) *Chem. Lett.*, **29**, 396.

143 Kang, Y.M. and Wan, B.Z. (1995) *Catal. Today*, **26**, 59.

144 Baiker, A., Maciejewski, M., Tagliaferri, S., and Hug, P. (1995) *J. Catal.*, **151**, 407.
145 Zheng, N. and Stucky, G.D. (2006) *J. Am. Chem. Soc.*, **128**, 14278.
146 Pratsinis, S.E. (1998) *Prog. Energy Combust. Sci.*, **24**, 197.
147 Strobel, R., Stark, W.J., Makler, L., Pratsinis, S.E., and Baiker, A. (2003) *J. Catal.*, **213**, 296.
148 Thybo, S., Jensen, S., Johansen, J., Johannessen, T., Hansen, O., and Quaade, U.J. (2004) *J. Catal.*, **223**, 271.
149 Hannemann, S., Grunwaldt, J.-D., Krumeich, F., Kappen, P., and Baiker, A. (2006) *Appl. Surf. Sci.*, **252**, 7862.
150 Strobel, R., Krumeich, F., Stark, W.J., Pratsinis, S.E., and Baiker, A. (2004) *J. Catal.*, **222**, 307.
151 Moser, W.R., Knapton, J.A., Koslowski, C.C., Rozak, J.R., and Vezis, R.H. (1994) *Catal. Today*, **21**, 157.
152 Johannessen, T. and Koutsopoulos, S. (2002) *J. Catal.*, **205**, 404.
153 Jensen, J.R., Johannessen, T., Wedel, S., and Livbjerg, H. (2003) *J. Catal.*, **218**, 67.
154 Mädler, L., Stark, W.J., and Pratsinis, S.E. (2003) *J. Mater. Res.*, **18**, 115.
155 Yanagisawa, T., Shimizu, T., Kuroda, K., and Kato, C. (1990) *Bull. Chem. Soc. Jpn.*, **63**, 988.
156 Inagaki, S., Fukushima, Y., and Kuroda, K. (1993) *J. Chem. Soc. Chem. Commun.*, 680.
157 Kresge, C.T., Leonowicz, M.E., Roth, W.J., Vartuli, J.C., and Beck, J.S. (1992) *Nature*, **359**, 710.
158 Zhao, D., Feng, J., Huo, Q., Melosh, N., Fredrickson, G.H., Chmelka, B.F., and Stucky, G.D. (1998) *Science*, **279**, 548.
159 Plyuto, Y., Berquer, J.-M., Jacquiod, C., and Ricolleau, C. (1999) *Chem. Commun.*, 1653.
160 Zhang, W.-H., Shi, J.-L., Wang, L.-Z., and Yan, D.-S. (2000) *Chem. Mater.*, **12**, 1408.
161 Lebeau, B., Fowler, C.E., Mann, S., Farcet, C., Charleux, B., and Sanchez, C. (2000) *J. Mater. Chem.*, **10**, 2105.
162 Zhang, L., Sun, T., and Ying, J.Y. (1999) *Chem. Commun.*, 1103.
163 Mehnert, C.P., Weaver, D.W., and Ying, J.Y. (1998) *J. Am. Chem. Soc.*, **120**, 12289.
164 Iwamoto, M. and Tanaka, Y. (2001) *Catal. Surv. Jpn.*, **5**, 25.
165 Gupta, G., Stowell, C.A., Patel, M.N., Gao, X., Yacaman, M.J., Korgel, B.A., and Johnston, K.P. (2006) *Chem. Mater.*, **18**, 6239.
166 Bronstein, L., Kramer, E., Berton, B., Burger, C., Forster, S., and Antonietti, M. (1999) *Chem. Mater.*, **11**, 1402.
167 Whilton, N.T., Berton, B., Bronstein, L., Hentze, H.-P., and Antonietti, M. (1999) *Adv. Mater.*, **11**, 1014.
168 Gupta, G., Shah, P.S., Zhang, X., Saunders, A.E., Korgel, B.A., and Johnston, K.P. (2005) *Chem. Mater.*, **17**, 6728.
169 Diaz, J.F., Balkus, K.J., Jr., Bedioui, F., Kurshev, V., and Kevan, L. (1997) *Chem. Mater.*, **9**, 61.
170 Lim, M.H., Blanford, C.F., and Stein, A. (1998) *Chem. Mater.*, **10**, 467.
171 Xie, Y., Quinlivan, S., and Asefa, T. (2008) *J. Phys. Chem. C*, **112**, 9996.
172 Freund, H.-J., Bäumer, M., Libuda, J., Risse, T., Rupprechter, G., and Shaikhutdinov, S. (2003) *J. Catal.*, **216**, 223.
173 Rupprechter, G. (2007) *Catal. Today*, **126**, 3.
174 Silvestre-Albero, J., Rupprechter, G., and Freund, H.-J. (2006) *Chem. Commun.*, 80.
175 Rupprechter, G. (2001) *Phys. Chem. Chem. Phys.*, **3**, 4621.
176 Narayanan, R. and El-Sayed, M.A. (2005) *J. Phys. Chem. B*, **109**, 12663.
177 Yoo, J.W., Hathcock, D.J., and El-Sayed, M.A. (2003) *J. Catal.*, **214**, 1.
178 Rioux, R.M., Song, H., Grass, M., Habas, S., Niesz, K., Hoefelmeyer, J.D., Yang, P., and Somorjai, G.A. (2006) *Top. Catal.*, **39**, 167.
179 Bratlie, K.M., Lee, H., Komvopoulos, K., Yang, P., and Somorjai, G.A. (2007) *Nano Lett.*, **7**, 3097.
180 Lee, H., Habas, S.E., Kweskin, S., Butcher, D., Somorjai, G.A., and Yang, P. (2006) *Angew. Chem. Int. Ed.*, **45**, 7824.
181 Vidal-Iglesias, F.J., Solla-Gullón, J., Rodríguez, P., Herrero, E., Montiel, V., Feliu, J.M., and Aldaz, A. (2004) *Electrochem. Commun.*, **6**, 1080.
182 Solla-Gullón, J., Vidal-Iglesias, F.J., Herrero, E., Feliu, J.M., and Aldaz, A. (2006) *Electrochem. Commun.*, **8**, 189.

183 Narayanan, R. and El-Sayed, M.A. (2005) *Langmuir*, **21**, 2027.

184 Narayanan, R. and El-Sayed, M.A. (2004) *J. Am. Chem. Soc.*, **126**, 7194.

185 Narayanan, R. and El-Sayed, M.A. (2004) *Nano Lett.*, **4**, 1343.

186 Narayanan, R. and El-Sayed, M.A. (2004) *J. Phys. Chem. B*, **108**, 5726.

187 Englisch, M., Jentys, A., and Lercher, J.A. (1997) *J. Catal.*, **166**, 25.

188 Santori, G.F., Casella, M.L., Siri, G.J., Adúriz, H.R., and Ferreti, O.A. (2002) *React. Kinet. Catal. Lett.*, **75**, 225.

189 Vannice, M.A. and Ben, B. (1989) *J. Catal.*, **115**, 65.

190 Serrano-Ruiz, J.C., López-Cudero, A., Solla-Gullón, J., Sepúlveda-Escribano, A., Aldazb, A., and Rodríguez-Reinoso, F. (2008) *J. Catal.*, **253**, 159.

191 Galvagno, S. and Capannelli, G. (1991) *J. Mol. Catal.*, **64**, 237.

192 Frank, M. and Baumer, M. (2000) *Phys. Chem. Chem. Phys.*, **2**, 3723.

193 Heemeier, M., Stempel, S., Shaikhutdinov, S., Libuda, J., Baumer, M., Oldman, R.J., Jackson, S.D., and Freund, H.-J. (2003) *Surf. Sci.*, **523**, 103.

194 Boudart, M. (1969) *Adv. Catal. Relat. Subj.*, **20**, 153.

195 Boudart, M. (1985) *J. Mol. Catal.*, **30**, 27.

196 Bond, G.C. (1991) *Chem. Soc. Rev.*, **20**, 441.

197 Che, M. and Bennett, C.O. (1989) *Adv. Catal.*, **36**, 55.

198 Semagina, N., Renken, A., and Kiwi-Minsker, L. (2007) *J. Phys. Chem. C*, **111**, 13933.

199 Abid, M., Paul-Boncour, V., and Touroude, R. (2006) *Appl. Catal. A Gen.*, **297**, 48.

200 Mohr, C., Hofmeister, H., and Claus, P. (2003) *J. Catal.*, **213**, 86.

201 Overbury, S.H., Schwartz, V., Mullins, D.R., Yan, W., and Dai, S. (2005) *J. Catal.*, **241**, 56.

202 Miller, J.T., Kropf, A.J., Zha, Y., Regalbuto, J.R., Delannoy, L., Louis, C., Bus, E., and van Bokhoven, J.A. (2006) *J. Catal.*, **240**, 222.

203 Abad, A., Corma, A., and Garcia, H. (2008) *Chem. Eur. J.*, **14**, 212.

204 Li, Y., Boone, E., and El-Sayed, M.A. (2002) *Langmuir*, **18**, 4921.

205 Cervantes, G.G., Cadete Santos Aires, F., and Bertolini, J.C. (2003) *J. Catal.*, **214**, 26.

206 Sarkany, A., Zsoldos, Z., Stefler, G., Hightower, J.W., and Guzci, L. (1995) *J. Catal.*, **157**, 179.

207 Phillipson, J.J., Wells, P.B., and Wilson, G.R. (1969) *J. Chem. Soc. A*, 1351.

208 Bates, A.J., Leszczynski, Z.K., Phillipson, J.J., Wells, P.B., and Wilson, G.R. (1970) *J. Chem. Soc. A*, 1351-2435.

209 Boitiaux, J.P., Cosyns, J., and Robert, E. (1987) *Appl. Catal.*, **32**, 145.

210 Silvestre-Alberto, J., Rupprechter, G., and Freund, H.-J. (2005) *J. Catal.*, **235**, 52.

211 Katano, S., Kato, H.S., Kawai, M., and Domen, K. (2003) *J. Phys. Chem. B*, **107**, 3671.

212 Silvestre-Albero, J., Rupprechter, G., and Freund, H.-J. (2006) *J. Catal.*, **240**, 58.

213 Kuhn, J.N., Huang, W., Tsung, C.-K., Zhang, Y., and Somorjai, G.A. (2008) *J. Am. Chem. Soc.*, **130**, 14026.

214 Tauster, S.J., Fung, S.C., and Garten, R.L. (1978) *J. Am. Chem. Soc.*, **100**, 170.

215 Ko, E.I. and Garten, R.L. (1981) *J. Catal.*, **68**, 223.

216 Resasco, D.E. and Haller, G.L. (1982) *Stud. Surf. Sci. Catal.*, **11**, 105.

217 Raupp, G. and Dumesic, J.A. (1984) *J. Phys. Chem.*, **88**, 660.

218 Pesty, F., Steinruck, H.-P., and Madey, T. (1995) *Surf. Sci.*, **339**, 83.

219 Poirer, G., Hance, B., and White, J.M. (1998) *J. Phys. Chem.*, **97**, 5965.

220 Datye, A., Kalakkad, D., and Yao, M. (1995) *J. Catal.*, **155**, 148.

221 Haruta, M., Kobayashi, T., Sano, H., and Yamada, N. (1987) *Chem. Lett.*, 405.

222 Milone, C., Ingoglia, R., Schipilliti, L., Crisafulli, C., Neri, G., and Galvagno, S. (2005) *J. Catal.*, **236**, 80.

223 Milone, C., Ingoglia, R., Pistone, A., Neri, G., Frusteri, F., and Galvagno, S. (2004) *J. Catal.*, **222**, 348.

224 Claus, P., Brückner, A., Mohr, C., and Hofmeister, H. (2000) *J. Am. Chem. Soc.*, **122**, 11430.

3
When Does Catalysis with Transition Metal Complexes Turn into Catalysis by Nanoparticles?

Johannes G. de Vries

3.1
Introduction

Catalysis is essential in the production of chemicals, particularly in the production of bulk chemicals, but increasingly also for the production of fine chemicals [1]. If we confine ourselves to metal-based catalysts, we distinguish three forms of catalysis: heterogeneous catalysis, homogeneous catalysis, and catalysis with nanoparticles. The latter class may be further subdivided in catalysis with soluble metal nanoparticles and catalysis with supported metal nanoparticles, although the latter is usually seen as heterogeneous catalysis.

3.1.1
Homogeneous Catalysis

The catalyst used in homogeneous catalysis is typically a transition metal complex in which the central metal atom is ligated by one or more organic ligands that bind via one or more coordinating atoms to the metal [2]. The metal can be in the zero oxidation state or it can be in a higher oxidation state, in which case it has counterions. The counterions may bind to the metal like a ligand; this occurs with halides and carboxylates. The counterion may also have a highly delocalized charge such as in BF_4^- or PF_6^- ; in that case, the counterion does not bind directly to the metal atom. This results in the formation of cationic complexes.

The advantages of homogeneous catalysis are the following:

1) The catalyst is in the same phase as the substrate, which leads to a very efficient catalysis that is not hindered by diffusion problems.
2) Every single metal atom is catalytically active and has the same catalytic performance.
3) Rate and selectivity of the catalyst can be altered by changing the metal, the counterion, and the ligand. Since there is a huge diversity of ligands accessible by organic synthesis, this property makes it almost always possible to develop

Selective Nanocatalysts and Nanoscience, First Edition. Edited by Adriano Zecchina, Silvia Bordiga, Elena Groppo.

an economic process by screening these variables. This process can be accelerated by the use of high-throughput experimentation.

4) Research into the mechanism is relatively easy because of the molecular nature of the catalyst, which allows the use of spectroscopic techniques such as NMR.

The most important disadvantage of homogeneous catalysis is the instability of the catalyst. All three classes of catalysts can be inhibited by reaction with, or adsorption on the surface of catalyst poisons. However, a deactivation mechanism that is unique for homogeneous catalysis is through the loss of ligands. This may happen simply by dissociation or by reaction of an external reagent with the ligand. Once the metal complex becomes unsaturated, it can start to dimerize by a number of different mechanisms. This agglomerization can continue to form multimetallic clusters. If the metal is in the zero oxidation state, this process can lead to the formation of nanoparticles and eventually to the formation of metal crystals that precipitate, leading to an effective halt of the catalysis.

Although many articles, particularly those describing the immobilization of homogeneous catalysts, claim that separating the homogeneous catalyst from the product is a major problem, this is in fact not true. In bulk chemical processes, the catalyst and the product are usually separated by distillation. This can become a problem if the molecular weight of the product is too high leading to high temperatures during the distillation process, which may lead to catalyst decomposition. However, the molecular weight of most bulk chemical products is quite low. In fact there are 21 different bulk chemical processes based on homogeneous catalysis, testifying to the fact that catalyst separation and recycle is not a major problem. In fine chemicals, the product is isolated either by distillation or crystallization. Since the catalyst is usually not recycled, there is no real problem.

3.1.2 Heterogeneous Metal Catalysis

A heterogeneous metal catalyst usually consists of crystals of a metal or agglomerates of more than one metal that are deposited on a solid carrier material which usually is a silica or aluminum oxide or active carbon, although other supports are also known [3].

The advantages of heterogeneous catalysis are the following:

1) The catalysts are highly robust, which allows their use at high temperatures, which may lead to very high rates.
2) The catalysts are not easily deactivated, which allows their use for prolonged periods of time.

The main disadvantage of heterogeneous catalysts is the limited number of parameters available for attenuating their activity and selectivity. Addition of other metals, variation in the size of the crystallites, and change in the carrier material are the only available variables. Another disadvantage is the fact that only the metal

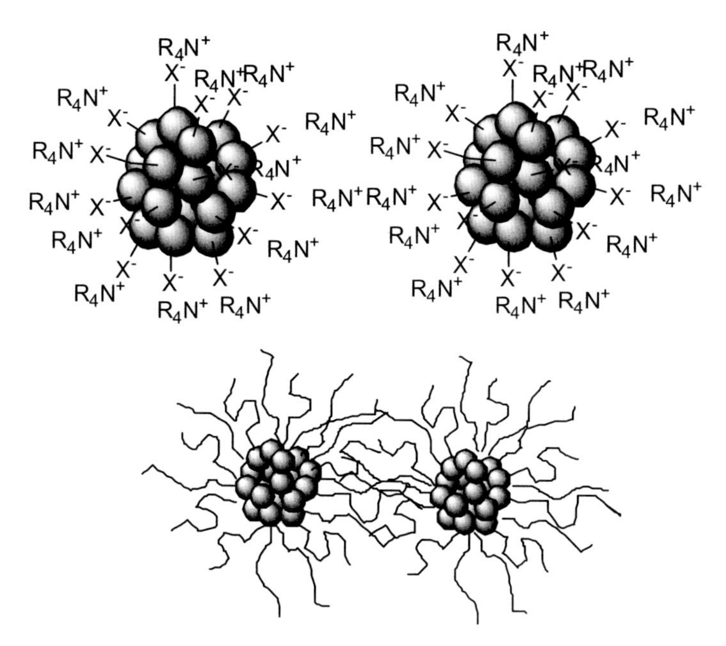

Figure 3.1 Stabilization of metal nanoparticles by tetra-alkylammonium halides or long-chain surfactants or amines.

atoms on the outside can possibly be active. Very often it is only the metal atoms at defects such as kinks and steps that are active.

3.1.3 Catalysis with Soluble Metal Nanoparticles

Nanoparticles are agglomerates of metals, usually in the zero oxidation state. They are prevented from growing to larger crystals by the presence of stabilizing ligands that form a steric or a charge barrier between them (Figure 3.1). Common stabilizers are tetra-alkylammonium halides or carboxylates, anionic surfactants, nitrogen, or phosphorus ligands carrying long alkyl chain or aryl rings or polar polymers, such as PVP. Their size is usually somewhere between 1 and 100 nm. Metal oxides and metal salts can also form nanoparticles. Nanoparticles can be prepared on purpose, for instance by reduction of a metal salt in the presence of a stabilizer. They can also be formed accidentally during catalysis.

The advantages of their use as catalysts can be described as follows:

1) The catalysts are soluble in the same reaction medium as the substrate; thus there are no diffusion limitations.
2) The stabilizing ligands tend to be bulk chemicals and are thus orders of magnitude cheaper than commonly used ligands in homogeneous catalysis.
3) The stabilizing ligands are usually not strongly bound and hence are easily displaced by the substrate.

4) Small-sized nanoparticles tend to be faster catalysts than the metal complex catalyst in the same reaction.

Metal nanoparticles have been used as catalyst in many types of reactions [4]. The disadvantages of the use of nanoparticle catalysts are their instability; particularly at higher temperatures, they tend to grow to larger size in a process known as Oswald ripening, which will end in their precipitation. On the other hand, if precipitation can be prevented *during* the reaction it can actually be used to good advantage to isolate the catalyst *after* the reaction.

Metal nanoparticles can be seen as the bridge between homogeneous and heterogeneous catalysis. They are soluble in the same medium as the substrate, they can react as a homogeneous catalyst for instance in the Heck reaction, but they also have a surface and thus can react as a heterogeneous catalyst as in hydrogenation reactions.

3.1.4
The Border between the Three Forms of Catalysis

Many cases are known where a catalyst in the form of a transition metal complex was converted into nanoparticles during a catalytic reaction. Often researchers are blissfully unaware of this fact. Several cases will be described in this chapter. Heterogeneous catalysts can also be solubilized. This is commonly referred to as leaching. Lesser known is that in many cases, it is in fact the leached metal that is responsible for the catalysis. This is true for most oxidation processes [5], but also for palladium-catalyzed C–C bond formation reactions. It is also possible that all or part of the metal from the heterogeneous catalyst is solubilized in the form of nanoparticles.

It is often not easy to distinguish between the three forms of catalysis. Finke devised a number of tests that can aid in making the distinction between catalysis by complexes and catalysis on surfaces [6]. He also stresses the fact that the distinction can never be made on the basis of a single test as each test has its own flaws.

1) *Visible evidence*: TEM will show the presence of nanoparticles. Light diffraction can also be used for this purpose.

2) *Kinetic evidence*: If nanoparticles are the active catalyst and are formed during the reaction an induction period may be apparent, often leading to a sigmoidal curve. The kinetics is sometimes irreproducible.

3) Quantitative poisoning studies (CS_2, or other sulfur-containing compounds, PPh_3,); Hg poisoning.

4) The identity of the true catalyst must be consistent with all the data.

This chapter will describe several cases where the catalysis operates at the borderline between homogeneous catalysis and catalysis by nanoparticles and also some cases where heterogeneous catalysts turn into nanoparticles catalysts.

3.2 Nanoparticles vs. Homogeneous Catalysts in C–C Bond-Forming Reactions

Most metal-catalyzed C–C bond-forming reactions catalyzed by transition metal catalysts proceed through a catalytic cycle in which the catalyst changes oxidation state, usually alternating between the zero and the plus two oxidation states. This is a typical situation in which catalyst destablization may occur. Whereas the +2 oxidation state usually results in quite stable complexes in view of the electron-donating nature of all ligands, the metal–ligand bond strength in the zero oxidation state may considerably be reduced leading to ligand dissociation and opening the way to clustering and possibly precipitation of metal crystals. This phenomenon is particularly prevalent with palladium.

3.2.1 The Heck–Mizoroki Reaction

In the Heck–Mizoroki reaction, a bond is formed between an olefin and an aromatic compound, which contains a leaving group [7]. In Heck's original version, these were mainly iodide and bromide [8]. Later versions were developed in which the leaving group could also be chloride, triflate, diazonium salts, iodonium salts, tosylate, acid chloride, anhydride, alkenyl esters, sulfonylchloride, or silanols [7] (Scheme 3.1).

Reetz and Beller independently have shown that the Heck reaction can also be catalyzed by preformed palladium nanoparticles that are stabilized by tetra-alkylammonium halides or by polar polymers such as PVP [9, 10]. The rate of these reactions was not higher than those catalyzed by palladium phosphine complexes and thus these results initially did not attract much attention.

The first sign that palladium nanoparticles were more prevalent in the Heck reaction came from the research of Reetz and coworkers on the Heck reaction using Jeffery conditions [11]. In this variant, no ligand is used–just palladium acetate, an inorganic potassium base, and a tetra-alkylammonium salt which was originally intended to aid the solubilization of the inorganic base in the reaction

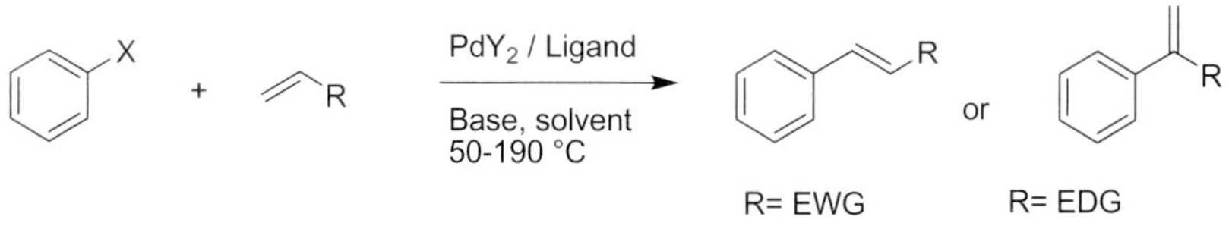

PdY_2 / Ligand

Base, solvent
50-190 °C

or

R= EWG

R= EDG

X= I, Br, Cl, $N_2^+X^-$, OSO_2R, COCl, $CO_2C(O)Ar$, $CO_2CH{=}CHR$,
R= H, alkyl, aryl, electron withdrawing group (EWG) or electron donating group (EDG)
Y= Cl, OAc, dba
Ligand = none, phosphine, phosphite, phosphoramidite
Base= Et_3N, $NaHCO_3$, K_2CO_3, KOAc, K_3PO_4

Scheme 3.1 The Heck–Mizoroki reaction.

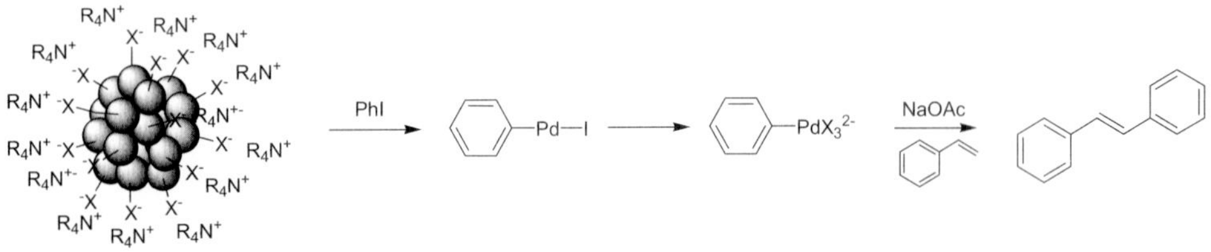

Scheme 3.2 Palladium nanoparticles as stoichiometric catalysts for the Heck reaction.

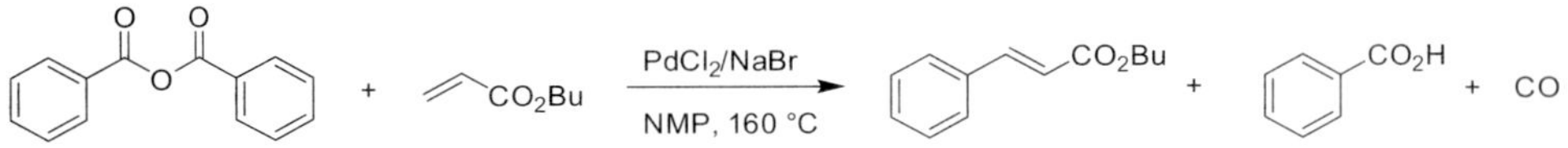

Scheme 3.3 Heck reaction on aromatic anhydrides.

medium [12]. Using TEM, Reetz showed the presence of palladium nanoparticles in these reactions [11]. He went one step further and reacted preformed palladium nanoparticles with an equivalent of iodobenzene (Scheme 3.2). Following this reaction with UV and ^{13}C NMR, he showed that the typical UV spectrum of the nanoparticles disappeared and that in the NMR the peaks of iodobenzene disappeared and at the same time a new set of peaks appeared which he attributed to an aryl palladium species, most likely $(PhPdI_3)^{2-}$. Adding styrene and NaOAc to this solution led to formation of the Heck product stilbene.

De Vries and coworkers have pioneered the use of aromatic anhydrides as arylating agents in the Heck reaction in an attempt to reduce the stoichiometric amount of salt waste that accompanies the formation of the Heck product [13]. In this reaction, CO and benzoic acid are the side products. The CO can be burned to CO_2 and the benzoic acid can be recycled back to the anhydride. This is one of the few Heck reactions that operate without base (Scheme 3.3).

One of the many other surprising elements regarding this reaction is the fact that it does not proceed in the presence of ligands, such as PPh_3. Later it was found that the presence of ligands prevents the decarbonylation reaction, which is a necessary step in the catalytic cycle [14]. However, the reaction is cocatalyzed by small amounts of chloride or better bromide salts. The maximum effect was obtained at a halide/palladium ratio of 4, which suggests an action at the level of the catalyst. Intrigued by these findings, the reaction was examined spectroscopically using TEM, EDX, EXAFS, and electrospray MS. TEM clearly showed the presence of nanoparticles, which was also confirmed by the EXAFS results that showed a Pd–Pd number of 8–12. In addition, the EXAFS showed the presence of multiple Pd–halide bonds and a single palladium–carbon bond. Most revealing was the electrospray MS, which showed the presence of a number of anionic monomeric and dimeric palladium species. One of these was $PhPdCl_2^-$ (in this reaction NaCl was used as additive). It is quite possible that this compound is present in solution as a dimeric species. These findings suggest that although a very large amount of the palladium is actually in the form of soluble nanoparticles, the actual catalysis proceeds via monomeric, or possibly dimeric, anionic species [15]. The role of the halide salt may be twofold: it stabilizes the colloids, thus preventing their further growth to palladium black, and they function as ligand for palladium in the actual catalytic cycle, which proceeds through anionic intermediates, in analogy with the proposals of Amatore and Jutand for the catalysis by palladium–phosphine complexes [16]. These findings led them to examine the Heck reaction under Jeffery conditions using TEM and ES-MS. They could confirm the presence of nanoparticles in the TEM. In the ES-MS, they detected the presence of $PhPdI_2^-$. In addition to that, a large peak attributable to PdI_3^- is visible [17]. Independent work performed by Evans using EXAFS showed the presence of large amounts of the dimeric species $Pd_2I_6^-$ in this reaction [18].

The above findings led them propose the following mechanism for these ligand-free Heck reactions (Scheme 3.4) [15, 17].

In the first phase of the reaction, Pd(II) is reduced to Pd(0) [19]. During the first 5 min of the reaction, they observed $([H_2O]PdOAc)^-$, which is a rare example of an anionic palladium(0) complex. This complex may undergo oxidative addition of aryl iodide to form $ArPdI_2^-$. This species and all other underligated species may well have a molecule of solvent (NMP, DMF) coordinated or they may be present as dimers [18]. The rest of the catalytic cycle proceeds along conventional lines via

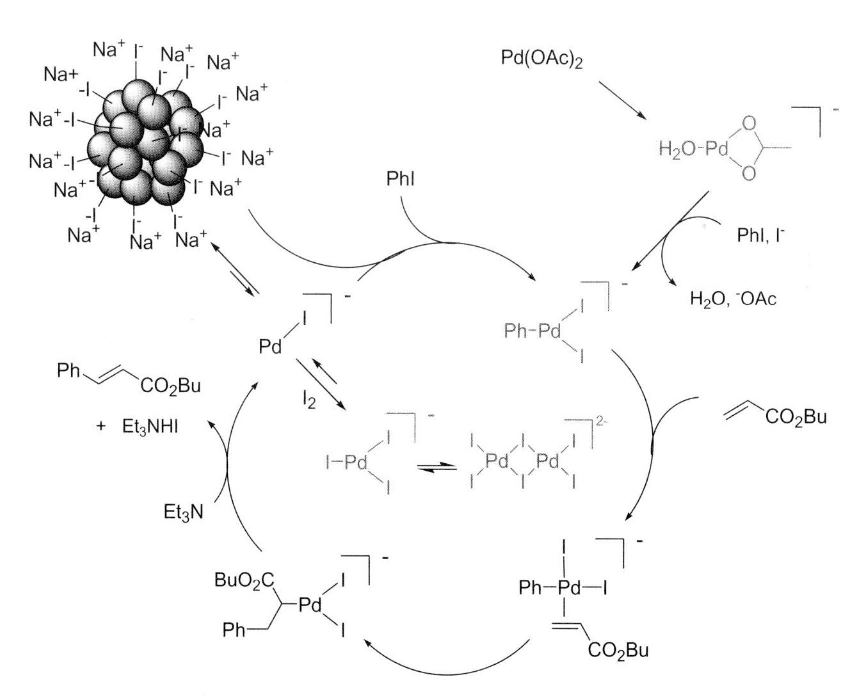

Scheme 3.4 Mechanism for the ligand-free Heck reaction. Intermediates in gray were observed by ES-MS.

olefin complex formation, olefin insertion, and beta-hydride elimination. However, all these intermediates are anionic. Hartwig recently showed that olefin insertion is much faster with these anionic complexes than with conventional phosphine-ligated neutral palladium complexes [20]. After beta-hydride elimination, the anionic palladium species left is highly underligated. At this stage, three pathways are possible: (i) reaction with I_2 to form PdI_3^- or more likely its dimer as found by Evans; (ii) formation of soluble nanoparticles; (iii) reaction with ArI. Since oxidative addition is fast with ArI, we would expect that in this case during the early stages of the reaction hardly any palladium colloids will be formed, depending on the substrate-to-catalyst (S/C) ratio. This is in agreement with the findings from Evans. At this point, it is unclear where the I_2 comes from; traces of oxygen may seem the most likely explanation [21]. However, Schmidt has proposed that it may arise from the formation of biphenyl [22]. This is a net reduction, which leads to the formation of Pd(II). Indeed, Schmidt has shown that the Heck reaction with ligand-less palladium actually is accelerated by the addition of sodium formate, which quickly reduces the formed Pd(II) back to Pd(0). At the end of the reaction, the nanoparticles will form rapidly, leading eventually to the formation of palladium black. Once the nanoparticles form, they can be solubilized again by reaction with aryl iodide under formation of $ArPdI_2^-$. The presence of the anion is essential to aid in the process of the solubilization.

Till recently, use of Jeffery conditions for the Heck reaction on aryl bromides was impossible. Here the oxidative addition of the aryl bromide is rate determining. The consequence is that all palladium will be present in the zero oxidation state and hence rapidly forms nanoparticles. The oxidative addition of the aryl halide will take out the palladium atoms at the rim of the nanoparticles in the form of $ArPdBr_2^-$. As this solubilization by oxidative addition is not fast enough, the Oswald ripening becomes the prevailing process, eventually leading to precipitation of palladium black. Scheme 3.5 shows this problem in a simplified form.

The DSM group tried a number of common stabilizers in order to prevent the further growth of the nanoparticles, but this was not terribly effective. In addition,

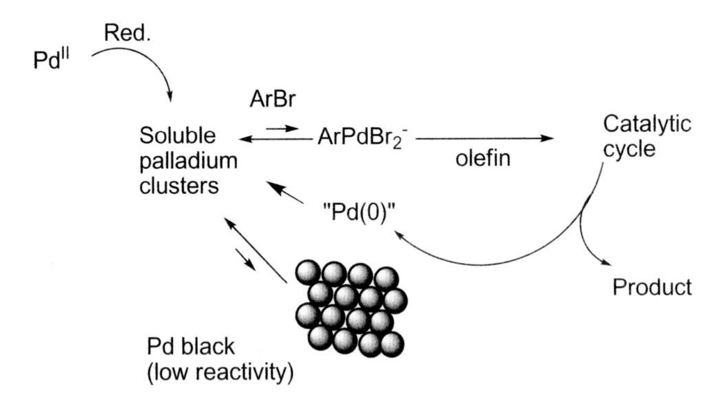

Scheme 3.5 Competition between oxidative addition of ArBr and Oswald ripening determines the size of the nanoparticles.

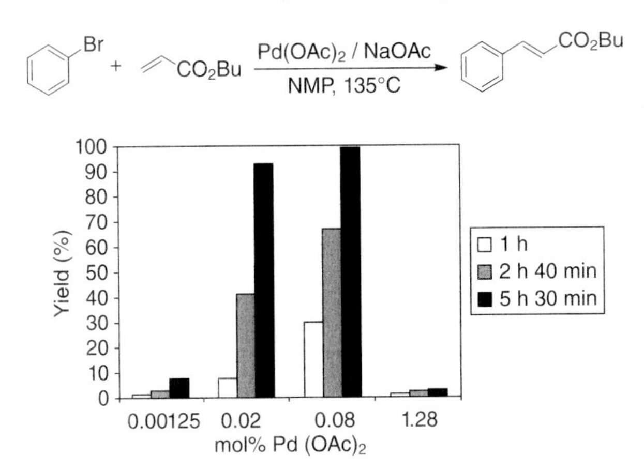

Figure 3.2 Effect of mol% of palladium on the rate of the ligand-free Heck reaction.

these additives defeat the purpose of the ligand-free Heck reaction, which is so convenient for synthetic purposes and even large-scale production in view of its easy workup. Looking again at Scheme 3.5, another solution to this problem presents itself. Whereas the Heck reaction is first order or possibly half order in palladium, the formation of the nanoparticles must follow a higher order. This suggests that lowering the palladium concentration might actually help in balancing the size reduction of the nanoparticles against their natural tendency to grow. Thus, four different palladium concentrations were tested in the ligand-free Heck reaction between bromobenzene and butyl acrylate. The rate of the reaction at these four concentrations is graphically depicted in Figure 3.2 [23].

At 1.28 mol%, the reaction is very slow, and the reason for this is clearly visible: almost immediately after the start of the reaction, palladium black started to form. However, at lower concentrations between 0.01 and 0.1 mol% of palladium, the reaction ran very smoothly without any precipitation of palladium during the reaction. These very low loadings were described by the authors and earlier by Beletskaya as "homeopathic palladium." Much lower amounts of catalyst are possible but need higher temperatures for a decent rate [24]. Comparing the rate at 0.08 and 0.02 mol%, it is clear that the turnover frequency (TOF) of the reaction actually increases by lowering the amount of palladium. This can be simply explained by the fact that the lower the palladium concentration the smaller the nanoparticles will be as their rate of formation is retarded. The smaller the nanoparticles, the more surface atoms will be available for catalysis and the higher the TOF. This phenomenon of increasing TOF with decreasing catalyst/substrate ratio is a telltale sign for the involvement of nanoparticles in these reactions. In fact, looking back at the many hundreds of publications describing new types of catalysts in the Heck reaction, this phenomenon is encountered quite frequently, suggesting that all high-temperature Heck reactions function via this mechanism [25]. We will discus some selected examples below.

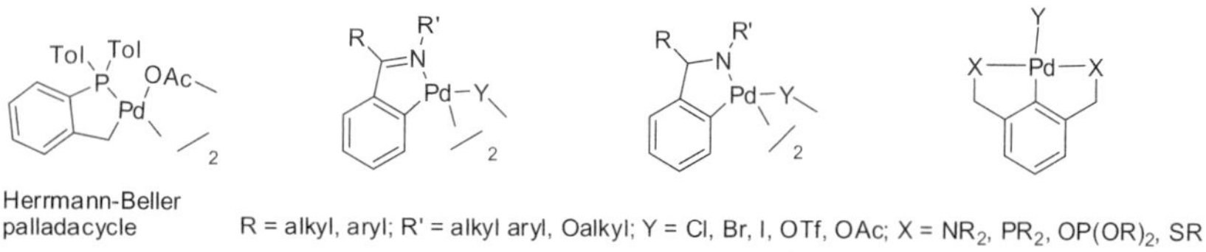

Herrmann-Beller palladacycle

R = alkyl, aryl; R' = alkyl aryl, Oalkyl; Y = Cl, Br, I, OTf, OAc; X = NR_2, PR_2, $OP(OR)_2$, SR

Figure 3.3 Palladacycles and pincer complexes used as catalysts in the Heck reaction.

The homeopathic palladium approach initially was not very successful for the conversion of aryl chlorides. The problem may be that here the oxidative addition step is even slower than that with the aryl bromides. Koehler found that very low amounts of $Pd(OAc)_2$ can indeed be used as catalyst for Heck reactions on aryl chloride if the reactions are carried out at higher temperatures (160 °C) [26]. In addition, he used tetra-alkylammonium bromide as stabilizer and finally he performs these reactions in air. The function of air is not entirely clear. It may actually oxidize the palladium on the outer rim of the nanoparticles and thus contribute to keep down the size of the nanoparticles. An alternative explanation would be the formation of palladium–oxygen complexes.

When Herrmann and Beller introduced the use of palladacycles as catalyst for the Heck reaction, this caused quite a bit of excitement and actually blew new life in an area that had been dormant for long [27]. Many research groups started to work in this area, and quite a few palladacyles and pincer complexes were reported that showed excellent activity in the Heck reaction (Figure 3.3) [28].

Immediately after their publication, a long debate ensued about the possible mechanism of the Heck reaction catalyzed by these complexes. Initial proposals by Herrmann and Beller, but also by Shaw, involved a Pd(II)/Pd(IV) cycle [27, 29]. It was shown by Hartwig that palladacycles are actually quite easily reduced to Pd(0) [30], which led to speculation about other possible mechanisms [31]. Interestingly an increase in TOF with increasing S/C ratio was already observed upon use of this catalyst in the Heck reaction. The story took an interesting twist when Nowotny and coworkers investigated the Heck reaction between iodobenzene and styrene using an immobilized palladacycle (Scheme 3.6) [32]. The obvious purpose of the study was to examine the recyclability of the immobilized catalyst. Disappointingly, the authors found that the solid catalyst lost all of its activity after the second run. However, the filtrate retained all the activity of the original catalyst. In addition, the second reaction catalyzed by the filtrate did not suffer from an induction period, whereas the one catalyzed by the virgin material did. This experiment clearly shows that the palladium becomes detached from the support, presumably via reduction to Pd(0). Thus, it is not the palladacycle itself that catalyzes the reaction and this clearly disproves the notion of a Pd(II)/Pd(IV) cycle. As ligand-free palladium must now be the catalyst, it is obvious that the mechanism is the same as in the ligand-free reactions described above.

Beletskaya in her work on nitrogen-containing palladacycles already commented that these were only the precursors of a Pd(0) species, based on the observation of induction periods and sigmoidal kinetic curves [33].

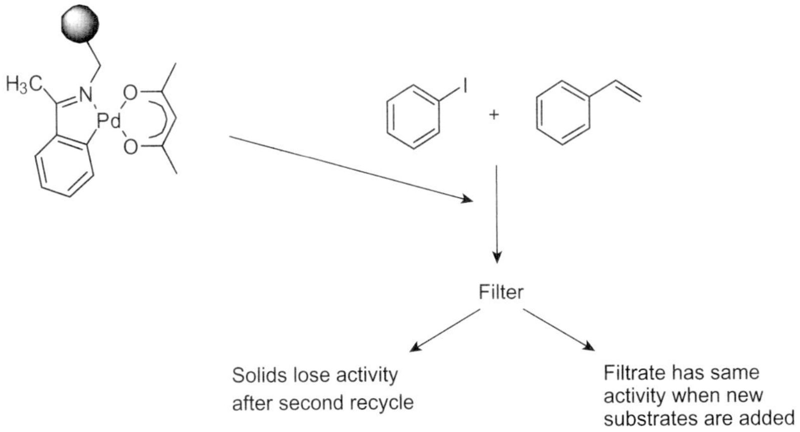

Filter

Solids lose activity after second recycle

Filtrate has same activity when new substrates are added

Scheme 3.6 Attempted recycle of an immobilized palladacycle reveals its instability.

De Vries and coworkers compared the rate of the Hermann–Beller palladacycle with their own homeopathic palladium method at the same S/C ratio. In these experiments, the catalysts were added to the hot solution containing all other ingredients. Under these circumstances, there hardly is any induction period in both reactions. This would seem to dispel the notion that palladacycles act as a slow release reservoir for Pd(0). Both reactions proceeded at about the same rate [23].

Attempts to create recyclable palladacycle or pincer type catalysts have been most revealing. Thus, Gladysz in an attempt to recycle an imine-based palladacycle carrying fluorous ponytails also found activity transferred to the nonfluorous phase and was able to prove the presence of Pd-colloids using TEM [34]. Several studies have appeared that show beyond doubt that PCP and SCS pincer complexes also decompose during the Heck reaction and lead to the formation of colloidal palladium [35]. Evidence was based on immobilization studies and on the application of the extensive Hg poisoning protocol developed by Finke, which proved the presence of palladium colloids [6].

Thus, the conclusion seems justified that all palladacycles and pincers decompose during the Heck reaction at high temperatures to form palladium colloids. A review by Jones and coworkers very neatly documents a lot of these cases [25].

Many reports have appeared on the use of heterogeneous palladium catalysts in the Heck reaction [25, 36]. Several different carrier materials have been used such as plain carbon, silica, or porous glass; the palladium has been built into the crystal lattice of zeolites, hydroxyapatite, and many other materials. Practically all reports showed that the catalyst could be reused several times and the authors have interpreted this as proof that the reaction takes place at the surface of the catalyst. Some researchers even bothered to do a hot filtration test to check for activity in the homogeneous phase. Most researchers reported that they did not find any activity; again seeming to confirm that the reaction really takes place at the surface of the catalyst and not in solution.

If one bothers to collect the results of all these paper in tabular form, a striking correlation is found: in all cases were the catalyst/substrate ratio was high the TOF was low, and reversely when the catalyst/substrate ratio was low the TOF was high [37]. This result is not easily explained on the basis of surface reactivity only and on the contrary is strongly reminiscent of the behavior of nanoparticle catalysis. Arai was the first to propose that it is indeed the leached palladium species that is responsible for the reaction in his work on the use of supported palladium catalysts in the Heck reaction of iodobenzene and methyl acrylate [38].

By using a three-phase test with an aryl iodide attached to a solid support, Davies was able to show that the Pd/C becomes active in the Heck reaction on butyl acrylate only after addition of a soluble monomeric aryl iodide or bromide [39]. Interestingly, he also found that the rate increased with increasing amounts of NaOAc. This suggests that the heterogeneous catalyst is solubilized by oxidative addition of the aryl halide and enters the catalytic cycle in the form of a soluble anionic species such as $(ArPd[OAc]Br)^-$ or $(ArPd[OAc]_n)^-$. Biffis reported more or less similar findings in the Heck reaction with Pd/Al_2O_3 (AO–Pd) and Pd deposited on an ion-exchange resin containing sulfonate groups (PS–Pd) [40]. Aryl bromide alone was sufficient to solubilize the palladium when AO–Pd was used, but ArBr and NaOAc where necessary with PS–Pd. A review on the use of heterogeneous catalysts in the Heck reaction in which Biffis and Zecca describe the importance of leaching is highly recommended reading [36].

Köhler performed extensive studies on a range of Pd/C catalysts that differed in Pd dispersion, Pd distribution, and oxidation state [41]. He found that most active systems were obtained with catalysts that had high dispersion, low degree of reduction, and uniform Pd impregnation. In addition, sufficient water content was also found to be important. He also concluded that palladium leaching correlates significantly with the reaction parameters. In a later paper, he found that Pd deposited on the zeolite NaY can even be used for the Heck reaction on aryl chlorides in the presence of tetrabutylammonium bromide [42]. Here again a good correlation was found between palladium in solution and the conversion of aryl bromide (Figure 3.6). During the reaction about one-third of the palladium goes into solution, but after the reaction is finished, less then 1 ppm of palladium is found in solution.

Thus, also in the case of heterogeneous palladium catalysts the palladium is dissolved via oxidative addition with aryl halide in the presence of anions to form $ArPdX_2^-$; after running the course of the Heck reaction, the palladium zero that is formed after the beta-hydride elimination remains solubilized in the form of nanoparticles till the end of the reaction when the rate of oxidative addition becomes slow. At this point, all the palladium reprecipitates as clearly shown in Figure 3.4.

Since it is known that $Pd(PPh_3)_4$ spontaneously forms nanoparticles within 1 day even at room temperature [44], it would seem safe to assume that at temperatures above 120 °C all palladium phosphine complexes also convert into nanoparticles. Thus, this type of catalyst is particularly unsuitable for high-temperature Heck reactions as the large excess of ligand with respect to the number of acces-

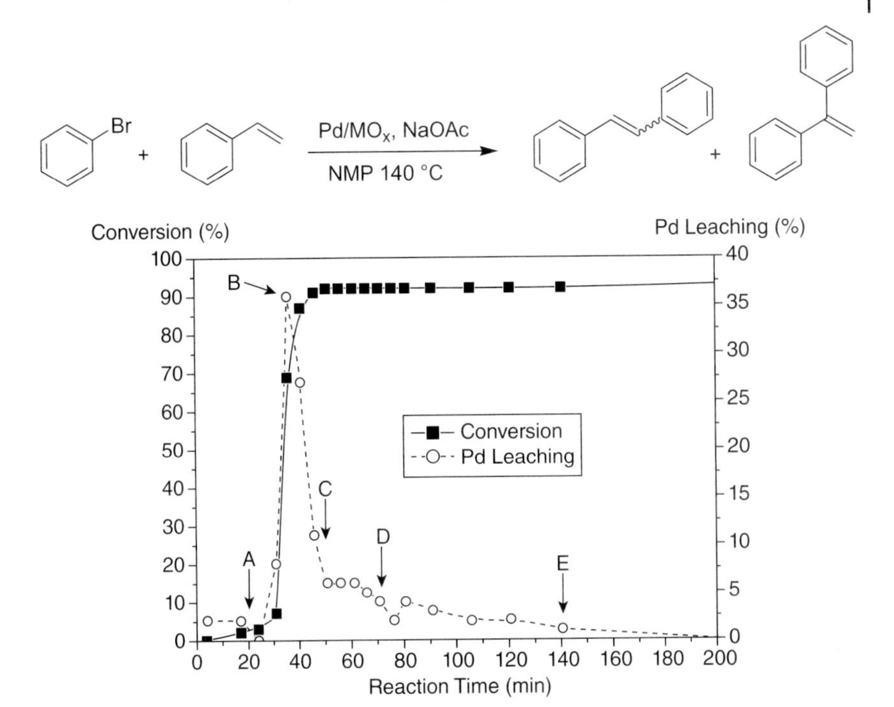

Figure 3.4 Time-dependent correlation of conversion and Pd leaching (percentage of the total Pd amount) in the Heck reaction of bromobenzene and styrene in the presence of Pd/TiO_2 (reaction conditions: 180 mmol bromobenzene, 270 mmol styrene, 216 mmol NaOAc, 0.2 mol% Pd catalyst, 180 mL NMP, 140 °C). Arrows A–E mark typical events during comparable experiments reported up to now (A:, Pd dissolution starting at reaction temperature; B: maximum amount of Pd in solution/highest reaction rate; C: substantial redeposition of Pd onto the support with increasing conversion; D: (far-reaching) completion of Pd redeposition; E: complete redeposition of even Pd traces by increased temperature/reducing agents) (reprinted from Ref. [43]).

sible palladium atoms will lead to a deactivated catalyst. Palladium NHC complexes are also known to form nanoparticles at high temperatures [45]. In conclusion, in view of the above we have to assume that the same mechanism of palladium nanoparticles and anionic intermediates is operative in all Heck–Mizoroki reactions at high temperatures, regardless of the type of palladium precursor.

3.2.2
The Kumada–Corriu Reaction

The Kumada–Corriu reaction is a cross-coupling reaction in which an alkyl or aryl Grignard reagent is coupled to an sp^2 carbon atom, usually in the form of an aryl or vinyl halide [46]. The reaction is catalyzed by palladium, nickel, and iron.

De Vries reported the use of ligand-free palladium in the Kumada reaction [47]. The reaction was not very selective and only a low yield of the coupling

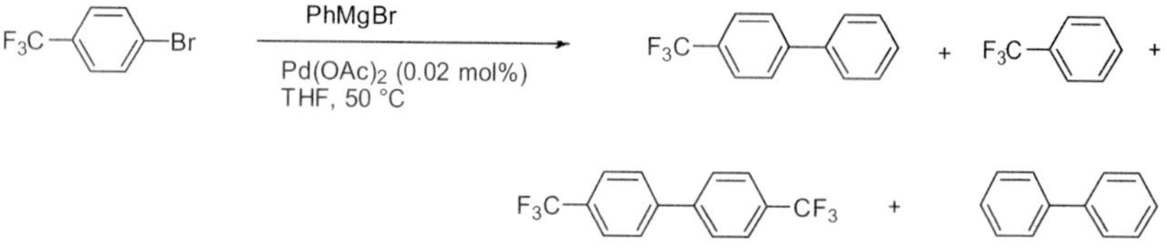

PhMgBr

$Pd(OAc)_2$ (0.02 mol%)
THF, 50 °C

Scheme 3.7 Kumada–Corriu reaction catalyzed by ligand-free palladium.

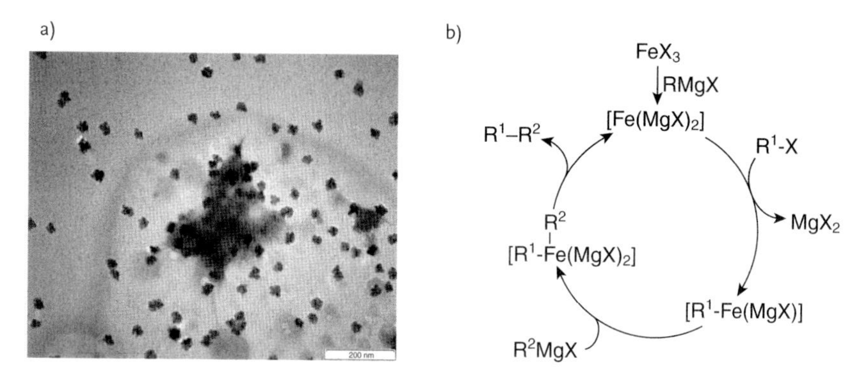

Figure 3.5 (a) TEM of Fe-nanoparticles that were used as catalyst for the Kumada-Corriu reaction by Bedford. (b) Mechanism proposed by Fürstner and coworkers for the iron-catalyzed Kumada–Corriu coupling.

product was obtained. Large amounts of the Grignard homo-coupling product, a small amount of the aryl bromide coupling product, and some reduction product were also obtained (Scheme 3.7). In fact, poor selectivity in the Kumada reaction catalyzed by palladium phosphine complexes may point to instability of the palladium complex used since, normally speaking, use of isolated palladium complexes leads to very high selectivities to the cross-coupling product.

Kochi has reported the use of iron salt as catalyst for cross-coupling reactions, mainly of alkenyl halides [48]. Recently, this chemistry has been taken up again and expanded by Fürstner [49], Cahier [50], Bedford [51], and others [52]. Excellent yields have been obtained in the cross-coupling of Grignards with aryl chlorides, tosylates, and triflates; surprisingly the bromides and iodides are coupled in poor yield, with large amount of the homo-coupling products also formed. Kochi in his first report already mentioned the formation of gray-black solutions upon addition of the Grignard reagent [48]. Bedford, examining the $FeCl_3$-catalyzed coupling between aryl-Grignards and secondary alkyl bromides, examined the dark solution at the end of the reaction using TEM and discovered the presence of iron nanoparticles (Figure 3.5a) [51]. He then showed that preformed iron nanoparticles, stabilized by PEG, are also very efficient catalysts for this reaction. There is still considerable controversy over the mechanism of the iron-catalyzed cross-coupling. Kochi originally proposed a Fe(I)/Fe(III) mechanism, whereas Bedford favored a

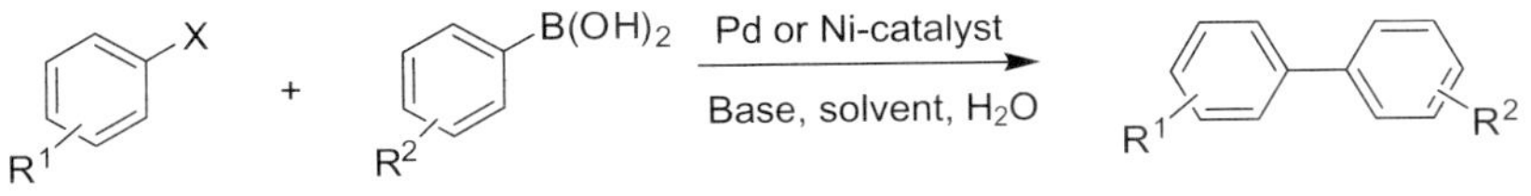

$$\text{R}^1\text{-C}_6\text{H}_4\text{-X} + \text{R}^2\text{-C}_6\text{H}_4\text{-B(OH)}_2 \xrightarrow[\text{Base, solvent, H}_2\text{O}]{\text{Pd or Ni-catalyst}} \text{R}^1\text{-C}_6\text{H}_4\text{-C}_6\text{H}_4\text{-R}^2$$

Scheme 3.8 The Suzuki reaction.

single-electron transfer (SET) mechanism based on the outcome of some radical clock-type experiments, which clearly showed that radical intermediates were involved. Fürstner, however, recently proposed a mechanism based on an Fe(−2)/Fe(0) cycle (Figure 3.5b) [53]. DFT calculations performed by Norrby and coworkers support an Fe(I)/Fe(III) mechanism [54].

3.2.3 The Suzuki Reaction

In the Suzuki reaction, the arylating agent is an arylboronic acid or derivative thereof. The reaction is catalyzed by palladium or nickel complexes [55] (Scheme 3.8).

El-Sayed and coworkers showed that the Suzuki reaction on aryl iodides can be catalyzed by palladium nanoparticles stabilized by poly-vinylpyrrolidine (PVP) in water [56]. However, they noted that during the reaction precipitation of palladium black occurred. Nacci and coworkers used palladium nanoparticles in tetraalkylammonium bromides as ionic liquid type solvent [57]. They were able to obtain excellent yields in the Suzuki reaction of aryl bromides and activated aryl chlorides. Many papers have since appeared on the use of different stabilizing agents for palladium nanoparticles used as catalyst in the Suzuki reaction. De Vries and coworkers showed that ligand-free palladium led to excellent results in the Suzuki reaction on aryl bromides with catalysts loadings between 0.01 and 0.05 mol% [47]. The authors presume that also in this case, nanoparticles are present that act as a reservoir of Pd(0). Trzeciak and coworkers used $Pd(OAc)_2$ or $PdCl_2$ immobilized on cyclohexyldiamine-modified glycidyl methacrylate polymer as catalyst in the Suzuki reaction. They examined the catalysts using TEM, SEM, EDX, and XPS. Palladium nanoparticles were formed under the conditions of the Suzuki–Miyaura reaction, the size of which depended on the type of anion. The palladium acetate derived catalyst was found to be fully reduced, leading to the formation of relatively small (2–5 nm) nanoparticles that were highly active. On the other hand, the palladium chloride derived catalyst was only partially reduced, and rather large nanoparticles were formed that had much less activity [58].

Rothenberg and coworkers prepared nanoparticles based on four different metals: Pd, Pt, Cu, and Ru [59]. They tested these nanoparticles and also mixed metal nanoparticles in the Suzuki–Miyaura reaction between iodobenzene and phenylboronic acid in DMF using K_2CO_3 as base (Table 3.1). As expected, palladium displayed the highest activity of the monometallic catalysts, affording quantitative yields after 4 h at 110 °C. No reaction was observed with the platinum

Table 3.1 Biphenyl yields and second-order rate constants obtained using mixed metal nanoparticle catalysts[a].

$B(OH)_2$ + I — Mixed metal NPs, DMF, K_2CO_3, 110 °C →

Entry	Catalyst composition	Yield (%)	k_{obs} (L/mol min^{-1})
1	Cu	62	3.2×10^{-3}
2	Pd	100	5.9×10^{-2}
3	Ru	40	2.0×10^{-3}
4	Cu/Pd	100	6.1×10^{-2}
5	Pd/Pt	94	9.7×10^{-3}
6	Pd/Ru	100	2.9×10^{-2}
7	Cu/Pd/Pt	92	2.5×10^{-2}
8	Cu/Pd/Ru	100	3.8×10^{-2}
9	Pd/Pt/Ru	81	7.3×10^{-3}
10	Cu/Pd/Pt/Ru	62	2.8×10^{-3}

a) Standard reaction conditions: 0.50 mmol iodobenzene, 0.75 mmol phenylboronic acid, 1.5 mmol K_2CO_3, 0.01 mmol catalyst (2 mol% total metal nanoclusters relative to PhI), 12.5 mL DMF, N_2 atmosphere, 110 °C.

clusters, but ruthenium and, surprisingly, copper clusters were found to be both active and stable. Of the bimetallic combinations, Cu/Pd was the most active, on par with pure Pd nanoparticles.

Leadbeater and coworkers reported a metal-free Suzuki reaction carried out in a microwave at >160 °C [60]. Later, these results were retracted as the authors found that the reaction was actually catalyzed by ppb amounts of palladium that were present as impurity in Na_2CO_3 that was used as the base [61]. This is of course in line with the earlier findings that the higher the substrate/catalyst ratio, the higher the TOF.

Gladysz used a palladacycle decorated with fluorous ponytails as catalyst in the Suzuki reaction. He found that catalyst activity transferred from the fluorous phase to the DMF phase [34]. This phase showed a characteristic orange-red color that suggested the presence of nanoparticles. In addition, upon repeated recycle the palladacycle catalyst completely lost its activity. TEM showed the presence of palladium nanoparticles. Similarly, Bedford developed a silica immobilized palladacycle, which was used as catalyst in the Suzuki reaction (Figure 3.6). The catalyst lost activity upon each recycle, and the authors were able to show that the palladacycle ligand is arylated leading to the formation of palladium nanoparticles [62].

Nájera compared the rate of the Suzuki reaction catalyzed by the acetophenone-oxime based palladacycle she developed previously with that of $Pd(OAc)_2$. Interestingly, the rates were very similar in the Suzuki reactions of aryl bromides, suggesting a mechanism which is the same for both catalyst precursors and presumably involves palladium nanoparticles. However, in the Suzuki reaction of unactivated aryl chlorides the palladacycle gave much better results, suggesting a different mechanism [63].

Not much mechanistic research has been performed on Suzuki reactions catalyzed by palladium nanoparticles. An interesting study was recently published by Fairlamb, Lee, and coworkers [64]. They studied the Suzuki reaction catalyzed by preformed palladium nanoparticles, stabilized by PVP. Using a range of nanoparticles of different size, they found that the normalized TOF did change as function of the total number of accessible palladium atoms; however, they did not change as a function of the number of defect sites (Figure 3.7). Next, they investigated the

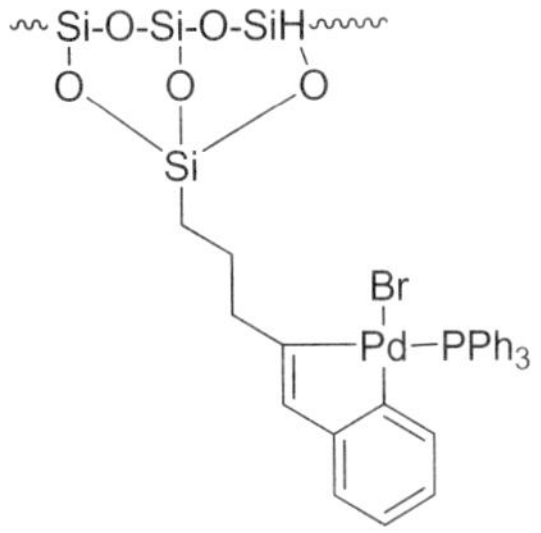

Si-O-Si-O-SiH
O O O
Si
Br
Pd–PPh_3

Figure 3.6 Bedford's immobilized palladacycle.

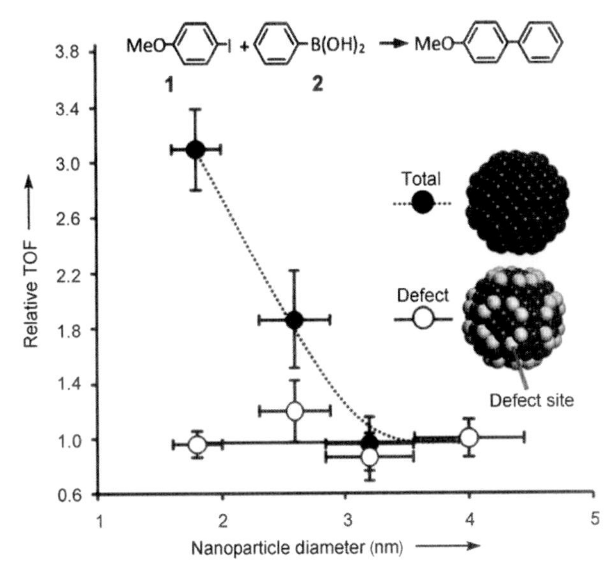

Figure 3.7 Structure-sensitive Suzuki coupling of 1 and 2 over size selected, cubeoctahedral, PVP-stabilized Pd nanoparticles. Turnover frequencies are normalized relative to the surface-atom densities of the largest nanoparticle: total surface atoms (•) or defect surface atoms ( ). The normalized cross-coupling rate should be independent of nanoparticle size if the correct active site has been identified (reprinted from Ref. [64]).

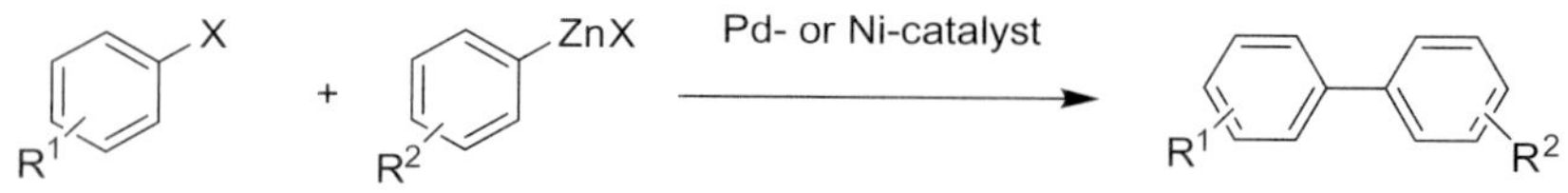

Scheme 3.9 The Negishi coupling.

change in size of the nanoparticles during the Suzuki reaction using EXAFS. They found no change, which led them to conclude that in contrast to the Heck reaction, the catalytic cycle actually takes place at the surface of the nanoparticles on the defect sites.

3.2.4 The Negishi Reaction

In the Negishi coupling, the arylating agent is an arylzinc halide (Scheme 3.9).

There are no reports on the use of preformed palladium nanoparticles as catalyst in the Negishi reaction. De Vries and coworkers showed that ligand-free palladium at 0.02 mol% actually was a good catalyst for Negishi reactions on aryl iodides and activated aryl bromides. Some homo-coupling side products were also formed [47].

Marder and coworkers also described the use of ligand-free palladium acetate as catalyst in the Negishi reaction using stoichiometric tetra-alkylammonium bromide as cocatalyst [65]. The reaction is very fast at room temperature and is completed in seconds; even at −20 °C the reaction was finished in just 30 min. At this temperature, the reaction kinetics shows the characteristic sigmoidal curve. The effect of the palladium concentration on the rate of the reaction was also investigated and again the TOF clearly increased at higher substrate catalyst ratios. They found the reaction was effectively retarded by the addition of only 0.5 eq. of PPh_3, which clearly indicate the involvement of nanoparticles in the catalysis.

3.2.5 The Sonogashira Reaction

In the Sonogashira reaction, a terminal alkyne is coupled to an aryl or alkenyl halide catalyzed by a catalytic system that consists of a homogeneous palladium catalyst, a copper(I) salt, and a secondary amine [66]. As the copper-salt also catalyzes the Glaser-type homo-coupling of the alkyne to form diynes, much work has been done on copper-free Sonogashira reactions, with great success.

Hyeon and coworkers examined the efficacy of both palladium nanoparticles as well as mixed Pd–Ni nanoparticles in the Sonogashira reaction [67]. The mixed nanoparticles were of the core/shell type with the core being mostly nickel and the shell mostly palladium (case c in Figure 3.8). As expected, the core/shell nanoparticles were more active on a palladium basis than the pure palladium nanoparticles as in the latter case most palladium was inside the nanoparticles

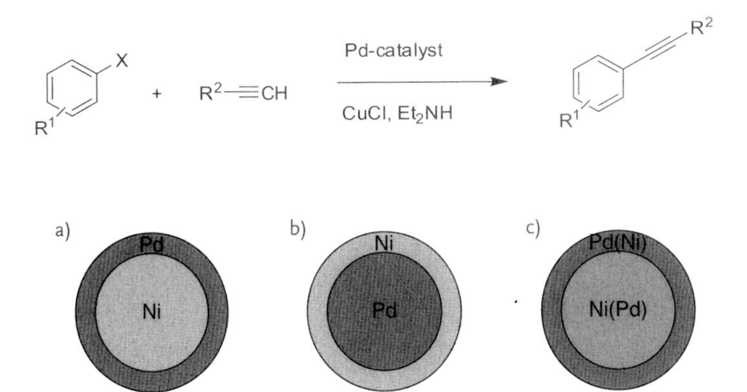

Figure 3.8 Mixed palladium-nickel nanoparticles as catalysts for the Sonogashira reaction.

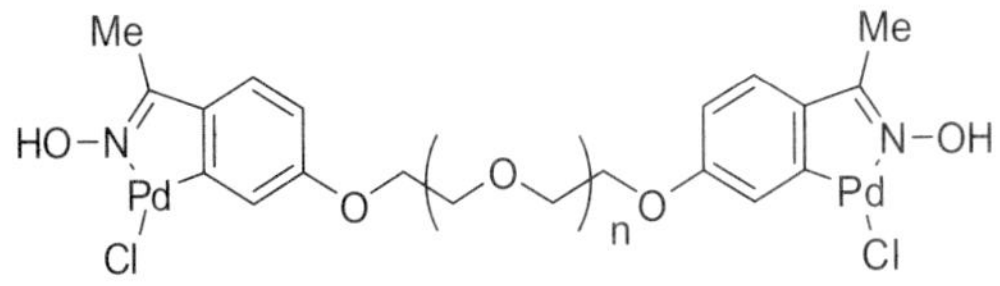

Figure 3.9 Nájera's palladacycle immobilized on PEG by Corma and coworkers.

and unavailable for catalysis. In a similar vain, Srinivasan and coworkers examined the effect of ultrasound on the rate of the Sonogashira reaction of aryl iodides and activated aryl bromides catalyzed by palladium nanoparticles in both acetone and ionic liquids [68]. They found that the reaction proceeded smoothly at 30 °C in high yield under the ultrasound conditions if 2 mol% of $PdCl_2$ was used as precatalyst. In acetone, the reaction was complete within 20 min, whereas in ionic liquids this took 2 h. However, the yields were somewhat higher in the ionic liquids. If no ultrasound was used at this temperature, the reaction did not proceed. However, if $PdCl_2$ was pretreated with ultrasound in the presence of Et_3N after which the reaction was carried out in silent mode, the yields were much lower. This shows that the ultrasound is important both to create an active catalyst, presumably by reducing the size of the Pd–Np's, but it also catalyzes the reaction itself, presumably caused by the local high temperatures and pressures in the cavitations.

A number of groups have reported the use of palladacyles and pincers as catalyst for the Sonogashira reaction. Corma and coworkers attached the palladacycle developed by Nájera and coworkers onto polyethylene glycol (Figure 3.9) [69]. They used this catalyst (5 mol%) in the Sonogashira reaction of *p*-bromoacetophenone and phenylacetylene at 150 °C. Although the catalyst could be reused up to 10 times, they did notice that palladium nanoparticles were formed. They also investigated the change in size of the nanoparticles over the repeated use and found

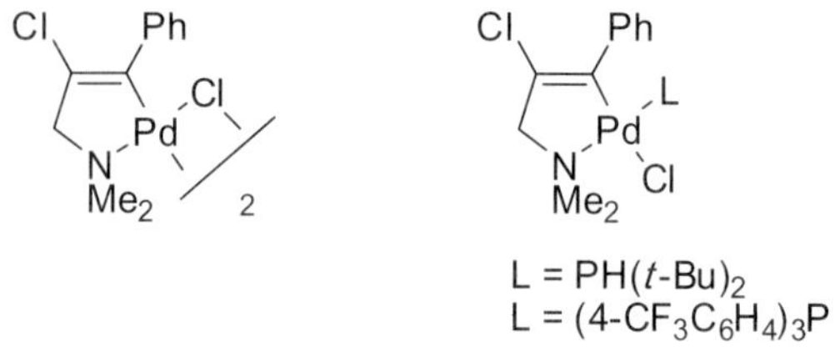

L = $PH(t\text{-}Bu)_2$
L = $(4\text{-}CF_3C_6H_4)_3P$

Figure 3.10 Dupont's palladacyles.

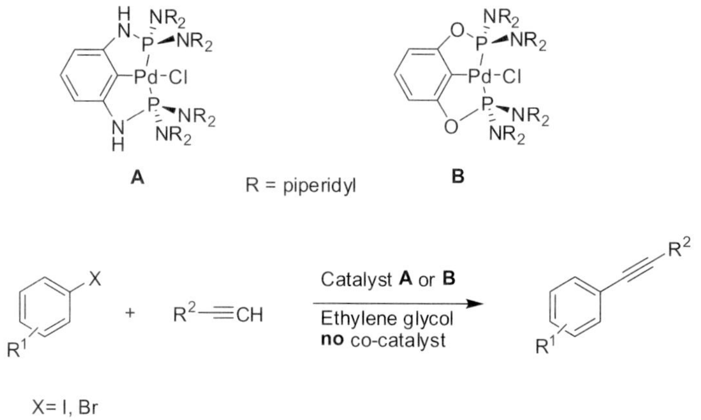

A R = piperidyl B

Catalyst **A** or **B**
Ethylene glycol
no co-catalyst

X= I, Br

Figure 3.11 Pincers as precatalysts in the Sonogashira reaction.

that the size of the nanoparticles increased only gradually, which they attributed to the stabilizing effect of the polyethylene glycol.

Dupont and coworkers investigated the use of three different palladacyles as catalyst for the Sonogashira reaction between alkynes and aryl iodides as well as activated aryl bromides (Figure 3.10) [70]. At temperatures around 120 °C, these palladacycles were extremely active even at S/C ratios as high as 5×10^5. They performed a series of tests to ascertain the true nature of the catalyst. They found that the catalysis was completely inhibited when 300 eq. of Hg was added. Also, in the Collmann test, which uses an immobilized substrate, they found full activity confirming that the catalyst is a soluble palladium species. Thus the catalyst clearly is a soluble Pd(0) species.

Bolliger and Frech explored the use of palladium PCP–pincer complexes as catalyst in the Sonogashira reaction (Figure 3.11) [71]. The catalysts were highly active at 140 °C in the copper-free Sonogashira coupling of aryl iodides at 50 ppm. Under similar conditions, aryl bromides were coupled using 100 ppm of catalyst. The kinetics of the reactions clearly showed sigmoidal curves. In addition, the TOF of the reactions increased at higher S/C ratios. However, based on a negative mercury test and the fact that the reaction was not accelerated by the addition of PEG or tetra-alkylammonium halides, the authors decided that nanoparticles were not

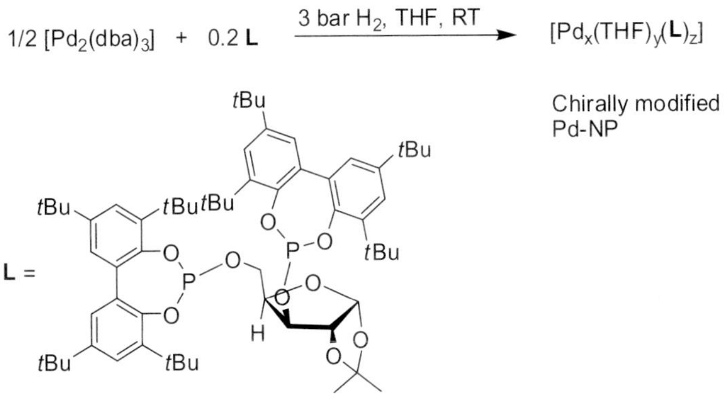

1/2 $[Pd_2(dba)_3]$ + 0.2 **L** —(3 bar H_2, THF, RT)→ $[Pd_x(THF)_y(\mathbf{L})_z]$

Chirally modified Pd-NP

Scheme 3.10 Preparation of chiral palladium nanoparticles.

involved. It should be noted, however, that the mercury test is rather tricky and needs a large excess of mercury as well as very good stirring.

3.2.6 Allylic Alkylation

Asymmetric allylic alkylation is a well-studied reaction that is catalyzed by several classes of transition metal complexes [72]. Much work has been performed with palladium complexes and enantioselectivity can reach up to 99% in selected cases.

Chung and coworkers showed that preformed palladium nanoparticles are in fact excellent catalysts for the allylation of a substituted malonic diester using allyl acetate [73]. Gómez and coworkers reported the asymmetric allylic alkylation of racemic 3-acetoxy-1,3-diphenylpropene with dimethyl malonate using palladium nanoparticles that were modified with a chiral bulky bisphosphite ligand (Scheme 3.10) [74]. They compared the activity and selectivity of this catalyst with the catalyst made from $(Pd[C_3H_5]Cl)_2$ and the same bulky bisphosphite ligand. Although with both systems the product was obtained with 97% ee, there was a striking kinetic difference. The colloidal catalyst on the one hand performed the reaction as a kinetic resolution with a maximum of 56% conversion after 24 h, which had not changed much after 165 h (Table 3.2, Entries 1 and 2). Consequently, the unconverted allylic acetate was obtained with 89% ee. The homogeneous catalyst on the other hand converted the substrate completely without any kinetic resolution in just 1.5 h. Addition of mercury or CS_2 totally inhibited the reaction with the colloidal catalyst, but not the reaction with the palladium complex.

On the other hand, Diéguez and coworkers prepared a series of palladium nanoparticles stabilized by five chiral sugar-based oxazolinyl–phosphite ligands, containing various substituents at the oxazoline and phosphite moieties [75]. These nanoparticles were applied in the Pd-catalyzed asymmetric allylic alkylation and Heck coupling reactions. A detailed study to elucidate the nature of the active species using a continuous-flow membrane reactor (CFMR), accompanied by TEM

Table 3.2 Allylic alkylation catalyzed by chiral nanoparticles.

Ph–CH(OAc)–CH=CH–Ph (*rac*-**I**) + $H_2C(CO_2Me)_2$ →[Pd-NP; BSA, KOAc, CH_2Cl_2, rt] Ph–CH($CH(CO_2Me)_2$)–CH=CH–Ph (**II**)

Entry	I/Pd/L	Time (h)	Conv. (%)	ee (II)	ee (I)
1	100:1:0.2	24	56	97 (*S*)	89 (*S*)
2	100:1:0.2	168	59	97 (*S*)	89 (*S*)
3	100:1:1.05	168	61	97 (*S*)	89 (*S*)

observations, classical poisoning experiments, and kinetic measurements was carried out. The membrane in the reactor had a cut-off of 700 Da allowing only molecular species to pass through. The allylic alkylation reaction with the chiral palladium nanoparticles was performed in the membrane reactor. Analysis of the solution collected during the first two reactor volumes pumped through the reactor showed a conversion of 10% with 19% enantioselectivity. They observed that the catalytic activity stopped after six reactor volumes. To the solution from the first two reactor volumes, 2 mg of KOAc were added and this sample was allowed to stir for a further 4 h. This led to an increase in conversion up to 32% while maintaining the same enantioselectivity. This is in agreement with the fact that the active species are molecular complexes that arise from leaching from the nanoparticles.

Indeed, in view of the mechanism of the reaction, in which an oxidative addition occurs on Pd(0), leaching of a monomeric palladium complex from the nanoclusters as in the Heck reaction seems highly likely.

3.3 Nanoparticles vs. Homogeneous Catalysts in Hydrogenation Reactions

While homogeneous hydrogenation usually is catalyzed by metal complexes in which the metal is oxidized, in heterogeneous catalysis and catalysis with nanoparticles the metal is in the zero oxidation state. If the transition metal complex somehow is reduced to a zero oxidation state complex, the transition to nanoparticles and heterogeneous catalysis is an easy one. Many examples have been documented, but often as a side remark noting that a black precipitate had formed at the end of the reaction.

3.3.1 Hydrogenation of Arenes

While the hydrogenation of benzo-fused heteroaromatic compounds and some heteroaromatic compounds such as furans and pyrroles using homogeneous tran-

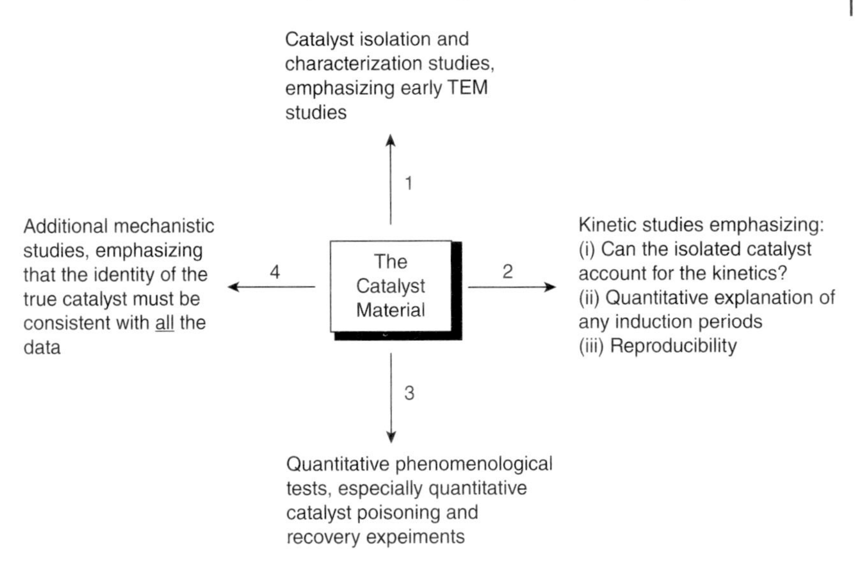

Scheme 3.11 Finke's methodology for distinguishing between homogeneous catalysis and nanocatalysis (or heterogeneous catalysis).

sition metal catalysts is well-documented, the hydrogenation of simple arenes is rare [76]. According to Finke, the only examples of well-established, monometallic, homogeneous catalysts for the hydrogenation of benzene are those developed by Rothwell and coworkers based on Nb(V) and Ta(V) hydrido complexes [77]. Finke has published extensively on hydrogenations catalyzed by homogeneous ruthenium complexes where the true catalysts turned out to be heterogeneous or in the form of soluble nanoparticles. In one case he examined the hydrogenation of benzene using ($Ru[C_6Me_6][OAc]_2$) as precatalyst. Based on the kinetics (sigmoidal curves), poisoning experiments and separately testing a black precipitate versus a red solution for activity, he established that in this case the true catalyst was heterogeneous in nature [77]. The Finke methodology is summarized in Scheme 3.11.

Dyson showed that a homogeneous ruthenium catalyst operating in a two-phase aqueous organic system actually increased in rate upon increasing pH (Scheme 3.12) [78]. He found the increase in rate correlated with the increased formation of ruthenium nanoparticles. Indeed many authors have noted that in aqueous two-phase system, the chances of forming colloids and/or heterogeneous metal are much increased [79].

One reason for the inability of monometallic transition metal catalysts to hydrogenate arenes is the assumption that a three metal ensemble that can be positioned ideally on the three double bonds of the benzene ring is necessary to reduce the first double bond [79]. For this reason, it was generally believed for a long time that triruthenium clusters are indeed truly homogeneous arene hydrogenation catalysts. Süss-Fink and coworkers had reported that $(Ru_3[\mu_2\text{-}H]_3[\eta^6\text{-}C_6H_6][\eta^6\text{-}C_6Me_6]_2[\mu_3\text{-}O])^+$ (Figure 3.12) is an excellent catalyst for the hydrogenation of benzene

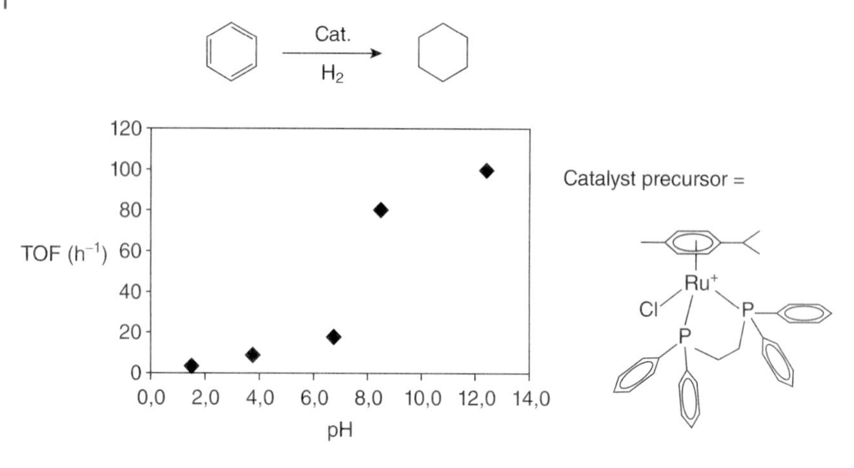

Scheme 3.12 Effect of pH upon rate in the hydrogenation of benzene to cyclohexene (reprinted with permission from Ref [78]).

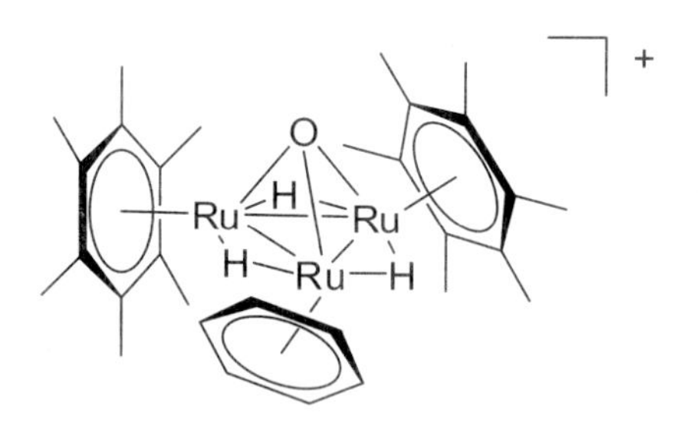

Figure 3.12 Triruthenium cluster.

to cyclohexene at 110 °C, 60 bar H_2, in water with a TOF of 289 h^{-1} and a total turnover (TTO) value of 740 after 2.5 h [80].

Furthermore, the catalyst could be reisolated at the end of the hydrogenation reaction in 95% yield, seemingly confirming its homogeneous nature. However, in a collaboration between the labs of Süss-Fink and Finke, it was shown that the missing 5% of catalyst was found back as a thin metal film on the walls of the reactor. The metal film proved to be a competent catalyst for the hydrogenation of benzene [81]. A series of experiments, including kinetics (sigmoidal curves), poisoning experiments, and the fact that H-D exchange which took place under the catalytic conditions did not involve the triruthenium cluster, showed that in this case it is probably ruthenium nanoparticles or heterogeneous ruthenium that is the real catalyst. Another interesting observation was the extremely fast hydrogenation of ethylbenzene, which was caused by contamination of this substrate with the hydroperoxide PhCH(Me)OOH, which accelerated the decomposition of the ruthenium cluster to form ruthenium nanoparticles.

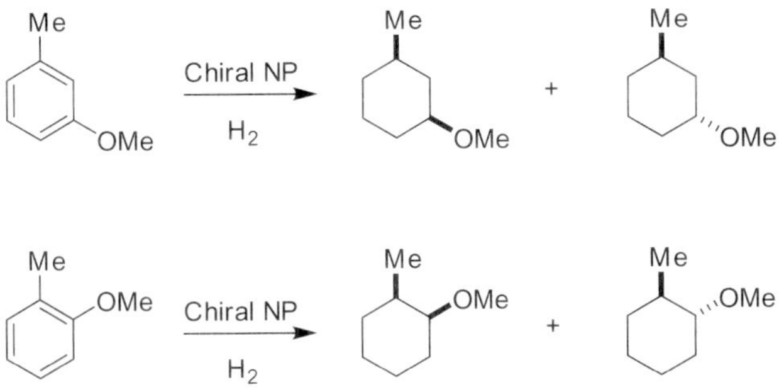

Scheme 3.13 Attempted asymmetric hydrogenation of substituted arenes.

3.3.2
Asymmetric Hydrogenation

Thus far there are no reports on the homogeneous asymmetric hydrogenation of substituted arenes, in line with Finke's suggestion that most homogeneous catalysts are incapable of aromatic hydrogenation. Philippot and coworkers examined the asymmetric hydrogenation of substituted anisoles using ruthenium, rhodium, and iridium nanoparticles stabilized by chiral 1,3-bisphophite ligands based on 1,3-pentanediol or on protected sugar derivatives [82]. The substrates were hydrogenated in good yields with excellent *cis*-selectivity. However, enantioselectivity in these reactions was barely measurable at mostly 6% (Scheme 3.13).

3.4
Platinum-Catalyzed Hydrosilylation

Many reports exist on the hydrosilylation reaction of olefins, catalyzed by homogeneous platinum catalysts [83]. In industry, this reaction is used in the synthesis of silane coupling agents and UV screeners. It is also utilized for the formation of three-dimensional networks via cross-linking between silicone hydride polymers with silicon vinyl polymers. Products are silicone rubbers, liquid injection molding compounds, paper-release coatings, and pressure-sensitive adhesives. Commonly used catalysts are Speier's catalyst (H_2PtCl_6 in isopropanol), $Pt(COD)_2$, and Karstedt's catalyst (Figure 3.13).

Although Chalk and Harrod had proposed a universally accepted mechanism along conventional lines involving discreet homogeneous monometallic complexes via oxidative addition of the silylhydride to the metal catalyst, there were a number of unexplained phenomena [84]. One of these was the catalytic effect of oxygen. The second one was the observation of an induction period, followed by an extremely fast reaction with TOFs in excess of $100\,000\,h^{-1}$. Marciniec and James found that when ruthenium phosphine complexes were used as catalyst, the oxygen served to oxidize the phosphine ligands [85]. Loss of ligands may

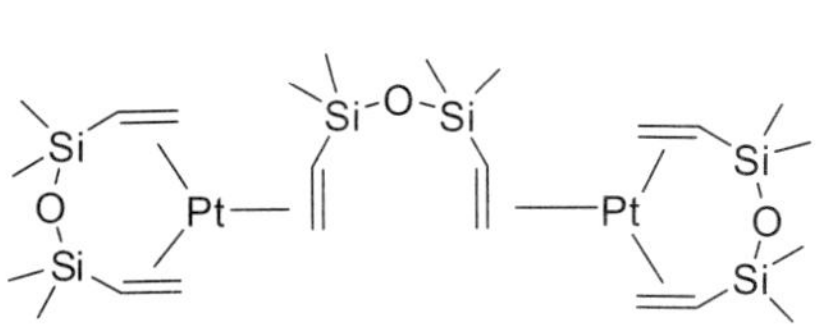

Figure 3.13 Karstedt's catalyst.

well be the start of nanoparticle formation although the authors did not mention this possibility. Lewis examined the reaction catalyzed by $Pt(COD)_2$ and found that upon addition of $(EtO)_3SiH$ a dark solution formed. TEM showed the presence of nanoparticles [86]. The preformed colloidal catalyst was actually faster than the catalyst precursor in the hydrosilylation reaction. Also in the case of Speier's catalyst, he was able to show the presence of nanoparticles. Interestingly, Speier's catalyst is almost an order of magnitude slower than $Pt(COD)_2$ in the hydrosilylation of $TMSCH{=}CH_2$ with $(EtO)_3SiH$. This was readily explained by the difference in size between the nanoparticles. Speier's catalyst was reduced to nanoparticles with an average size of 83 nm, whereas $Pt(COD)_2$ was reduced to much smaller particles, readily explaining the difference in reactivity. Lewis also showed that addition of mercury completely inhibited the catalyst, again confirming the fact that the platinum nanoparticles were indeed the effective catalyst. They noticed that in the absence of oxygen, the catalyst solutions turned darker over time with catalysis slowing down. Thus the oxygen seemed to play a role in keeping the size of the nanoparticles small, possibly as a ligand.

However, in a remarkable turnaround Lewis and coworkers distanced themselves from this mechanism based on the involvement of nanoparticles in spite of all the evidence [87]. This was mainly based on a study of the mechanism of the hydrosilylation reaction catalyzed by Karstedt's catalyst in which they made use of EXAFS, SAXS, and UV-vis. Regardless of the hydrosilane olefin ratio, they found that the catalyst during the reaction is a monomeric platinum compound containing one silicon and carbon in the first coordination sphere. The final platinum compound is a function of the ratio between the olefin and the hydrosilylating agent. At excess olefin, the platinum "end product" contains only platinum carbon bonds, whereas in the presence of excess hydrosilylating agent the platinum end product is multinuclear and contains only platinum silicon bonds. They find that oxygen functions by disrupting multinuclear platinum species that are formed when poorly coordinating olefins are used. The mechanism they proposed is depicted in Scheme 3.14.

3.5 Conclusions

It is clear from the many examples cited above that distinguishing between a mechanism in which a monomeric metal complex is the catalyst and a mechanism

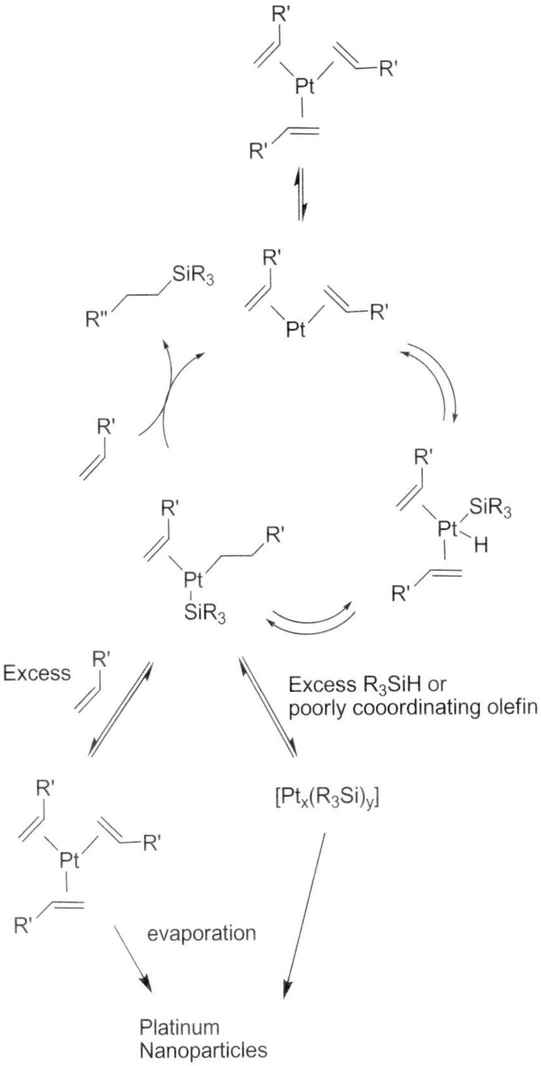

Scheme 3.14 Proposed mechanism for the platinum-catalyzed hydrosilylation.

in which metal nanoparticles are the catalyst is not always easy. The obvious (visible) presence of one form does not rule out the catalytic activity of the other form. Kinetics helps; the observation of sigmoidal curves is often a tell-tale sign of catalysis by nanoparticles or even a heterogeneous catalyst. Another clue is the observation of an increase in TOF with increasing substrate/catalyst ratio. EXAFS is a very powerful technique as it allows the observation of metal species during the catalytic reaction. In particular, it can be used to determine their oxidation state and the number of other atoms the average metal atom is surrounded with.

Thus the metal–metal number, which is an indication of nanoparticle formation, can be easily determined. This can even be followed over time. Poisoning studies can help to validate that the catalysis takes place at a surface rather than via a homogeneous complex. Observation of nanoparticles via TEM by itself should not be considered as conclusive evidence, but their absence is rather a strong evidence for a mechanism via a monometallic complex.

Nanoparticle catalysis has some obvious advantages as well as disadvantages when compared to catalysis with discreet metal complexes. Obvious advantage is the fact that no expensive ligands are used, which in addition simplifies the purification of the product. At the end of the reaction, the nanoparticles usually precipitate, which makes catalyst separation from the product quite easy. Many new developments are expected from this field.

References

1 Horvath, I.T. (ed.) (2003) *Encyclopedia of Catalysis*, vol. 1–6, John Wiley & Sons, Inc., New York.

2 van Leeuwen, P.W.N.M. (2004) *Homogeneous Catalysis. Understanding the Art*, Kluwer Academic Publishing, Dordrecht.

3 Ertl, G., Knözinger, H., Schüth, F., and Weitkamp, J. (eds) (2008) *Handbook of Heterogeneous Catalysis*, vol. 1–8, 2nd edn, Wiley-VCH Verlag GmbH, Weinheim.

4 Astruc, D. (ed.) (2008) *Nanoparticles and Catalysis*, Wiley-VCH Verlag GmbH, Weinheim.

5 Arends, I.W.C.E., and Sheldon, R.A. (2001) *Appl. Catal. A Gen.*, **212**, 175.

6 Widegren, J.A. and Finke, R.G. (2003) *J. Mol. Catal. A Chem.*, **198**, 317–341.

7 (a) Oestreich, M. (ed.) (2009) *The Mizoroki-Heck Reaction*, Wiley-VCH Verlag GmbH, Weinheim; (b) Beletskaya, I.P. and Cheprakov, A.V. (2000) *Chem. Rev.*, **100**, 3009.

8 (a) Heck, R.F. and Nolley, J.P., Jr. (1972) *J. Org. Chem.*, **37**, 2320; (b) Heck, R.F. (1982) *Org. React.*, **27**, 345.

9 Reetz, M.T., Breinbauer, R., and Wanninger, K. (1996) *Tetrahedron Lett.*, **37**, 4499.

10 Beller, M., Fischer, H., Kühlein, K., Reisinger, C.P., and Herrmann, W.A. (1996) *J. Organomet. Chem.*, **520**, 257.

11 Reetz, M.T. and Westermann, E. (2000) *Angew. Chem. Int. Ed.*, **39**, 165.

12 (a) Jeffery, T. (1996) *Advances in Metal-Organic Chemistry*, vol. 5 (ed. L.S. Liebeskind), JAI Press, Greenwich, CT, USA, p. 153; (b) Jeffery, T. (1996) *Tetrahedron*, **52**, 10113.

13 Stephan, M.S., Teunissen, A.J.J.M., Verzijl, G.K.M., and de Vries, J.G. (1998) *Angew. Chem. Int. Ed. Engl.*, **37**, 662.

14 Jutand, A., Negri, S., and de Vries, J.G. (2002) *Eur. J. Inorg. Chem.*, 1711.

15 de Vries, J.G. (2006) *Dalton Trans.*, 421–429.

16 Amatore, C. and Jutand, A. (2000) *Acc. Chem. Res.*, **33**, 314.

17 de Vries, A.H.M., Parlevliet, F.J., Schmieder-van de Vondervoort, L., Mommers J.H.M., Henderickx, H.J.W., Walet, M.A.N., and de Vries, J.G. (2002) *Adv. Synth. Catal.*, **344**, 996.

18 Evans, J., O'Neill, L., Kambhampati, V.L., Rayner, G., Turin, S., Genge, A., Dent, A.J., and Neisius, T. (2002) *J. Chem. Soc. Dalton Trans.*, 2207.

19 Reetz has shown that acetate itself can function as reductant: Reetz, M.T., and Lohmer, G. (1996) *Chem. Commun.*, 1921. In the case of benzoic anhydride the authors detected the formation of chloroacrylate esters, which points in the direction of a Wacker type mechanism for the reduction [13]. This latter mechanism has also been proposed by Heck [8b].

20 Carrow, B.P. and Hartwig, J.F. (2010) *J. Am. Chem. Soc.*, **132**, 79.

21 Schmidt, A.F., Al-Halaiqa, A., Smirnov, V.V., and Kurokhtina, A.A. (2008) *Kinet. Catal.*, **49**, 638.

22 Schmidt, A.F. and Smirnov, V.V. (2003) *J. Mol. Catal. A Chem.*, **203**, 75.

23 de Vries, A.H.M., Mulders, J.M.C.A., Mommers, J.H.M., Henderickx, H.J.W., and de Vries, J.G. (2003) *Org. Lett.*, **5**, 3285.

24 Arvela, R.K. and Leadbeater, N.E. (2005) *J. Org. Chem.*, **70**, 1786.

25 Phan, N.T.S., Van Der Sluys, M., and Jones, C.W. (2006) *Adv. Synth. Catal.*, **348**, 609.

26 Kleist, W., Pröckl, S.S., and Köhler, K. (2008) *Catal. Lett.*, **125**, 197.

27 Beller, M., Fischer, H., Herrmann, W.A., Öfele, K., and Broßmer, C. (1995) *Angew. Chem. Int. Ed. Engl.*, **34**, 1848; (b) Herrmann, W.A., Broßmer, C., Reisinger, C.-P., Riermeier, T.H., Öfele, K., and Beller, M. (1997) *J. Am. Chem. Soc.*, **119**, 1357.

28 (a) Dupont, J., Pfeffer, M., and Spencer, J. (2001) *Eur. J. Inorg. Chem.*, 1917; (b) Beletskaya, I.P. and Cheprakov, A.V. (2004) *J. Organomet. Chem.*, **689**, 4055; (c) Dupont, J., Consorti, C.S., and Spencer, J. (2005) *Chem. Rev.*, **105**, 2527; (d) Bedford, R.B., Cazin, C.S.J., and Holder, D. (2004) *Coord. Chem. Rev.*, **248**, 2283.

29 Shaw, B.L. (1998) *New J. Chem.*, 77.

30 Louie, J. and Hartwig, J.F. (1996) *Angew. Chem. Int. Ed. Engl.*, **35**, 2359.

31 (a) Beller, M. and Riermeier, T.H. (1998) *Eur. J. Inorg. Chem.*, 29; (b) Böhm, V.P.W. and Herrmann, W.A. (2001) *Chem. Eur. J.*, **7**, 4191; (c) Rosner, T., Bars, J.L., Pfaltz, A., and Blackmond, D.G. (2001) *J. Am. Chem. Soc.*, **123**, 1848.

32 Nowotny, M., Hanefeld, U., van Koningsveld, H., and Maschmeyer, T. (2000) *Chem. Commun.*, 1877.

33 Beletskaya, I.P., Kashin, A.N., Karlstedt, N.B., Mitin, A.V., Cheprakov, A.V., and Kazankov, G.M. (2001) *J. Organomet. Chem.*, **622**, 89.

34 (a) Rocaboy, C. and Gladysz, J.A. (2002) *Org. Lett.*, **4**, 1993; (b) Rocaboy, C. and Gladysz, J.A. (2003) *New J. Chem.*, **27**, 39.

35 (a) Yu, K., Sommer, W., Richardson, J.M., Weck, M., and Jones, C.W. (2004) *Adv. Synth. Catal.*, **347**, 161; (b) Eberhard, M.R. (2004) *Org. Lett.*, **6**, 2125; (c) Bergbreiter, D.E., Osburn, P.L. and Frels, J.D. (2005) *Adv. Synth. Catal.*, **347**, 172.

36 Biffis, A., Zecca, M., and Basato, M. (2001) *J. Mol. Catal. A Chem.*, **173**, 249.

37 Dams, M. (2004) Dissertation. Catholic University of Leuven, Belgium.

38 Zhao, F., Bhanage, B.M., Shirai, M., and Arai, M. (2000) *Chem. Eur. J.*, **6**, 843.

39 Davies, I.W., Matty, L., Hughes, D.L., and Reider, P.J. (2001) *J. Am. Chem. Soc.*, **123**, 10139.

40 Biffis, A., Zecca, M., and Barsato, M. (2001) *Eur. J. Inorg. Chem.*, 1131.

41 Pröckl, S.S., Kleist, W., Gruber, M.A., and Köhler, K. (2004) *Angew. Chem. Int. Ed.*, **43**, 1881.

42 Köhler, K., Heidenreich, R.G., Krauter, J.G.E., and Pietsch, J. (2002) *Chem. Eur. J.*, **8**, 622.

43 Djakovitch, L., Köhler, K., and de Vries, J.G. (2008) *Nanoparticles and Catalysis* (ed. D. Astruc), Wiley-VCH Verlag GmbH, Weinheim, p. 303.

44 Ye, E., Tan, H., Li, S., and Fan, W.Y. (2006) *Angew. Chem. Int. Ed.*, **45**, 1120.

45 Inés, B., SanMartin, R., Moure, M.J., and Domínguez, E. (2009) *Adv. Synth. Catal.*, **351**, 2124; (b) Karimi, B.B. and Enders, D. (2006) *Org. Lett.*, **8**, 1237.

46 (a) Tamao, K., Sumitani, K., and Kumada, M. (1972) *J. Am. Chem. Soc.*, **94**, 4374; (b) Corriu, R.J.P. and Massé, J.P. (1972) *J. Chem. Soc. Chem. Commun.*, 144; (c) Ackermann, L. (2009) *Modern Arylation Methods*, Wiley-VCH Verlag GmbH, Weinheim.

47 Alimardanov, A., Schmieder-van de Vondervoort, L., de Vries, A.H.M., and de Vries, J.G. (2004) *Adv. Synth. Catal.*, **346**, 1812.

48 (a) Tamura, M. and Kochi, J.K. (1971) *J. Am. Chem. Soc.*, **93**, 1487; (b) Neumann, S.M. and Kochi, J.K. (1975) *J. Org. Chem.*, **40**, 599.

49 Sherry, B.D. and Fürstner, A. (2008) *Acc. Chem. Res.*, **41**, 1500.

50 (a) Cahiez, G., Chavant, P.Y., and Metais, E. (1992) *Tetrahedron Lett.*, **33**, 5245; (b) Cahiez, G. and Marquais, S. (1996) *Tetrahedron Lett.*, **37**, 1773; (c) Cahiez, G. and Avedissian, H. (1998) *Synthesis*,

1199; (d) Duplais, C., Bures, F., Korn, T., Sapountzis, I., Cahiez, G., and Knochel, P. (2004) *Angew. Chem. Int. Ed.*, **43**, 2968.

51 Bedford, R.B., Betham, M., Bruce, D.W., Davis, S.A., Frost, R.M., and Hird, M. (2006) *Chem. Commun.*, 1398.

52 (a) Bolm, C., Legros, J., Le Paih, J., and Zani, L. *Chem. Rev.*, 2004, **104**, 6217; (b) Shinokubo, H. and Oshima, K. (2004) *Eur. J. Org. Chem.*, 2071; (c) Fürstner, A. and Martin, R. (2005) *Chem. Lett.*, **34**, 624; (d) Correa, A., Mancheno, O.G., and Bolm, C. (2008) *Chem. Soc. Rev.*, **37**, 1108; (e) Czaplik, W.M., Mayer, M., Cvengros, J., and Jacobi von Wangelin, A. (2009) *ChemSusChem*, **2**, 396.

53 Fürstner, A., Martin, R., Krause, H., Seidel, G., Goddard, R., and Lehmann, C.W. (2008) *J. Am. Chem. Soc.*, **130**, 8773.

54 Kleimark, J., Hedström, A., Larsson, P.-F., Johansson, C., and Norrby, P.-O. (2009) *ChemCatChem*, **1**, 152.

55 Miyaura, N. and Suzuki, A. (1995) *Chem. Rev.*, **95**, 2457.

56 Li, Y., Hong, X.M., Collard, D.M., and El-Sayed, M.A. (2000) *Org. Lett.*, **2**, 2385.

57 (a) Calo, V., Nacci, A., Monopoli, A., and Montingelli, F. (2005) *J. Org. Chem.*, **70**, 6040; (b) Calo, V., Nacci, A., and Monopoli, A. (2006) *Eur. J. Org. Chem.*, 3791.

58 Borkowski, T., Trzeciak, A.M., Bukowski, W., Bukowska, A., Tylus, W., and Kępiński, L. (2010) *Appl. Catal. A Gen.*, **378**, 83.

59 Thathagar, M.B., Beckers, J., and Rothenberg, G. (2002) *J. Am. Chem. Soc.*, **124**, 11858.

60 (a) Leadbeater, N.E. and Marco, M. (2003) *Angew. Chem. Int. Ed.*, **42**, 1407; (b) Leadbeater, N.E. and Marco, M. (2003) *J. Org. Chem.*, **68**, 5660.

61 Arvela, R.K., Leadbeater, N.E., Sangi, M.S., Williams, V.A., Granados, P., and Singer, R.D. (2005) *J. Org. Chem.*, **70**, 161.

62 Bedford, R.B., Cazin, C.S.J., Hursthouse, M.B., Light, M.E., Pike, K.J., and Wimperis, S. (2001) *J. Organomet. Chem.*, **633**, 173.

63 (a) Alacid, E., Alonso, D.A., Botella, L., Nájera, C., and Pacheco, M.C. (2006) *Chem. Rec.*, **6**, 117; (b) Alonsa, D.A. and Nájera, C. (2010) *Chem. Soc. Rev.*, **39**, 2891.

64 Ellis, P.J., Fairlamb, I.J.S., Hackett, S.F.J., Wilson, K., and Lee, A.F. (2010) *Angew. Chem. Int. Ed.*, **49**, 1820.

65 Liu, J., Deng, Y., Wang, H., Zhang, H., Yu, G., Wu, B., Zhang, H., Li, Q., Marder, T.B., Yang, Z., and Lei, A. (2008) *Org. Lett.*, **10**, 2661.

66 (a) Sonogashira, K. (2004) *Metal-Catalyzed Cross-Coupling Reactions*, vol. 1 (eds F. Diederich and A. de Meijere), Wiley-VCH Verlag GmbH, Weinheim, p. 319; (b) Chinchilla, R. and Nájera, C. (2007) *Chem. Rev.*, **107**, 874.

67 Son, S.U., Jang, Y., Park, J., Na, H.B., Park, H.M., YunJ, H.J., Lee, J., and Hyeon, T. (2004) *J. Am. Chem. Soc.*, **126**, 5026.

68 Gholap, A.R., Venkatesan, K., Pasricha, R., Daniel, T., Lahoti, R.J., and Srinivasan, K.V. (2005) *J. Org. Chem.*, **70**, 4869.

69 Corma, A., García, H., and Leyva, A. (2006) *J. Mol. Catal.*, **240**, 87.

70 Consorti, C.S., Flores, F.R., Rominger, F., and Dupont, J. (2006) *Adv. Synth. Catal.*, **348**, 133.

71 Bolliger, J.L. and Frech, C.M. (2009) *Adv. Synth. Catal.*, **351**, 891.

72 Trost, B.M. and Crawley, M.L. (2003) *Chem. Rev.*, **103**, 2921.

73 Park, K.H., Son, S.U., and Chung, Y.K. (2002) *Org. Lett.*, **4**, 4361.

74 (a) Jansat, S., Gómez, M., Philippot, K., Muller, G., Guiu, E., Claver, C., Castillón, S., and Chaudret, B. (2004) *J. Am. Chem. Soc.*, **126**, 1592; (b) Favier, I., Gómez, M., Muller, G., Axet, M.R., Castillón, S., Claver, C., Jansat, S., Chaudret, B., and Philippot, K. (2007) *Adv. Synth. Catal.*, **349**, 2459.

75 Diéguez, M., Pàmies, O., Mata, Y., Teuma, E., Gómez, M., Ribaudo, F., and van Leeuwen, P.W.N.M. (2008) *Adv. Synth. Catal.*, **350**, 2583.

76 Bianchini, C., Meli, A., and Vizza, F. (2007) *Handbook of Homogeneous Hydrogenation*, vol. 1 (eds J.G. de Vries and C.J. Elsevier), Wiley-VCH Verlag GmbH, Weinheim, p. 455.

77 Widegren, J.A., Bennett, M.A., and Finke, R.G. (2003) *J. Am. Chem. Soc.*, **125**, 10301.

78 Daguenet, C. and Dyson, P.J. (2003) *Catal. Commun.*, **4**, 153.

79 Dyson, P.J. (2003) *Dalton Trans.*, 2964.

80 Süss-Fink, G., Therrien, B., Vieille-Petit, L., Tschan, M., Romakh, V.B., Ward, T., Dadras, M., and Laurenczy, G. (2004) *J. Organomet. Chem.*, **689**, 1362.

81 Hagen, C.M., Vieille-Petit, L., Laurenczy, G., Süss-Fink, G., and Finke, R.G. (2005) *Organometallics*, **24**, 1819.

82 Gual, A., Godard, C., Philippot, K., Chaudret, B., Denicourt-Nowicki, A., Roucoux, A., Castillón, S., and Claver, C. (2009) *ChemSusChem*, **2**, 769.

83 Marciniec, B. (2005) *Coord. Chem. Rev.*, **249**, 2374.

84 Chalk, A.J. and Harrod, J.F. (1945) *J. Am. Chem. Soc.*, **87**, 16.

85 Gulinski, J., James, B.R., and Marciniec, B. (1995) *J. Organomet. Chem.*, **499**, 173.

86 (a) Lewis, L.N. and Lewis, N. (1986) *J. Am. Chem. Soc.*, **108**, 7228; (b) Lewis, L.N. (1990) *J. Am. Chem. Soc.*, **112**, 5998.

87 Stein, J., Lewis, L.N., Gao, Y., and Scott, R.A. (1999) *J. Am. Chem. Soc.*, **121**, 3693.

4
Capsules and Cavitands: Synthetic Catalysts of Nanometric Dimension

Giuseppe Borsato, Julius Rebek Jr., and Alessandro Scarso

4.1
Introduction on Supramolecular Catalysis

Catalysis is at the deep core of living systems; chemical transformations take place in nature at a relatively narrow range of temperatures and pressures and often in quite dilute systems. Thanks to their incredible rate accelerations, a number of different catalysts can steer reaction pathways through multiple routes and maintain selectivity for both reagents and products. In humans and in other evolutionary developed living organism, enzymes are known to catalyze organic transformations with rate accelerations that are many orders of magnitude greater that the uncatalyzed reaction [1] and they display unique features related to both substrates and products. In fact, enzymes can regulate chemical transformations with a great level of precision that is expressed in terms of chemo-, regio-, stereo-, and enantioselectivity of the given process. What at a first glance could seem like a sort of miracle, is more basically described with the concept originally expressed by Pauling [2] that enzymes are perfect hosts for reactions transition states. As a consequence, the enzyme stabilizes the preferred pathway and transforms the selected reagent into the selected product with exclusion of several other competitive pathways (chemo-, regio-, stereo-, and enantio-control) that, under the same experimental conditions, bear transition states that are characterized by much higher activation energies. It is, therefore, natural to consider molecular recognition as the mother concept that lies at the base of the discussion, and consequently supramolecular chemistry as the interdisciplinary subject that unravels subnanometric interactions.

This new discipline that crosses the boundaries between different pure sciences was elegantly defined as the chemistry beyond the molecule [3], and with this concept it is intended the bottom-up approach that allows nanofabrication of a plethora of thermodynamically controlled functional molecular devices made of several units. Reversibile self-correction and self-recognition are the peculiar features of supramolecular objects that can be characterized by different geometries, symmetries, and forms widening the borders of practical chemical structures [4].

Selective Nanocatalysts and Nanoscience, First Edition. Edited by Adriano Zecchina, Silvia Bordiga, Elena Groppo.

Another aspect of supramolecular chemistry, more and more attracting the interest of scientists coming from different disciplines, is its intrinsic advantage compared to common organic or inorganic synthesis to provide access to highly organized large complex structures. Starting from simple subunits, mixing the components in the correct proportions, and leaving the weak intermolecular forces to do the work, the structure emerges through self-assembly saving time and effort. The chemist's ambition to mimic the catalytic efficiency of enzymes, together with the growing ability to manipulate supramolecular assemblies, has prompted many chemists to embark on the adventure of creating synthetic or "artificial" enzymes. This idea of manipulating transition states is a consequence of the ability to design supramolecular sensors [5] for analytical purposes through molecular recognition exploiting a proper use of weak intermolecular forces.

Synthetic enzyme [6] is the name associated to manmade catalysts designed and intended to mimic one or more of the specific features observed in enzymatic catalysis. In terms of substrate or product selectivity, synthetic catalysts can often compete with naturally occurring ones, but in terms of catalytic activity only synthetic metalloenzymes can rival with natural enzymes as selective catalysts in aqueous media [7].

Even though the comparison with enzymes is sometimes rough, there is no doubt that the study of catalytic processes with synthetic models, allows the controlled isolation of a specific aspect of a complex problem. This focus on a restricted portion of a chemical transformation represents the only possible approach that, by subsequent improvement with a bottom-up approach, will lead one day to the preparation of true synthetic enzymes. These can be classified into several classes, all with nanoscience as a common background [8]. In any case, there will always be a conceptual difference between covalently bound homogeneous and heterogeneous catalysts and self-assembled supramolecular catalysts. The latter combine the ease of preparation, through attractive supramolecular interactions, to the catalytic efficiency provided by the presence of either suitable organic functional groups or metal centers. Moreover, the field of supramolecular catalysis [9, 10] in its broader sense is a very wide area of interest [11] as it spans from typical examples of hosting catalytic species like cavitand and capsules described in this contribution, to the preparation of libraries of new ligands for metal catalysis via self-assembling exploiting hydrogen bonding or metal–ligand interactions [12]. The common underlying strategy is based on the preparation of a limited number of subunits decorated with peripheral functionalities that are mixed in solution in the proper stoichiometry allowing the formation of reciprocal attractive interactions. These subunits self-assemble into well-defined structures that are dynamic but survive long enough to express their catalytic activity toward specific substrates [13]. The most common example is the use of micelles associated with homogeneous catalysts. Micelles are composed of amphiphilic molecules as subunits that because of hydrophobic effect [14] self-assemble in water into pseudospherical aggregates acting as nanometric reactors capable of dissolving and concentrating substrates either in the interior or on the surface of the assembly. Selectivity at all

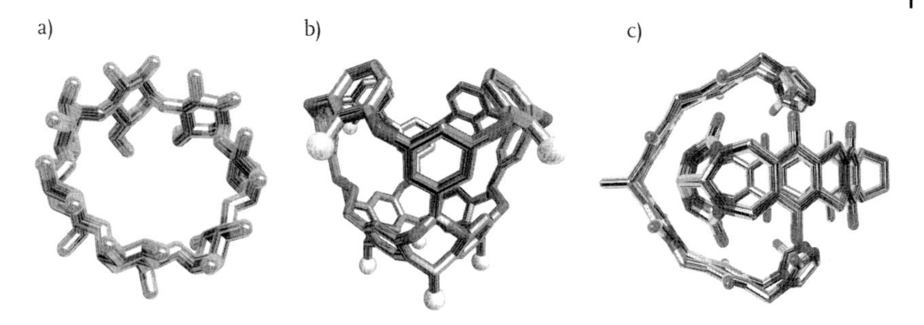

Figure 4.1 Examples of supramolecular catalysts: (a) cyclodextrin as a typical tube-like catalyst, (b) cavitand as a vase like catalyst, and (c) supramolecular self-assembled capsule as a spherical catalyst.

levels is often enhanced because of the local polarity gradients and the anisotropy of the assembly compared to ordinary solvents [15].

Often other catalytic systems like polymers, cyclodextrins, cucurbiturils, protein cages, and many other covalent systems are referred to as supramolecular nanoreactors where the assembly is related to their folding into active conformations. The shapes assumed by the nanocatalysts range from tube-like species like cyclodextrins with openings on both ends that allow the cavity to be accessed from two opposed directions, to vase-like or conical cavitands when the cavity has only one open end, to globular or capsular assemblies where the catalyst is completely sheltered inside the cavity and access occurs through partial dissociation, unfolding, or gating through the walls of the capsule (Figure 4.1).

Tube-like molecular and supramolecular catalysts are characterized by a generally fast guest uptake by virtue of the presence of two openings. Hydrophobic effect is the dominating driving force for binding while the challenge is the achievement of the proper orientation of the bound guest. There are several examples of successful supramolecular catalysts based on such structures where additional functionalities have been added and the topic has been extensively covered in the recent years [13, 16].

The aim of the present chapter is to focus the attention of the reader mainly to cavitands and capsules that more closely resemble enzymatic catalytic sites. In these, the pocket hosting the substrate surrounds a large portion of its surface providing several weak intermolecular attractive forces that, together with shape and steric requirements and the juxtaposition of catalytic functionalities, allow an extremely high level of chemo-, regio-, stereo-, and enantioselectivity typical and unique for natural catalyst.

Molecular and supramolecular cavitands are open-ended structures composed of one or more units characterized by a shape reminiscent of a vase or a pocket. They feature a tapered bottom and walls that limit the movement of the guest imparting a structure that ranges from a calix/pyramid to a cone. Because of the presence of an open end, the guest uptake and release is usually quite rapid and this better matches the requirements of a high turnover.

Molecular and supramolecular capsules are closed nanometric containers which can completely surround a guest [17]. For unimolecular capsules, more commonly named carcerands, the in-out exchange of the guest is a rather slow process that occurs via sneaking through the slits on the surface of the molecular container which breathes due to thermal motion. For supramolecular capsules held together by weak intermolecular attractive forces, the guests exchange more rapidly through portals opened by partial dissociation of a limited number of weak intermolecular interactions or by complete dissociation and reassociation of one or more subunits.

While cavitands are usually unimolecular organic structures, supramolecular capsules are assemblies held together by relatively weak forces, usually hydrogen bonds, metal–ligand coordination, and hydrophobic effects. The chemical nature of the attractive interactions involved in holding the capsule is sometimes directly involved in the catalytic activity, while in other cases they act as simple spectators and the effect of the host is to favor compartmentalization. Notable advantage of capsules over cavitands is their tighter and prolonged binding that permits the reversible interaction with elusive guests such as gases at mild ambient experimental conditions [18, 19]. The following discussion aims at focusing the attention of the reader on the aspects of the complementarity between the supramolecular catalyst and the nature of the reaction to be catalyzed; in fact, the interactions involved provide stabilization of the transition state of the reaction more closely mimicking enzyme action. The topic is gaining attention by the scientific community and several aspects have been recently and brilliantly covered in perspectives, highlights, and review articles [20–22].

4.1.1
Weak Intermolecular Forces

The forces that are responsible of intermolecular attractive interactions are generally small in terms of energy involved and much weaker compared to covalent bonds (170–450 kJ/mol). Their energies are metal–ligand interactions (40–130 kJ/mol) [23], ionic interactions (2–10 kJ/mol) [24], hydrogen bonding (8–80 kJ/mol) [25], cation-π [26], anion-π [27], CH-π [28], and the recently outlined halogen bond (0.2–3.0 kJ/mol) [29]. Nevertheless, thanks to their cooperative action, they are responsible for phenomena that span from the subnanoscopic catalytic efficiency of enzymes to the macroscopic aggregation states of matter.

To further stress this point which is essential to understand enzymatic catalysis, it is worth noting that weak interactions can influence greatly the selectivity of reactions and simple organometallic systems were developed to demonstrate this. Recent examples showed that a new era of asymmetric catalysis is on the horizon by simply using chiral counteranions in place of chiral cationic complexes; the only prerequisite is a tight contact in the ion pair and this is favored in the presence of noncoordinating solvents. Asymmetric induction through such kind of ionic interactions has been shown by different research groups using chiral, enantiopure atropoisomeric phosphate esters in the intramolecular cyclization of allene

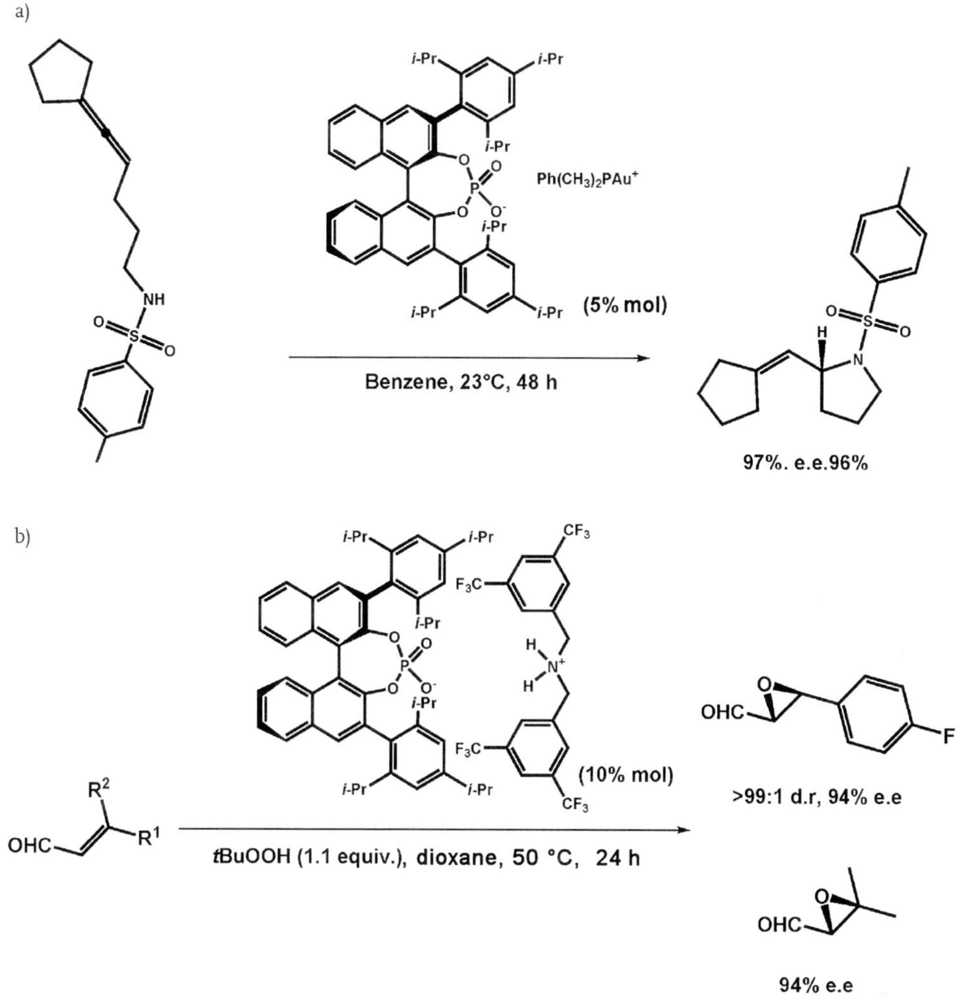

Figure 4.2 Enantioselective catalytic reactions where asymmetric induction is provided by chiral counteranionic species: (a) intramolecular cyclization and (b) enal epoxidation.

sulfonamides with Au(I) achiral complexes (Figure 4.2a) [30], as well as in the epoxidation of enals with achiral amines via iminium cation formation (Figure 4.2b) [31].

4.1.2
Compartmentalization and Catalysis

The presence of extensive surface interaction between catalyst and substrate is a pivotal prerequisite both for selectivity and catalytic activity of synthetic enzymes.

The rigidity of the catalyst is crucial for these purposes, and recent studies have more precisely defined this issue. In particular, it was demonstrated that substrate binding should not be too loose because this would reduce catalytic activity, but also not too tight; the best results are observed with a catalyst that maintains a cavity that provides proper substrate orientation, offering additional intrareceptor attractive interactions that reinforce molecular recognition. This results in good activation and selectivity without limiting dynamic exchange or turnover capability [32].

Another important aspect of host–guest supramolecular interaction is worth mentioning: in encapsulation phenomena where the host completely surrounds the guest, a rather general phenomenon related to the packing coefficient (PC) [33] was observed. This is defined as the ratio between the volume of the guest and the volume of the cavity. The PC was observed to be generally about 0.55 for system characterized by good binding, while $PC < 0.45$ or > 0.60 were observed for guests that did not show specific interactions with the host. This value is also commonly observed in organic solvents, and as long as solvation effects are dominant, like those observed in capsules, about half of the available space can be filled with the guest. The remaining free space allows for the appropriate thermal motion of the guest molecules. Accordingly, simple solvation effects are not enough to impart strong catalyst–substrate interactions; instead, the presence of specific attractive interactions within the cavity is the best solution to the demand for selective recognition (Figure 4.3).

As long as bimolecular reactions are concerned, the main effect displayed by supramolecular self-assembled catalysts is to favor gathering of the substrates in close proximity and in the proper orientation to promote functional group reciprocity. The overall effect is to transform the intermolecular chemical reaction to something more like an intramolecular one. The effective molarity defined as $EM = k_{intra}/k_{inter}$ is the ratio between the intramolecular and the intermolecular reaction rates, and provides a value, expressed as molar (M) concentration, that represents the hypothetical concentration of one of the reactants required for the intermolecular reaction to proceed with a pseudo first-order specific rate equal to that of the intramolecular reaction [34]. The higher the EM value of a catalyst, the more efficient is the catalyst. This provides a measure that depends solely on the way the catalyst acts as a template holding simultaneously the reactants in a reciprocal arrangement that is suitable for the reaction to occur.

Affinity and selectivity are two important recurring concepts in supramolecular chemistry that are generally thought to be inter-related, that is, the higher the binding, the better the selectivity. A recent review by Schneider on the subject concludes that this statement is true as long as the binding is driven by one single kind of weak interactions, but it fails when different contributions add together to make the overall affinity and selectivity. The best system would be a host where a primary interaction is responsible for most of the binding, and a secondary interaction accounts for the selectivity [35]. The same concepts are also strictly related to catalysis; in fact the affinity expressed by catalysts in binding transition states is responsible for the acceleration imparted, while the preference in binding for one

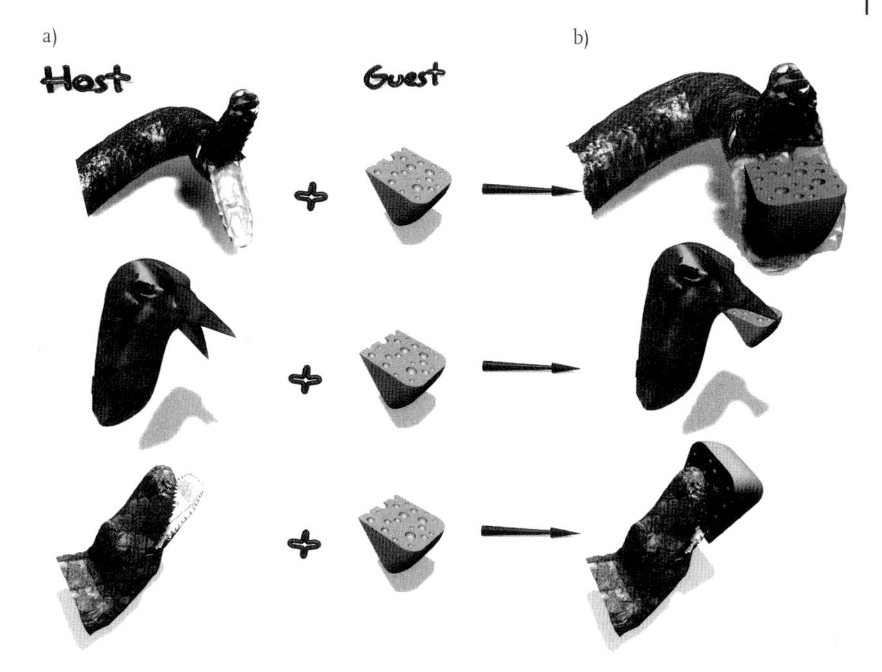

Figure 4.3 Schematic representation of the evolution of synthetic supramolecular receptors: (a) loose binding of a flexible host that changes its conformation to better complement the guest (snake's mouth); (b) tight binding of a rigid host that is intrinsically modeled to complement the guest (bird's bill); (c) additional secondary attractive interactions present within the binding pocket of a rigid host enhance the selectivity (teeth in the crocodile's mouth).

over several possible transition states (selectivity) influences the selectivity of the catalyst at all levels. While in several chemical transformations activity and selectivity have been demonstrated not to be related [36], the ideal development of catalysts with primary interactions (activity) and secondary interactions (selectivity) for transition states would provide real synthetic enzymes. Cavitands and capsules, thanks to their large surface interaction with substrates, may be the most promising synthetic catalysts on the way to enzyme mimicking. In fact such catalysts with enhanced host properties provide a sort of solvation of reagents, and they hold such solvation long enough to potentially impart high activity and selectivity [37]. Exchange rates for guests are important in catalysis, and the in/out mechanisms involved are numerous [38]. The reversible formation of the assembly is a prerequisite for capsules to promote catalytic turnover. Moreover, controlling these dynamic processes would theoretically allow the regulation of catalytic activity through something reminiscent of allosteric effects. In nature, these involve signaling molecules that bind at remote sites and switch enzymes between on and off states.

4.1.3
Cavitands and Self-Assembled Capsules as Synthetic Enzymes

The vase-shape molecular containers and the self-assembled chemical structures developed in the recent years displaying impressive host–guest chemistry are many. The present work is not intended to represent an exhaustive gallery of the different kinds and features of supramolecular catalysts; rather the aim of the authors is to provide the reader with selected examples among the most representative cavitands and self-assembled capsules that have become landmark contributions in the development of synthetic enzymes and in the development of enzyme mimics.

The chemical structure of the hosts is reported together with a molecular model and the cartoon which represents the host, and will be used subsequently to simplify the discussion of catalytic activities. In the realm of synthetic receptors, cavitands represent one of the most important actors; in fact, their vase-like shape complements the convex guest and the high surface interactions between host and guest often impart high selectivity. This is a key feature that can open the door to efficient synthetic enzymes. In the following section, three main examples of cavitands are described and discussed later in their catalytic performances. The first (Figure 4.4a) is a unimolecular species developed by Rebek and collaborators where intramolecular hydrogen bonds between adjacent amide residues maintain the vase-like shape that forms a hydrophobic pocket able to accommodate suitable guests [39]. The hydrogen bond seam on the rim of the cavitand is characterized by a dynamic behavior that interconverts two cycloenantiomers as observed when chiral enantiopure guests were bound with formation of two diastereoisomeric complexes. The preferred guests for this cavitand are cationic organic species where attractive cation–π interactions exist with the walls of the host. The seam is broken during the uptake and release of guests, that is, the cavitand folds around its target much as an enzyme folds around its substrate.

Another landmark cavitand which is self-assembled from multiple subunits was introduced by Fujita and collaborators [40]. This was based on the metal–ligand coordination between four 2,4,6-tris(3-pyridyl)-1,3,5-triazine walls and six Pd(II) metal centers $[M_6L_4]^{12+}$ (Figure 4.4b). The pyramidal structure is water soluble and provides a polycationic chamber that beautifully complements anionic species. The interior is hydrophobic and favors, through the hydrophobic effect, the binding of congruent neutral guests as well.

A third supramolecular structure bearing several metal–ligand interactions characterized by a closed shape, which can be considered in between a cavitand and a capsule, was developed by Reek and collaborators and is based on the employment of ligands bearing one soft donor atom and three hard donor atoms, P and N, respectively [41]. Such ligands can bind three rigid porphyrin- or salen-type walls and one soft catalytic metal complex, thus providing a self-assembled cavitand with a catalytic center in the middle (Figure 4.4c). This assembly is not very deep, and substrate binding is driven by the soft metal center while the walls of the cavitand impart proper size and shape selectivity basically operating as steric constraints.

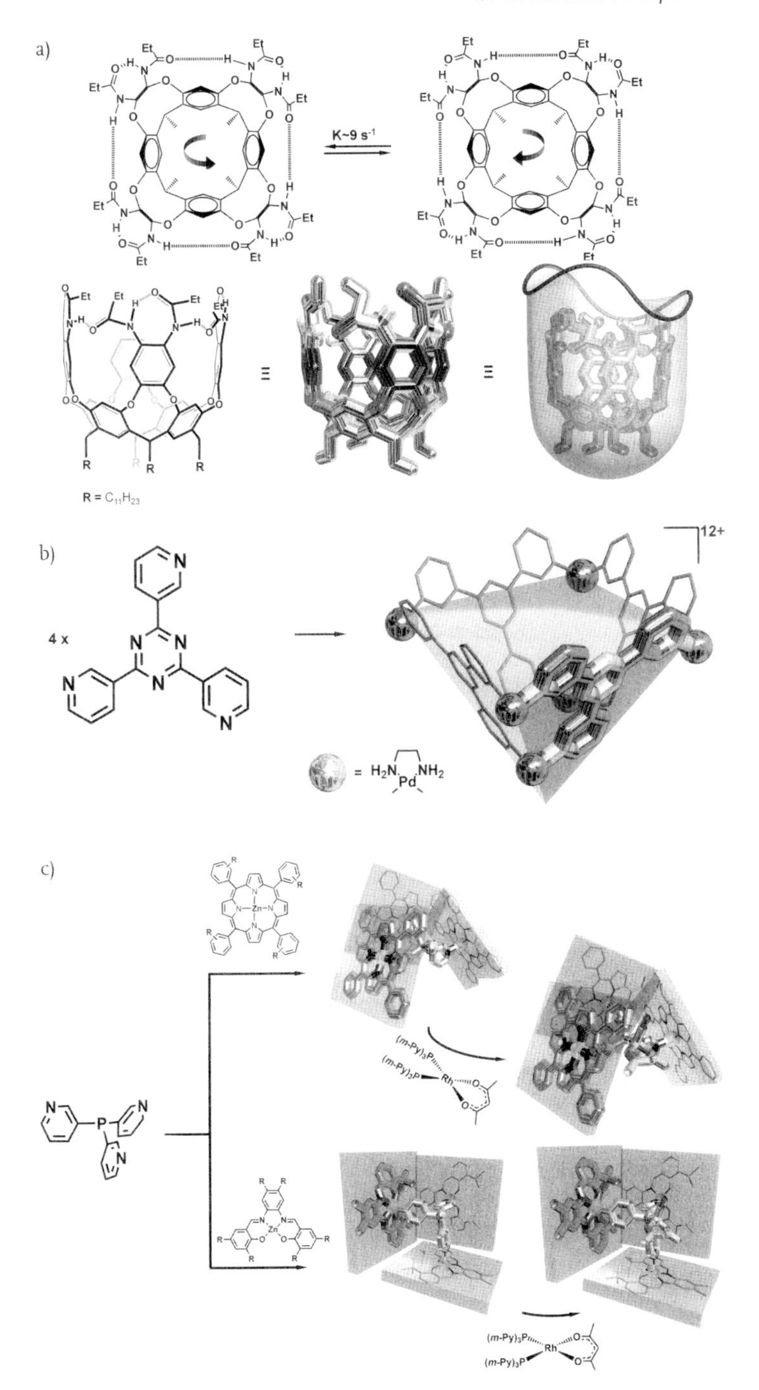

Figure 4.4 (a) Hydrogen-bonded, self-folding cavitand; (b) metal–ligand self-assembled cavitand; (c) metal–ligand self-assembled structures (shielded catalyst).

Self-assembled molecular capsules represent a further step in the mimicking of natural receptors, as they more or less completely surround their target(s). These can be divided into three families, hydrogen-bonded capsules, metal–ligand assembled capsules, and hydrophobically assembled capsules. The first family was introduced about 15 years ago by Rebek. The hydrogen-bonded self-assembled capsules range from the smallest dimeric "tennis ball" to the large hexameric resorcin[4]arene based capsules [42, 43]. Two most interesting examples that display unique catalytic activities are reported in Figure 4.5. The first is the cylindrical dimeric capsule which features a resorcin[4]arene base to which four aromatic walls with imide functionalities on the top were attached. Two modules self-assemble through a network of eight bifurcated hydrogen bonds and result in a cavity of about 425 Å^3 that can be filled with up to three guest molecules, with preference for neutral or anionic species. The second capsule is the slightly larger dimeric "softball" compared to "tennis ball" where glycoluril subunits are implemented into a polycyclic rigid structure bearing bridged bicyclic centerpiece. This provides a concave surface with hydrogen-bonding units at the extremities that self-assemble into a spherical dimer with a cavity of about 400 Å^3 that can accommodate usually up to two guests [44]. In both capsules, guest exchange occurs through opening of flaps of the dimeric assembly. Slow in-out exchange of guests on the chemical shift timescale was observed by NMR investigations for most of the guests. Dimerization is controlled by solvents; those that compete well for hydrogen bonds such as alcohols readily dissociate the capsules, whereas those that do not (aromatics, $CHCl_3$, and the like) show assemblies with association constants of $>10^5\,M^{-1}$.

The second family of capsules is based on metal–ligand coordination for holding together the subunits. This strategy is characterized by a general higher directionality and tighter binding [45, 46] that allows the preparation of capsules based on a higher number of subunits [47] and that, thanks to a general higher solubility in water, can take advantage of the hydrophobic effect. These favor coencapsulation and better control of molecular interactions by the hollow surface of the coordination cages [48]. Two most interesting capsules in terms of catalytic activity are those developed by the groups of Raymond and Fujita. The first is a tetrahedral structure $[M_4L_6]^{12-}$ where catecholamide residues attached to a naphthalene core self-assemble with trivalent hard metal centers (Figure 4.6a) [49]. The structure thus obtained is characterized by a polyanionic nature which imparts solubility in water. The bidentate ligands and the octahedral geometry of the metal centers provide chiral Δ or Λ configuration to each corner, and the mechanical coupling between the corners imparted by the rigid ligands results in the exclusive formation of homochiral enantiopure Δ,Δ,Δ,Δ or Λ,Λ,Λ,Λ capsules with a chiral cavity. The in-out exchange was investigated in detail and in most cases it occurs through a nonrupture mechanism where guests squeeze through apertures created by deformation of the capsule leaving the host completely assembled [50, 51]. Such capsules provide a hydrophobic cavity of about 300–500 Å^3 that can accommodate organic guests in water. Organometallic species can be encapsulated as well; they must have proper size, shape, and preferentially a positive charge in order to better

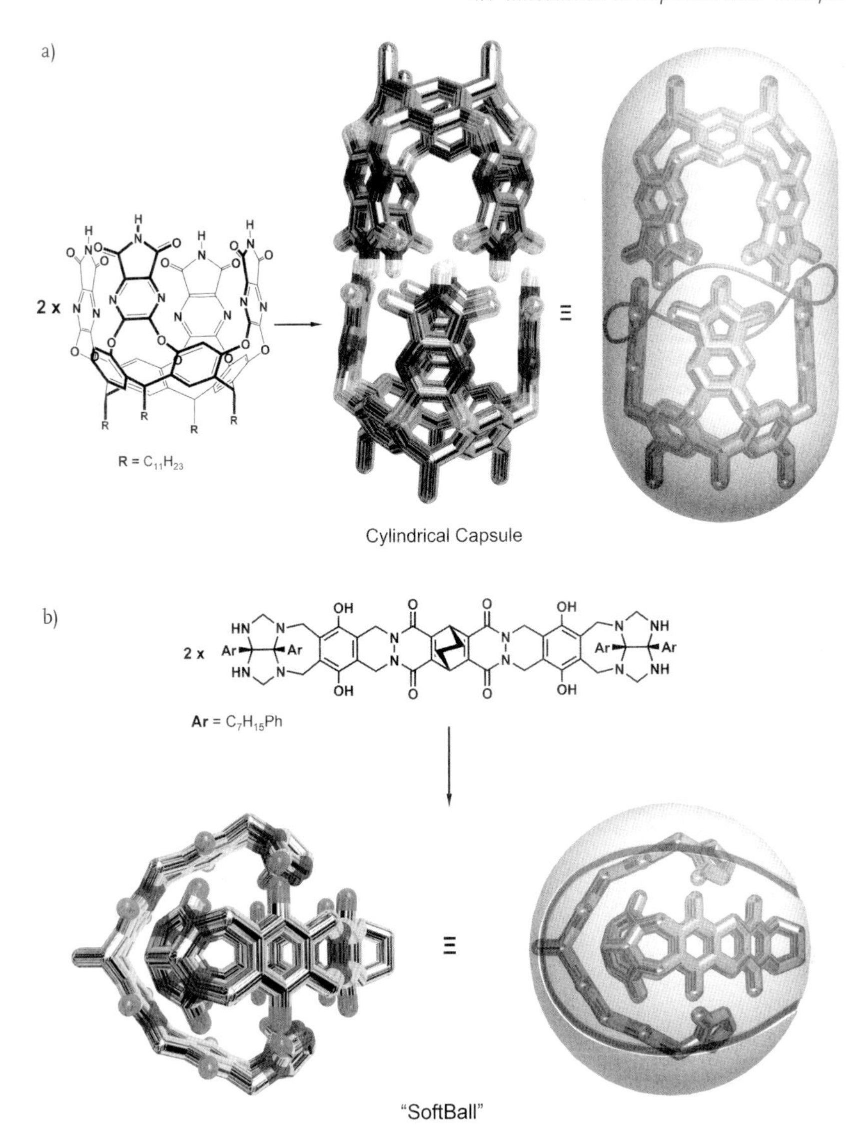

Figure 4.5 Chemical structures, stick molecular models, and cartoon representations of (a) cylindrical dimeric capsule; (b) softball dimeric capsule. In the modeled structures, some peripheral substituents have been omitted for clarity.

complement the anionic corners and to interact with the aromatic surfaces of the inner space of the capsule through cation–π interactions.

The second important metal–ligand self-assembled capsule was developed by Fujita and collaborators, and is a direct variation of the cavitand reported in Figure 4.4b. By substitution of the 2,4,6-tris(3-pyridyl)-1,3,5-triazine ligand with the 2,4,6-tris(4-pyridyl)-1,3,5-triazine ligand and coordination to Pd(II) corners, a

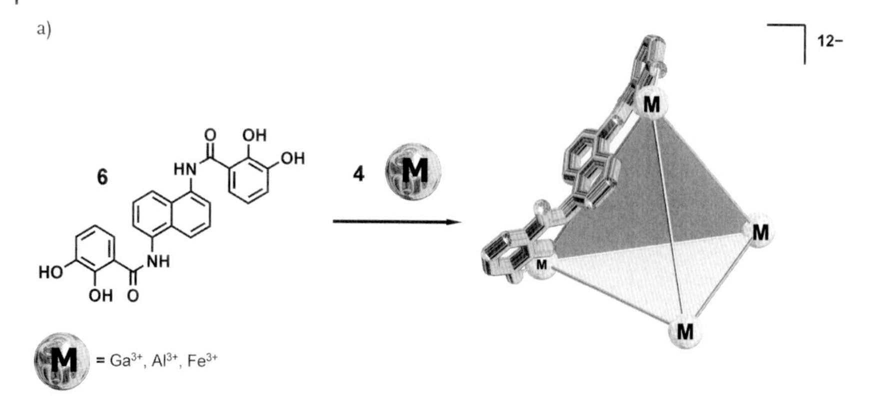

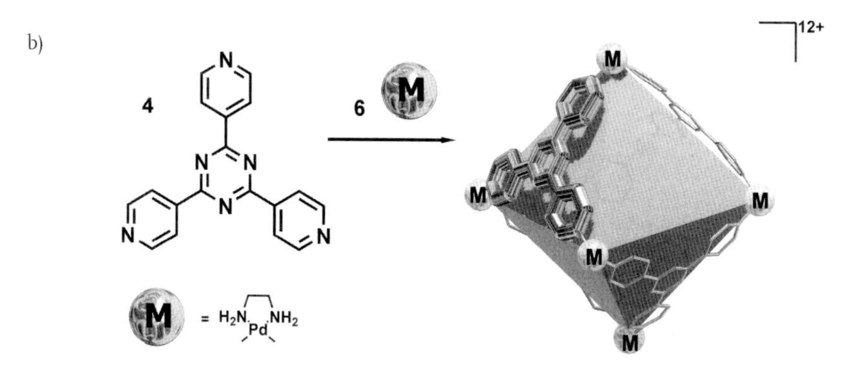

Figure 4.6 Chemical structures, stick molecular models, and cartoon representations of (a) tetrahedral $[M_4L_6]^{12-}$ capsule; (b) octahedral $[M_6L_4]^{12+}$ capsule. In the modeled structures, some substituents have been omitted for clarity.

$[M_6L_4]^{12+}$ assembly (Figure 4.6b) results. The octahedral structure is similar to the parent cavitand in terms of guest selection, and it prefers anionic and neutral species. The cavity can accommodate up to two molecular guests with one of the highest pairwise selectivities ever observed for molecular capsules.

The third kind of capsule is a recent discovery of the group of Gibb [52] that developed a deepened cavitand based on resorcin[4]arene scaffold to which two further aromatic belts were attached with overall eight carboxylic groups that impart solubility in water. The apolar cavity of the structure together with the use of water as solvent allows the binding of apolar guests. These are buried in the interior of the cavity with consequent dimerization of the host aided also by π–π stacking of aromatic residues that adorn the rim of the cavity. The guests in this case can be accommodated in an overall available volume of about 500–800 Å^3. The guests act as templates for the formation of the assembly, which is a process completely driven by the hydrophobic effect. Guest exchange is a rather slow

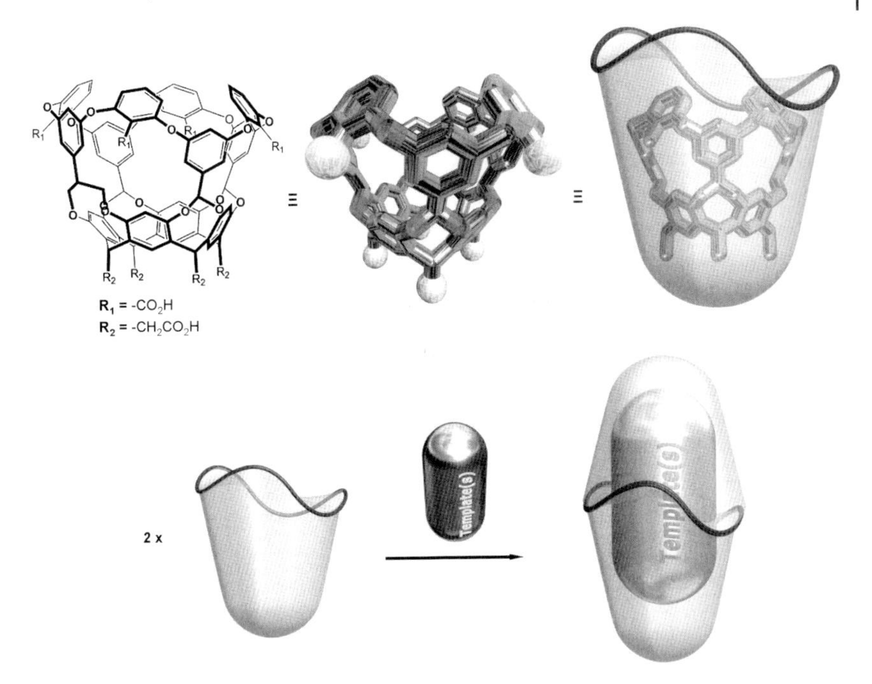

Figure 4.7 Chemical structure, stick molecular model, and cartoon representation of the dimeric octa-acid capsule held together by hydrophobic effect with templating guest. In the modeled structures, some substituents have been omitted for clarity.

process that takes place in the order of $0.1–1\,s^{-1}$ and is governed by the assembly and disassembly process (Figure 4.7).

4.2 Compartmentalization of Reactive Species in Synthetic Hosts as Supramolecular Catalysts

In the macroscopic world, compartmentalization is a common feature of living systems; in fact every living organism is composed of several isolated organs, each one separated from the rest of the body and deployed to accomplish a precise function. Looking closer, it can be seen that the cell membrane isolates the ingress and egress of the cell content from the bulk of the solution, and it regulates the exchange of all the chemical entities by means of active or passive control. Going down further to a nanometric view, in the active site of enzymes the substrate is isolated from the bulk solution and it resides transiently into a new environment that is highly complementary to it and, more precisely, much more complementary to the transition state of the reaction. This is a major difference with common organic or organometallic catalysis where the substrate interacts with the catalyst with a limited portion of its surface, while the rest remains in contact with the bulk solution. Better proximity between reactants, stabilization of intermediate species, new reaction pathways are all common features of catalysis observed ubiquitously in homogeneous organic/inorganic/organometallic as well as in heterogeneous and enzymatic catalysis. But only in the latter case thanks to the very large surface interactions between substrate and catalyst, selectivity issues reach the highest levels. These are directly related to the discrimination among possible reaction pathways that can be very similar and separated by small activation barriers: the more intimate the contact between substrate and catalyst, the higher the control on the pathway of the reaction and the selectivity of the process. Often the effect of the enzyme is to stabilize a highly reactive intermediate species for a sufficiently long time to drive it to the formation of a certain product. The same species could not survive in the bulk solution where it would proceed to decomposition or some other unproductive destiny. Segregation in a suitable environment is, therefore, one of the simplest and most effective tools observed in nature to solve such problems, and stabilization of reactive species by means of complexation in preorganized cavities is a commonplace in biological catalysis. Obviously, the in–out motion of the intermediate dictates the average lifetime of the species. Chemists have been studying such phenomena for a long time, initially isolating reactive species within covalent assemblies where the unstable species can be created but where they must reside because of the lack of suitable exits. Landmark examples provided by Cram and Warmuth are the stabilization of cyclobutadiene, *o*-benzyne, and cycloheptatetraene in covalent hemicarcerands thus preventing them from dimerization or reaction with the bulk-phase species [53–55]. Further steps toward the mimic of reversible uptake and stabilization of reactive intermediates by synthetic hosts was accomplished using self-assembled capsules, in particular those held together by metal–ligand bonds which are characterized by higher stability. Examples are the enclathration of two molecules of *cis*-4,4′-dimethylazobenzene within an octahedral capsule where the N=N double bond does not undergo photoisomerization under irradiation even after few weeks at room temperature. The geometric constrain provided by the host and the slow in-out exchange causes the guest to reside within the cavity (Figure 4.8a) providing the sources of longevity [56].

Organic reactive species like diazonium and tropilium cations [57] once sheltered within the tetrahedral metal–ligand self-assembled capsule were stable in solution for time periods much longer than those in the bulk solvent. The lifetime for the reactive encapsulated guests in some cases can increase up to a hundred times, as observed for the organometallic species [CpRu(*cis*-1,3,7-octatriene)]BF_4 [[58] that, once sheltered within the capsule, it is stable for several weeks in aqueous solution, while in a weakly coordinating solvent like CD_2Cl_2 it survives only for 20 h at room temperature (Figure 4.8c). Hydrogen-bonded capsules are usually more labile, but the proper combination of host and guest allows in some cases highly stable adducts like in the case of benzoyl peroxide which can be easily accommodated within the cylindrical capsule as in Figure 4.8d [59]. Protected by the nanoenvironment, the peroxide is stable for at least 3 days at 70 °C, while in

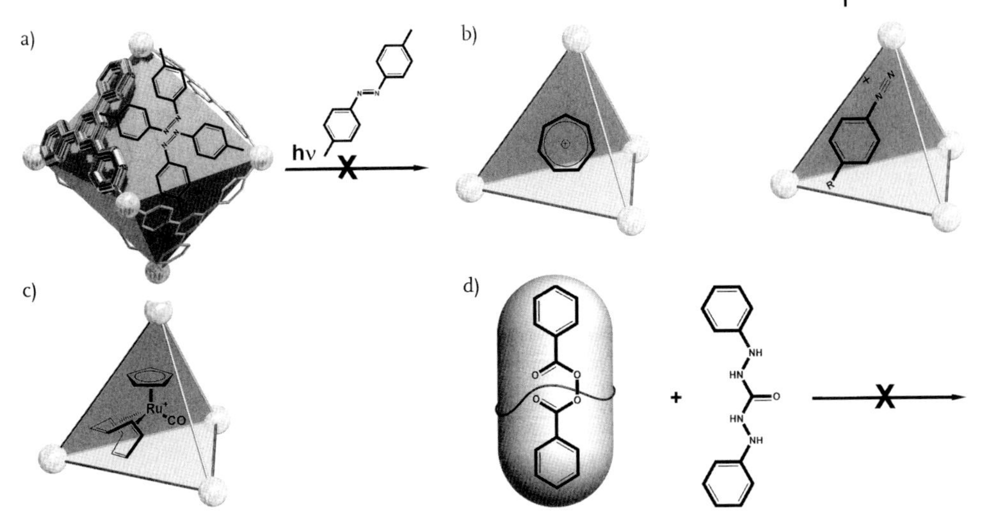

Figure 4.8 (a) Stabilization of a *cis*-diazoaryl derivative under photoirradiation by encapsulation; (b) stabilization of tropilium and diazonium cations; (c) stabilization of reactive Ru-alkene species CpRu(COD)CO; (d) encapsulation of dibenzyl peroxide thwarts its homolysis and decomposition.

solution it can survive no longer than 3 h. Only the addition of competitive species like *p*-[*N*-(*p*-tolyl)]toluamide or hydrogen-bonding entities like DMF or acetic acid liberates the guest that returns to express its oxidizing power toward phosphines or 1,5-diphenylcarbazide as common reducing agents (Figure 4.8d).

Protection of molecules by encapsulation together with the size and shape selectivity typical of capsular hosts are responsible for unique reaction profiles observed in the simple coupling of an aromatic carboxylic acid and aniline with dicyclohexylcarbodiimide (DCC) as coupling agent (Figure 4.9) [60]. While the two different acids react with similar rates and profiles in the bulk, when the cylindrical capsule is present in solution both reactions are slower in rate but a marked difference in reactivity is observed. The longer acid reacts with a rate that is approximately five times slower than the shorter and the kinetic profile is characterized by an induction period. The capsule is initially filled with DCC and the coupling is initiated by the residual carbodiimide present in solution. As soon as the shorter product is formed, this being a better guest than DCC (4-methylbenzamide ~ dicyclohexylurea > DCC), it displaces the latter that moves to the bulk solution where it mediates the formation of more products, giving rise to the amplification effect observed. The same is not observed with the longer reagent that is not a suitable guest for the capsule. The capsule does not influence the reaction between reagents and DCC, but it regulates the rate at which they encounter each other.

A step further in the comprehension of the phenomena involved in nature for stabilizing reactive intermediates came from the employment of properly tailored

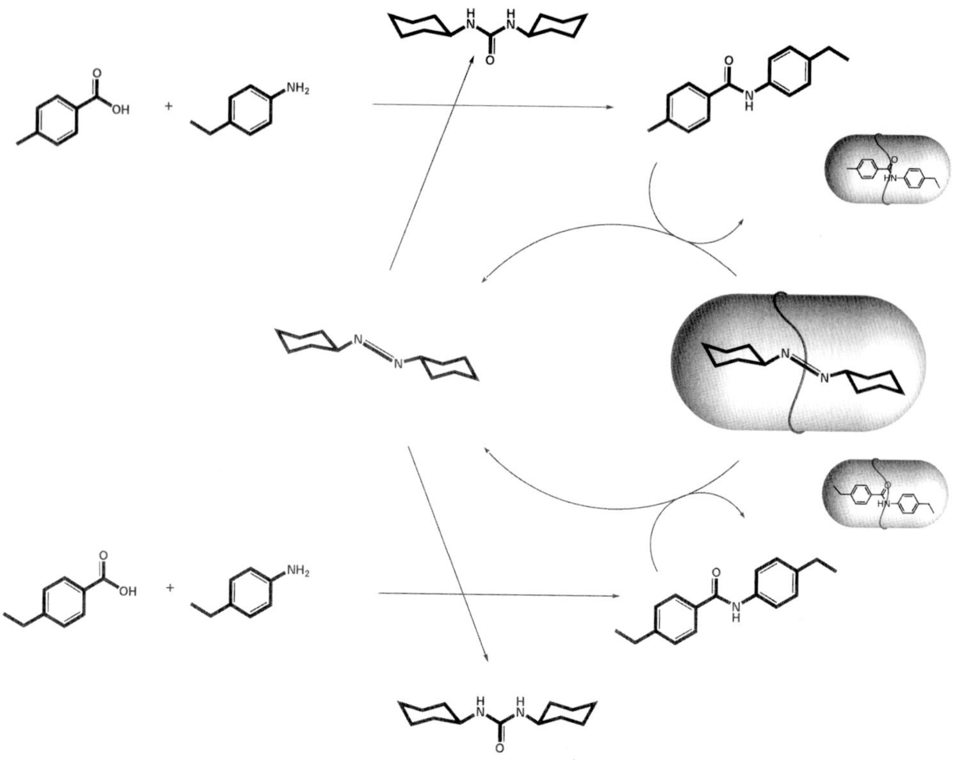

Figure 4.9 A self-assembled capsule can control the rate of reactions by sequestering selectively the reagents depending on their size, shape, and complementarity with the capsule's cavity. The shorter 4-methylbenzoic acid compared to the 4-ethyl substituted analog reacting with the aniline leads to a shorter amide product that better binds into the capsule's cavity. It liberates more coupling agent and this accelerates the reaction with an autocatalytic or chain reaction kinetic profile.

cavitands endowed with introverted functionalities like acids and aldehydes. Such constricted nanoenvironments, together with the employment of suitable precursors characterized by adequate sizes and shapes, allowed the direct observation of *O*-acyl isourea or *N*-acyl isourea [61] derivatives transiently present in the cavity (Figure 4.10a). Analogously, hemiaminals have never been observed except in particular cases with exotic reagents and cannot survive in the bulk solution. When isolated within the self-folding cavitand, they could live long enough to be fully spectroscopically characterized (Figure 4.10b) [62, 63]. The same strategy was employed to fully characterize by means of NMR and IR spectroscopy the intermediate isoimide products derived by interaction between an introverted acid and an isonitrile. Such species can be produced under mild conditions in the cavitand and protected from the attack of external nucleophiles [64].

Highly energetic conformations can be as difficult to stabilize as reactive intermediates. In nature, substrates can adopt highly energetic conformations once

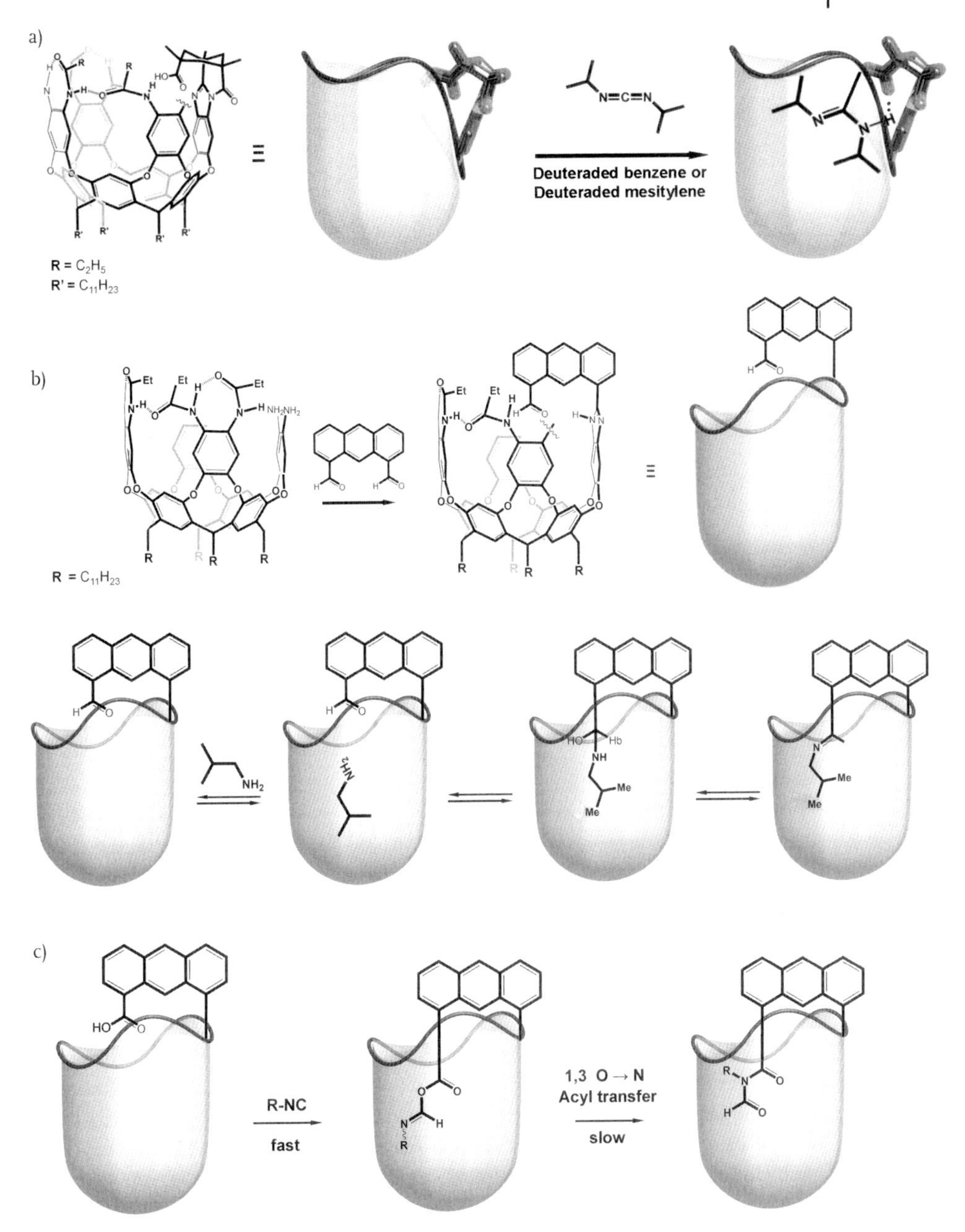

Figure 4.10 Examples of stabilization of reactive intermediates by isolation in cavitands: (a) *O*-acyl isourea or *N*-acyl isourea derivatives derived from the reaction between alkyl carbodiimide and an introverted carboxylic acid; (b) hemiaminals derived by the reaction between primary amines and an introverted aldehyde; (c) isoimide intermediates characterized within a cavitand.

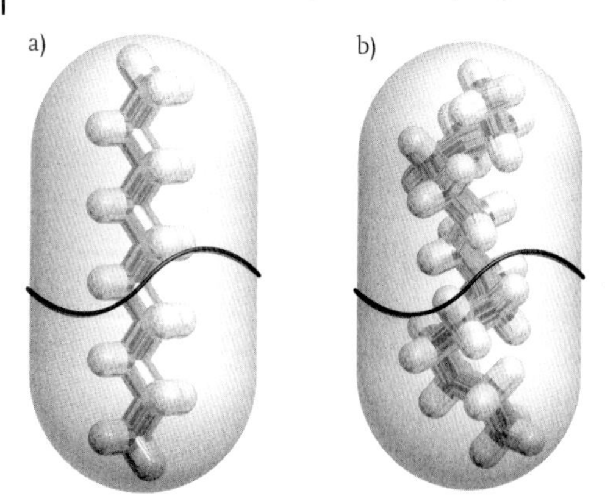

Figure 4.11 The cylindrical capsule can encapsulate (a) *n*-decane in its extended conformation, while (b) *n*-tetradecane, which is too long to be hosted in the same conformation, assumes a helical conformation where the strain energy derived by the presence of several *gauche* interactions is compensated by better CH–π and dispersion attractive interaction between the guest and the aromatic walls of the host.

buried in the enzymes' cavity that are not present when the species is free in solution. Often such conformations are responsible for peculiar selectivities and products observed. A noteworthy example of this phenomenon is found in the dimeric hydrogen-bonded cylindrical capsule. The semirigid aromatic walls that define the cavity of the capsule can accommodate linear alkanes characterized by precise length. For *n*-decane, the guest is bound in its elongated anti conformation but *n*-tetradecane can be hosted too, but only in a helical conformation where the enthalpic cost of several gauche interactions is overcome by better attractive CH–π interactions together with entropic solvent effects (Figure 4.11) [65].

4.2.1
Cavitands and Capsules as Synthetic Enzymes

Supramolecular synthetic catalysts operate both on a thermodynamic and on a kinetic level for a given chemical process. The following sections will initially describe those systems where the catalyst influences the equilibrium position of reversible reactions, while later examples will discuss the effects on reaction rates and product selectivity for irreversible reactions.

4.2.1.1 Reversible Reactions

Enzymes are the best catalysts known, their catalytic sites evolved to bind substrates in the proper orientation and contain suitable chemical tools to operate on the substrate and to stabilize the transition state of the reaction. Often they influence tautomeric equilibria largely favoring one species over the others by means of juxtaposition of complementary groups as well as favorable polarity and solva-

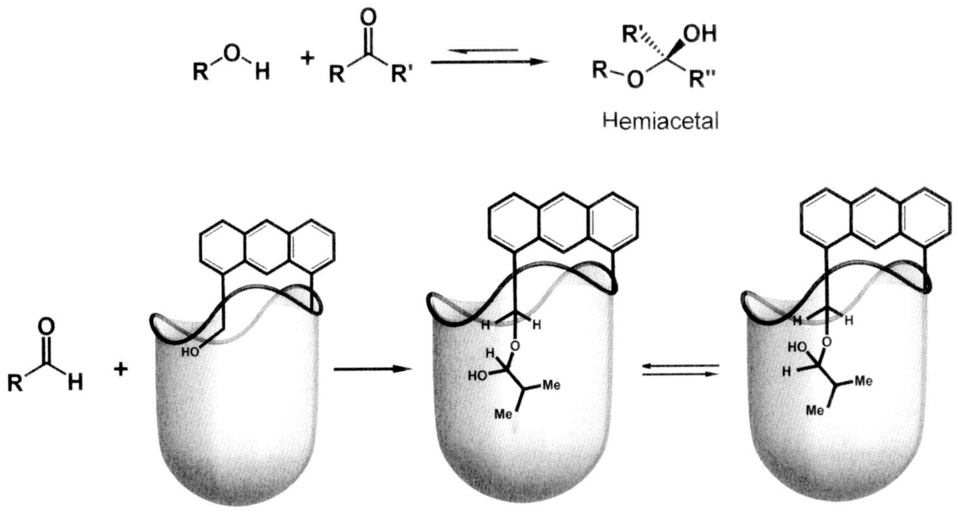

Hemiacetal

Figure 4.12 Hemiacetal stabilization within a cavitand bearing an introverted alcohol. The size of the aldehyde influences the equilibrium position with differences up to one order of magnitude between butanal and 2-methylpropanal.

tion effects. Similar behavior has been observed with cavitands and capsules, thanks to their cavities characterized by apolar surfaces lined by aromatic residues.

4.2.1.1.1 Cavitand Catalysts

A cavitand bearing an introverted alcohol (Figure 4.12) was used in the investigation of the reversible hemiacetal formation reaction between the alcoholic residue and carbonyl compounds, with the aim of investigating the stabilization effect imparted by the cavity (Figure 4.12) [66]. The addition of aldehydes to such an introverted alcohol in a noncompetitive solvent like mesitylene-d_{12} was followed by NMR, and rapid formation of the hemiacetal products was observed. This was confirmed by the presence of two diastereoisomeric species arising from the combination of the new stereogenic center present on the hemiacetal carbon and the clockwise or counterclockwise arrangement of the seam of amide hydrogen bonds that adorn the rim of the cavity. In these cases, enhancement in the order of 13- to 138-fold of the equilibrium toward the formation of the hemiacetal was observed for *n*-butyl and *i*-butyl aldehyde, respectively. More unexpected was the observation of hemiketal formation with $>10^5$-fold enhancement for the reaction of the introverted alcohol with cyclohexanone. The other possible combination of introverted aldehyde with free alcohols like propanols and butanols did not provide good results probably because of the small size of the guests. Exceptions were observed with *N,N*-dimethylethanolamine, *trans*-1,2-cyclohexanediol, and *trans*-1,2-cyclopentanediol with equilibrium enhancements from 5000-fold to

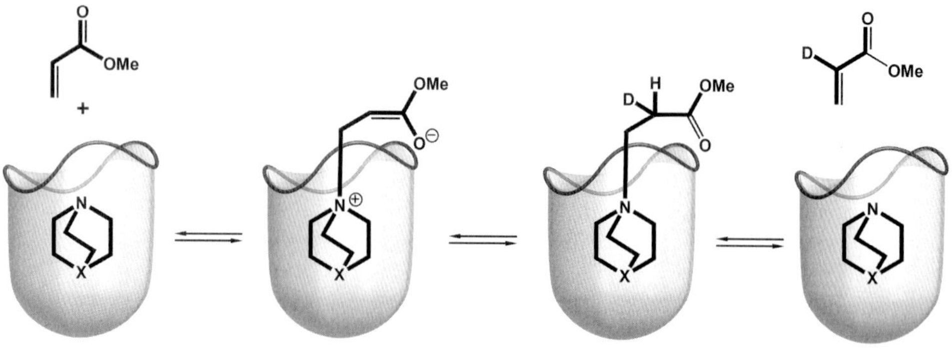

X= N
X= CH

Figure 4.13 Up to 1400-fold rate accelerations were provided by cavitand in the H/D exchange at the α position of methyl acrylate.

>10^5-fold that, especially in the latter two cases, are the result of the presence of extra intramolecular hydrogen-bonding acceptors that favor the stabilization of the hemiacetal.

The cavitand was also a suitable host for DABCO leading to the formation of a complex where the amine protrudes from the cavity; the nitrogen looks out into the bulk solvent like a moray looks out of its shelter (Figure 4.13) [67]. Such species in the presence of methyl acrylate in acetone-d_6 led to the substitution of the α-hydrogen of the acrylate with deuterium from the solvent with a mechanism similar to the Bayliss–Hillman reaction. The reaction with stoichiometric amount of cavitand was complete in 3 days, while without cavitand or with acetanilide a mimic of the cavitands wall, no reaction occurred. With quinuclidine as base, the acceleration observed was 1400-fold with methyl vinyl ketone and 800-fold with phenyl vinyl ketone. Stabilization of the enolate intermediate by the seam of amide hydrogen bonds that adorns the rim of the cavity appears to be the source of the enhanced rates (Figure 4.13).

4.2.1.1.2 Self-Assembled Capsule Catalysts

Self-assembled capsular nanoreactors can be characterized by rather slow exchange rates for ingress and egress of the reagents/guests but this allows sufficiently long contact times between the species and the inner surface of the cavity, ensuring stabilization and modification of chemical equilibria. It is, therefore, quite common to observe rather different equilibrium constants for reversible transformations when these occur within supramolecular cavities. Compared to the same reaction in the bulk solvent, amplifications of uncommon isomers can be driven by steric or electronic interactions with the inner surface of the cavity. For elongated guests in the self-assembled dimeric cylindrical capsule as those in Figure 4.14, the equilibrium between the Schiff base and the corresponding 1,4-dihydro-2*H*-3,1-benzoxazine ring forms was greatly altered. In the capsule, an increase in the

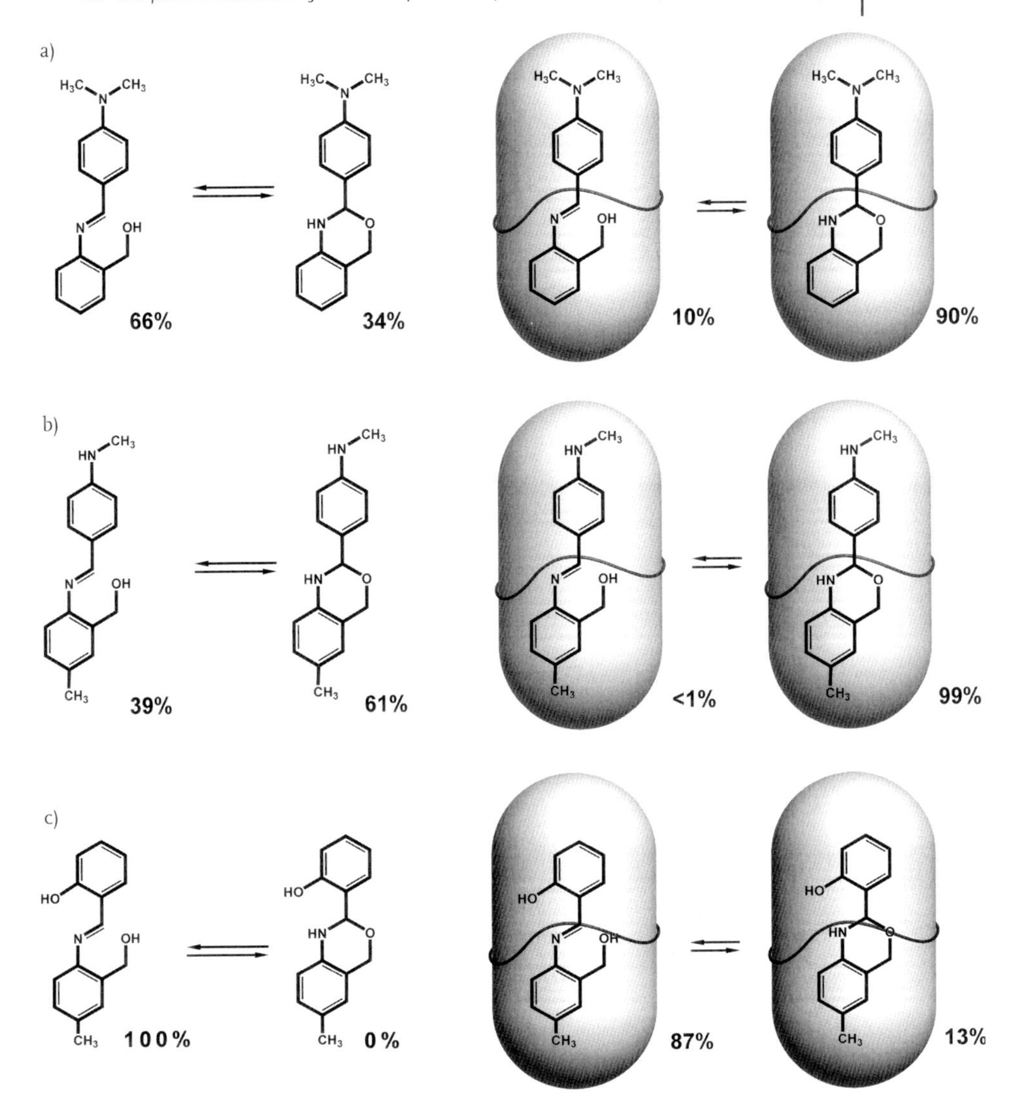

Figure 4.14 The cylindrical dimeric capsule shifts the equilibrium of the intramolecular cyclization to favor the closed benzoxazine isomer.

K_{in} of about 18-fold in favor of the ring form compared to K_{out} was found [68]. The interconversion is thought to occur while in the cavity, aided by the polar belt of hydrogen bonds that holds the capsule. In some cases, cyclic species that have never been observed in solution were detected in sufficient amounts to be characterized by NMR (Figure 4.14c). These cases provide further evidence that encapsulating unstable species is a viable method to access short-lived chemical entities.

Analogous behavior was observed for the tautomeric equilibrium $K = [\text{enol}]/[\text{ketone}]$ of a single long β-ketoester. In solution, a $K_{out} = 0.43$ was calculated while in the cavity K_{in} becomes 3.57, a value that has never been observed before in any solvent [69]. The ninefold increase in the concentration of the enol form can be ascribed to solvation effects due to the positioning of the enol proton in close proximity of the middle of the capsule where several hydrogen-bonding donors and acceptors are available. The encapsulation of smaller β-ketoesters left enough room to observe coencapsulation phenomena and the formation of two possible orientations of the guests in the capsule (Figure 4.15), the so-called social isomerism. The ratio between the social isomers was naturally influenced by the co-guest present, but more remarkably for both social isomeric forms the equilibrium constant between enol and ketone form was influenced by the co-guest as a unique example of a single molecule solvation effect. As far as the β-ketoester is concerned, replacing CH_3CH_2OH (52 Å^3) with a bigger CCl_3Br (114 Å^3) as the co-guest caused variations in $K_{CH_3\,middle}$ from 0.47 to 0.10 and $K_{CH_3\,bottom}$ from 2.54 to 0.28, while in the

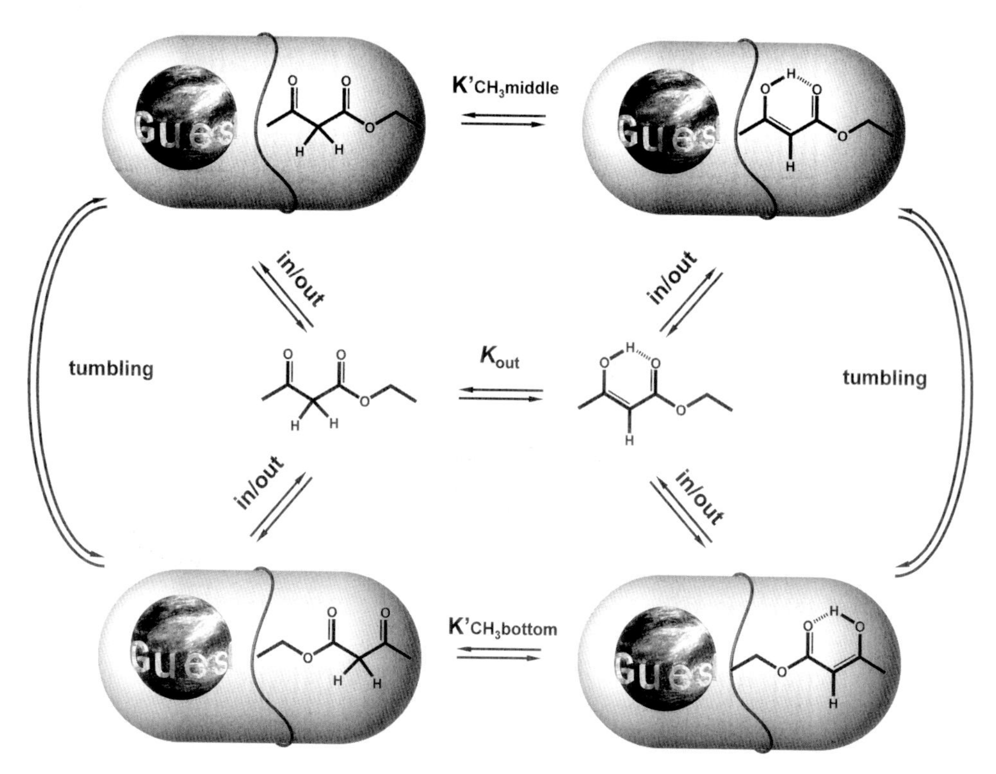

Figure 4.15 The dimeric cylindrical capsule places keto–enol partners of ethyl acetacetate into different arrangements called social isomers characterized by the acetyl group toward the bottom or the middle of the cavity. The lining of the cavity influences the equilibrium position, and a secondary guest plays another role in the keto/enol ratio related to its size.

bulk K_{out} remained 0.31–0.33. The co-guest's volume had a critical effect on pushing the tautomers away from the center of the capsule, thus enhancing attractive as well as repulsive interactions whose solvation effects were evident at a subnanometric level. This clearly mimics quintessential features of enzymes like guest solvation and extremely precise positioning for catalytic activity.

Metal/ligand self-assembled molecular capsules are usually water soluble and, apart from other features, they can exploit hydrophobic effect to steer equilibria toward uncommon positions. The tetrahedral assembly reported in Figure 4.16a combines a hydrophobic cavity with the presence of extended aromatic surfaces that nicely complements alkylammonium species through several attractive CH–π as well as cation–π interactions. This property was used to study the effect of the self-assembled capsule on the basicity of several encapsulated amines and phosphines isolated from the bulk solvent water [70]. Several mono- and diamines as well as phosphines proved to be encapsulated solely in the protonated form. As confirmed by mass analysis and different NMR techniques, a marked substrate selectivity occurs with a range of sizes that showed optimum binding. For instance, tetramethyldiamines separated by methylene units were good guests in the protonated cyclic form as long as the number of methylenes is between one and six, while with larger diamines no encapsulation occurred. With monoamines, tertiary ones like triethylamine showed broad signals while the larger tris-propyl derivative afforded sharp signals as expected for good encapsulation. An analogous trend was observed with the corresponding secondary amines. A thermodynamic cycle was applied to the system to determine the basicity shift imparted by the encapsulation phenomenon because of the difficulty observed in the encapsulation of neutral amines. The process was found to be largely entropically driven as expected by the presence of hydrophobic effect that governs such kind of host–guest interactions. The final effect on the basicity of the amine was a general shift of about 4 pK_a units, thus providing a clear example of a synthetic self-assembled host that mimics enzymes in altering acidity and basicity of the substrates.

The stabilization of reactive species once buried and sheltered in the deep cavity of enzymes is one of the most beautiful and effective tools used by nature to perform reactions that are incompatible with the aqueous medium. The self-assembled capsule showed several imitations of this phenomenon; one remarkable example was the stabilization in the cavity of reactive cationic species derived by the reversible addition of phosphines to carbonyl compounds (Figure 4.16b) [71]. Trialkylphosphines could be encapsulated in the protonated form into capsule, while in the presence of acetone in solution a new species was detected in the cavity which was identified as the phosphonium cation $[Me_2C(OH)PEt_3]^+$. Such species are not stable in the bulk water but could exist in the self-assembled capsule thanks to attractive interactions between the cationic guest and the hydrophobic aromatic cavity and the slow in-out exchange rate that ensured sufficiently long lifetimes.

The tetrahedral capsule showed a similar effect on the iminium equilibrium reaction between ketones and cyclic secondary amines [72]. The reaction in aqueous solution provides negligible concentration of the cationic species, but

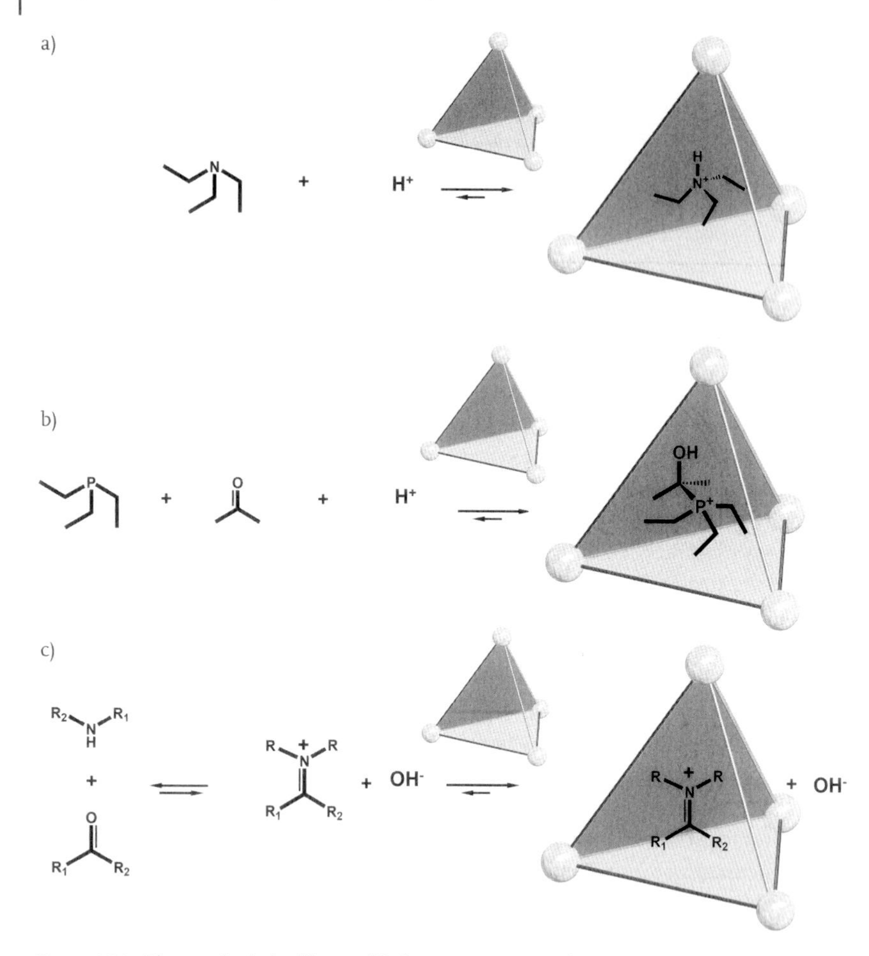

Figure 4.16 The tetrahedral self-assembled capsule provides a suitable cavity that stabilizes cationic species, shifting equilibria toward their formation. Examples are (a) protonated amines or more elusive and unstable structures such as (b) phosphonium and (c) iminium cations.

inside the self-assembled capsule the elusive iminium adduct could be isolated and characterized (Figure 4.16c). The size and shape of both ketones and amines are crucial for the proper filling of the cavity. A remarkable substrate selectivity was observed with competition experiments that underlined the preference for azetidine compared to pyrrolidine with 2-heptanone (74% vs. 26%, respectively) and 4-heptanone compared to 3- and 2-isomers with azetidine (74% vs. 18% and 8%, respectively). The high substrate and product selectivity displayed by the capsule discriminates both the position of the carbonyl group and the length of the chain. These are volumetric restrictions imparted by the cavity and are all typical features that are commonly encountered in enzymatic catalysis.

4.2.2 Irreversible Reactions

In the following section, irreversible reactions catalyzed by cavitands as well as capsules are described. The examples are divided into promoters that accelerate the reaction but, due to lack of turnover ability, must be employed in stoichiometric amounts compared to reagents, and truly catalytic systems that enable from few to several catalytic cycles and where catalysts are employed in a substoichiometric amount compared to substrates.

4.2.2.1 Cavitand Catalysts

4.2.2.1.1 Stoichiometric Catalysts

Unimolecular vase-like shaped cavitands are useful molecular containers which can efficiently bind reagents allowing the proper orientation and stabilization of the transition states. The first generation of cavitands was represented by simple molecular containers and was followed by a second generation where chemical functionalities, specifically selected to impart the desired catalytic activity, were incorporated. The cavitand reported in Figure 4.4a is an example of the former group, characterized by a π-electron rich cavity that greatly complements cationic organic species and is intrinsically a good candidate to stabilize transition states where positive charges are developed in the route from reagents to products. This cavitand was employed in the Menschutkin reaction, the quaternarization of tertiary amines with alkyl halides (Figure 4.17) [73]. Quinuclidine is a suitable guest in acetone with association constant K_a 40 M^{-1} and, in the presence of alkyl halides, rapid nucleophilic SN_2 reaction occurred between the reagents facilitated by the cavitand. The role of the host was to stabilize the incipient formation of cationic species thorough cation–π interactions as well as carbonyl–cation interactions. The acceleration compared to the uncatalyzed reaction was remarkable, with butyl chloride K_{acc}/K_{ctrl} 1300-fold and simple analogues of the wall of the cavitand displaying no activity, confirming the importance of preorganization of catalytic tools around the substrate to favor efficient stabilization of transition states. The steric

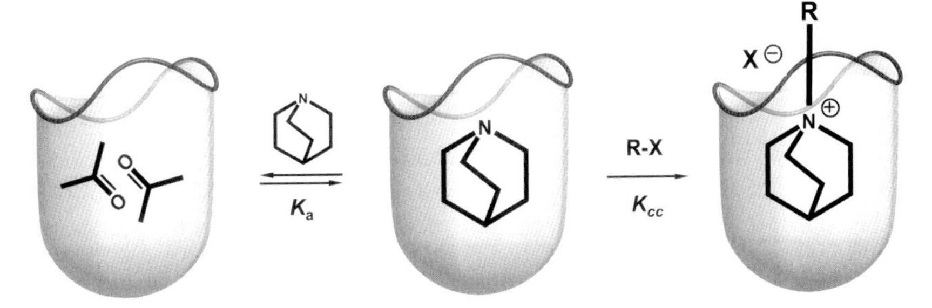

Figure 4.17 Rate accelerations for the reaction of quinuclidine with alkyl halides mediated by cavitand.

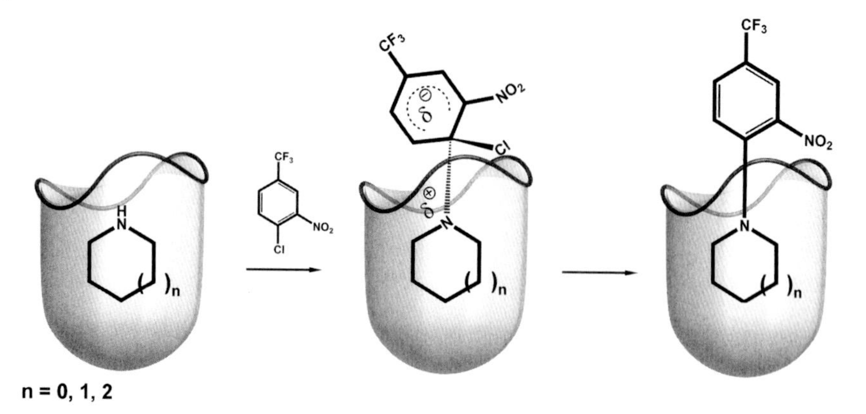

Figure 4.18 Rate acceleration imparted by cavitand for the nucleophilic aromatic substitution of electron poor chlorobenzenes via stabilization of the Meisenheimer intermediate.

properties of the alkyl halide had a remarkable effect on acceleration; in fact with bulkier electrophiles, butyl mesylate or tosylate acceleration K_{acc}/K_{ctrl} was in the order of 150- to 100-fold, while a more congested substrate like *t*-butyl bromide did not react at all. The products were all cationic species which hampered reagent binding; the system accelerated the reaction but was unable to provide turnover.

Analogous behavior was played by the same cavitand toward the nucleophilic aromatic substitution of cyclic secondary amines to electron-poor aromatic substrates bearing halogen-leaving groups (Figure 4.18) [74]. Amines are suitable guests for the cavitand with K_a in the range 14–50 M^{-1} under slow exchange on the NMR timescale. Upon addition of the aromatic substrate, rapid formation of the S_NAr product was observed with acceleration up to 100–400 times greater than the background uncatalyzed reaction. The cationic Meisenheimer intermediate and the developing charge present in the transition state of the reaction were stabilized by the juxtaposition of several amide groups. In bulk solution, the presence of four equivalents of an aromatic bis-imide wall of the cavitand had no effect on the reaction. Once again, product inhibition occurred and the cavitand had to be used in stoichiometric amount.

The same cavitand scaffold was more recently endowed with several chemical tools, both organic or organometallic, with the aim to catalyze different reactions. Modification of one of the aromatic walls with Kemp's triacid led to the synthesis of an introverted acid which can be easily converted into the corresponding methyl ether. The acid catalyst was tested in the intramolecular cyclization of ω-hydroxy-12,-epoxides. It provided the five-membered ring tetrahydrofurane or six-membered ring tetrahydropyrane alcohols depending on the attack of the hydroxyl moiety on the carbon atoms of the epoxide (Figure 4.19) [75]. The reaction was remarkably accelerated by the presence of the cavitand where the introverted Brønsted acid acted as an organocatalyst. The vase-like shape of the cavitand favored coiling of the reagent and greatly influenced the selectivity between the two isomeric cyclic

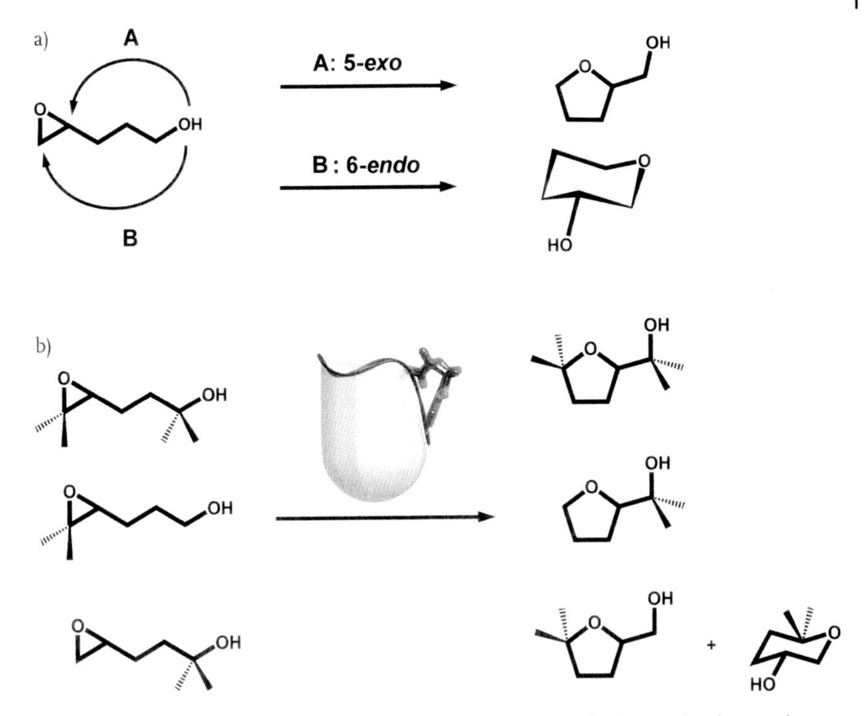

Figure 4.19 (a) The intramolecular cyclization of ω-hydroxy-1,2-epoxides provides five-membered ring tetrahydrofurane or six-membered ring tetrahydropyrane alcohols via 5-exo or 6-endo intramolecular attack. (b) The reaction catalyzed by the introverted acid cavitand favors the tetrahydrofurane derivative.

products that was also sensitive to the chemical structure, shape, and steric hindrance of the substrates employed. In fact, substrates bearing dimethyl groups attached to the epoxide led exclusively to the furanyl alcohols with greater selectivity compared to the reaction performed with simple acid catalysis that provided a 83 : 17 mixture of products. Moreover, 50- and 100-fold accelerations of the reaction compared to a carboxylic acid as catalyst were observed, confirming the cooperative effect of the presence of the acid residue together with the restricted space available within the cavitand. The cavity favored coiling and limited the number of possible conformers while enriching the system in those productive for the nucleophilic attack. Conversely, the substrate bearing dimethyl groups close to the alcohol nucleophile reacted much faster (with a 300-fold acceleration) but delivered a 1 : 1 ratio of the two regioisomeric products. This suggested a looser binding in the cavity and a higher degree of freedom due to diminished CH–π interactions.

The functionalization of the above-mentioned cavitand was further improved with the preparation of a species where one of the walls of the structure was replaced by a phosphite–oxazoline ligand connected to a Pd(II) center (Figure 4.20) [76]. This preparation followed the same strategy reported above and consisted of combining a catalytic site with a binding pocket, in this case to promote and

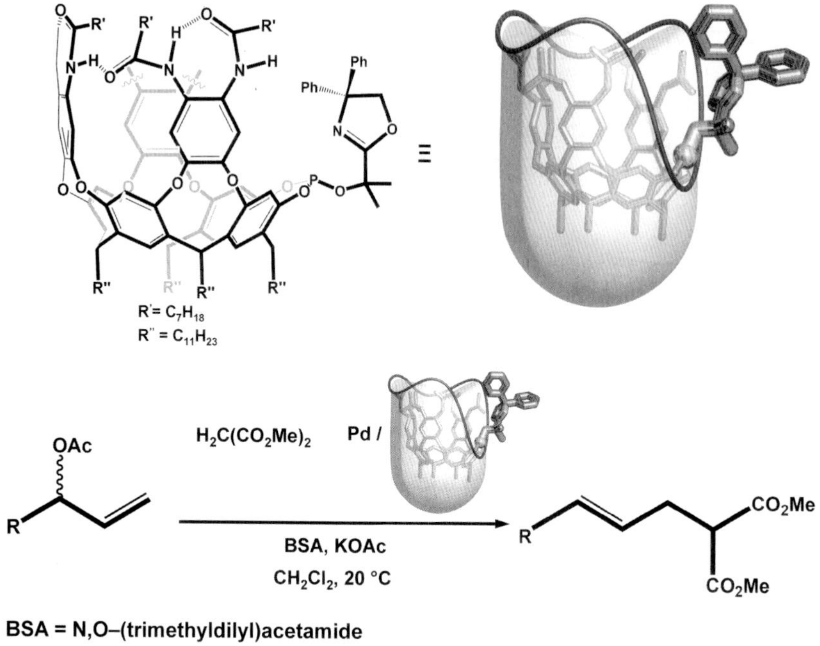

Figure 4.20 Allylic alkylation of various substrates with dimethylmalonate is more precisely modulated by cavitand-based phosphite–oxazoline ligand than with the free ligand. Marked substrate selectivity results through recognition of the R-group of the allyl acetate with the cavitand.

control allylic alkylation reactions. In this reaction, which is now a well-established chemical transformation whose mechanistic aspects have been thoroughly investigated, the nucleophilic attack takes place preferentially at the allyl terminus, *trans* to the Pd–P bond favoring reaction at the unsubstituted allyl end. The reactions between allylic substrates and dimethyl malonate as nucleophile in the presence of the simple complex Pd(II)–phosphite–oxazoline showed no differences as the R-residue of the substrate was varied. In contrast, when phosphite–oxazoline cavitand was employed the reactivity was much lower and marked differences in terms of reaction times and yield were observed with different R-groups. This effect is likely due to the interactions between the R-residues and the cavity of the catalyst that discriminates between substrates characterized by different size, shape, and bulkiness. The relative stabilization of the different cationic Pd-allyls is a crucial point in this reaction and was probed by competition experiments where combinations of two substrates were reacted with dimethyl malonate in the presence of the simple phosphite–oxazoline ligand and with the cavitand. Analysis of the system by means of mass spectrometry confirmed that the solution system showed equimolar formation of different cationic species no matter what was the reagent present, while in the cavitand the cyclohexyl substituent was preferred over *i*-Pr and 3-heptyl. This system represents a good example of subtle substrate specificity, and sheds light on the importance of proper tailoring of binding sites in order to better mimic and understand the peculiar selectivity observed.

As long as metal–ligand assemblies are concerned, one important example is based on the water-soluble square-pyramidal cavitand reported in Figure 4.4b that, because of the presence of extended electron-poor aromatic surfaces, can easily complement hydrophobic electron-rich aromatic guests. This self-assembled cavitand was exploited as catalyst for the synthesis of silanols from trialkoxysilanes, a reaction that leads to a wide range of possible products – monomers, dimers, trimers, and several oligomers [77]. The reaction of 2-naphthyltrimethoxysilane in water in the presence of the cavitand at room temperature led to the binding of two molecules of the reagent in the cavity, and once heated at 100 °C their complete and selective conversion into the dimeric silanol as the sole product was observed within 1 h (Figure 4.21a). Analogously, *m*-biphenyltrimethoxysilane provided the corresponding dimer in 84% yield. This is a remarkable example of cavity-directed synthesis where the hydrophobic effect together with strong π–π interactions favors the close proximity between reactants and limit possible synthetic pathways to those that provide suitable intermediates for the nanometric reaction chamber. The inevitable side effect of such good complementarity between the cavitand and one of the possible products was product inhibition and the lack of catalytic turnover of the system.

Olefins photodimerization reactions can also be catalyzed by this remarkable cavitand [78]. For example, the cavitand in water in the presence of naphthoquinone formed initially a host–guest species bearing two molecules of guest bound

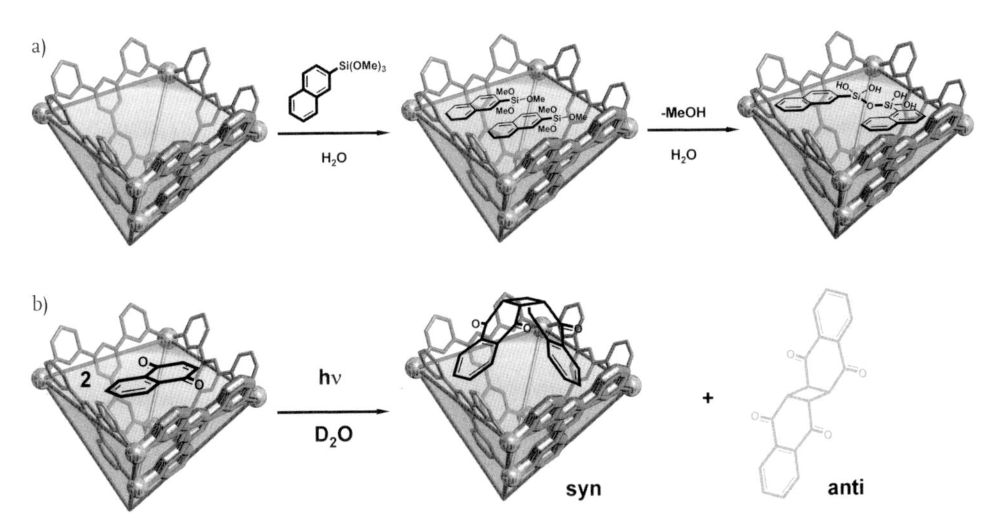

Figure 4.21 Self-assembled water-soluble cavitand provides an apolar cavity suitable to promote (a) formation of only dimers of 2-naphthyltrimethoxysilane and prevents further condensation and (b) imparts extreme *syn* selectivity in the photodimerization of two molecules of naphthoquinone.

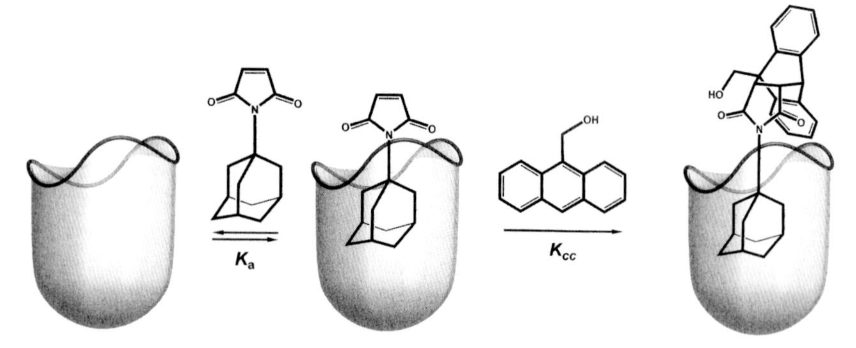

Figure 4.22 A hydrogen-bonded cavitand, in substoichiometric amounts, enables acceleration of the Diels–Alder reaction between maleimides and dienes and show catalytic turnover capacity

in the cavity. Irradiation at room temperature then led to the [2+2] photomediated dimerization of alkenes with formation of the product characterized by *syn* stereochemistry in >98% yield. In organic solvents, the same combination of reagents usually provides the *anti*-isomer exclusively (Figure 4.21b).

4.2.2.1.2 **Catalytic Systems**

With capsules and cavitands, the presence or absence of turnover activity by a potential catalyst is often a matter of the right combination of reagents, reaction, and products. A stoichiometric catalyst can become a turnover catalyst when tested on a different reaction, in particular when the products are worse guests than the reagents through steric and geometric constrains. This is the case of the hydrogen-bonded cavitand when used to promote the Diels–Alder reaction between dienophiles like maleimides and dienes like polycyclic aromatic substrates (Figure 4.22) [79]. The key was the use of substituted maleimide dienophiles that were good guests for the cavitand with the formation of hydrogen bonds between the carbonyl groups of the reagent and the imide seam of hydrogen bond of the cavitand. This was sufficient to activate the substrate that, in the presence of 9-anthracenemethanol (which does not compete for complexation), reacted readily at room temperature with substoichiometric amounts of cavitand with a Michaelis–Menten kinetic profile typical of enzymatic catalysis. The rate accelerations were on the order of 20–60-fold over the background reaction with a stark shape selectivity for the reaction. Specifically, maleimides bearing R-groups too long to allow hydrogen bonding to the amide rim of the cavitand showed limited acceleration. Isomeric adamantyl reagents showed that substrate selectivity could be fine-tuned. While the 1-adamantyl derivative was truly catalytic, the longer substrate provided a more tightly bound product. The solvent was as important as all the other variables: changing from mesitylene, which was not a suitable guest for the cavitand, to benzene, which on the contrary is well accommodated in the pocket, caused the complete inactivation because of the competition between solvent and substrate for the cavitand.

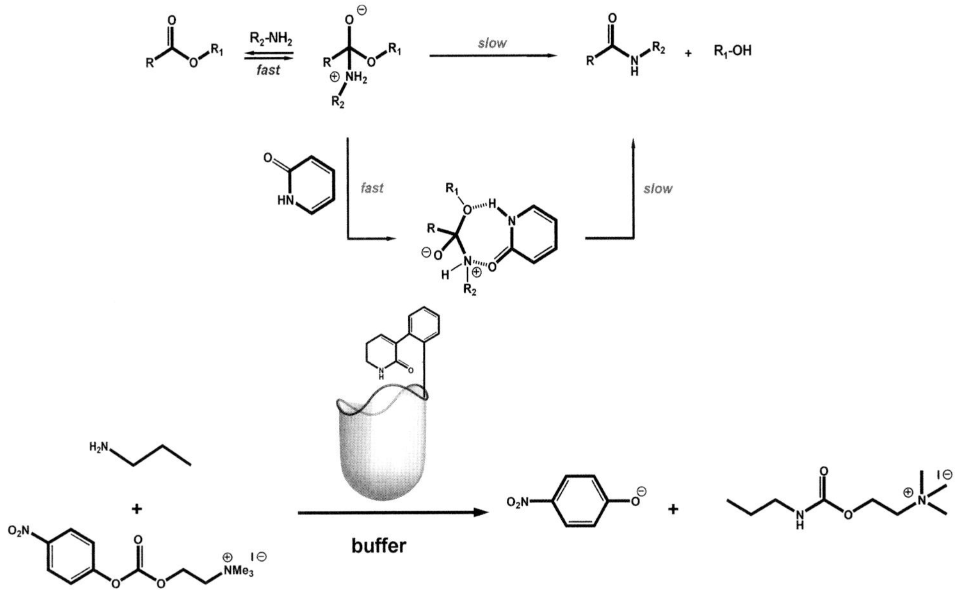

Figure 4.23 Pyridone functionalized cavitand presents a catalytic group that accelerates hydrolysis of *p*-nitrophenyl choline carbonate exploiting the binding affinity of the cavity for the alkylammonium residue present in the substrate.

Pyridone is well known as bifunctional organocatalyst, especially suited for the breakdown of tetrahedral intermediates. Such species was incorporated into the hydrogen-bonding cavitand with the aim of providing a binding pocket for cationic guests in close proximity to a catalytic site active in the aminolysis of carboxylic esters, especially for choline esters and carbonates (Figure 4.23) [80]. The reaction of the cavitand with *p*-nitrophenyl choline carbonate PNPCC and *n*-propylamine was 16 times faster in the presence of two equivalents of cavitand, while with the same amount of free pyridone only a slight acceleration was observed (10% with respect to the background reaction). With a lower amount of catalyst (10 mol%) turnover occurred but with only a modest 2.2-fold acceleration. No catalysis was observed with the bulky tritylamine or when a carbonate lacking the trimethylammonium residue was employed.

Inorganic complexes can be attached to the cavitand scaffold as well. In the example reported in Figure 4.24, a Zn-salen moiety was used to provide a metal center as a catalytic site placed on the rim in close proximity to a binding site [81]. The deep aromatic space provided by the cavitand is well known to complement choline derivatives as a consequence of proper space filling and attractive cation–π interactions. This allowed the employment of the Zn-salen cavitand as a highly active catalyst for anhydrides and carbonates in their reaction with nucleophiles, alcohols, and water to yield the corresponding esters or acids. In the first reaction, PNPCC was placed in the presence of Zn-salen cavitand in TFA/DIEA buffered

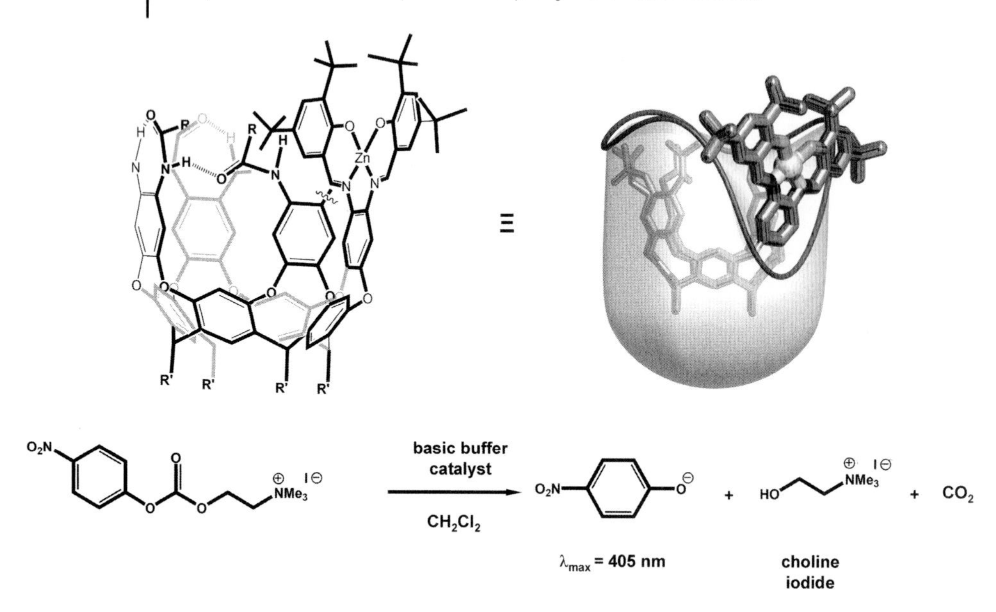

Figure 4.24 Zn-salen endowed cavitand promotes efficient hydrolysis of an organic carbonate under catalytic conditions.

CH_2Cl_2 and the reaction was monitored by UV-vis technique at 405 nm to follow the released *p*-nitrophenate. With a stoichiometric amount of catalyst, the reaction was 50 times faster compared to the background reaction and with 20% mol catalyst loading, the reaction was 12 times faster. Again the presence of the cavity was crucial to gain proper acceleration. Conversely, the reaction performed sluggishly in the presence of simple Zn-salen or with a carbonate lacking the ammonium residue. It is worth noting that the released free choline seemed to be a weak competitor with PNPCC for the cavity and therefore allowed catalytic turnover.

The same cavitand in the presence of choline as reagent and with acetic anhydride as acylating agent showed good acceleration for the esterification reaction [82]. Up to 720-fold rate increases resulted when the catalyst was used in stoichiometric amounts (Figure 4.25) and 320-fold when 40% mol was employed. The presence of both the binding site and the Zn(II) catalytic site was demonstrated to be crucial for activity, since the reaction performed with a cavitand lacking the Zn(II) center was not active at all and the reaction with the simple Zn-salen lacking a cavitand was 23 times slower. The reaction with more sterically demanding anhydrides showed slower rates as expected. With the more bulky substrate triethyl choline, the Zn-salen cavitand was only three times faster than the simple Zn-salen complex; this confirmed the high substrate selectivity imparted by the vase-like shape of the catalyst.

The water-soluble, metal–ligand assembled cavitand reported in Figure 4.26 showed true catalysis in the Diels–Alder reaction between 9-hydroxymethylanthracene and *N*-phenyl-maleimide. The formation of the cycloadduct at positions 9 and 10 of the anthracene reagent was observed with very good

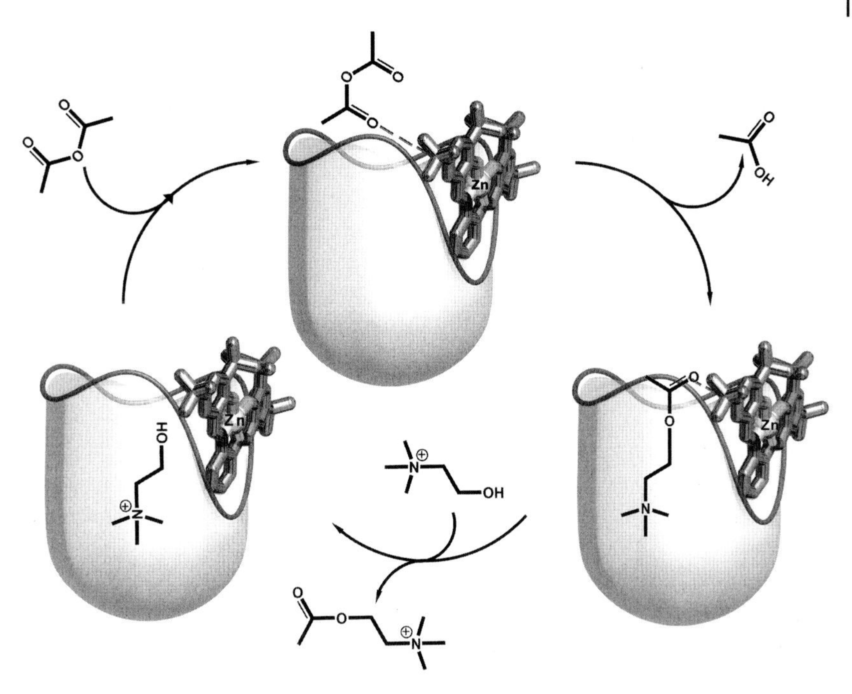

Figure 4.25 The reaction between choline and acetic anhydride is positively accelerated by catalytic amounts of cavitand endowed with a Zn-salen wall.

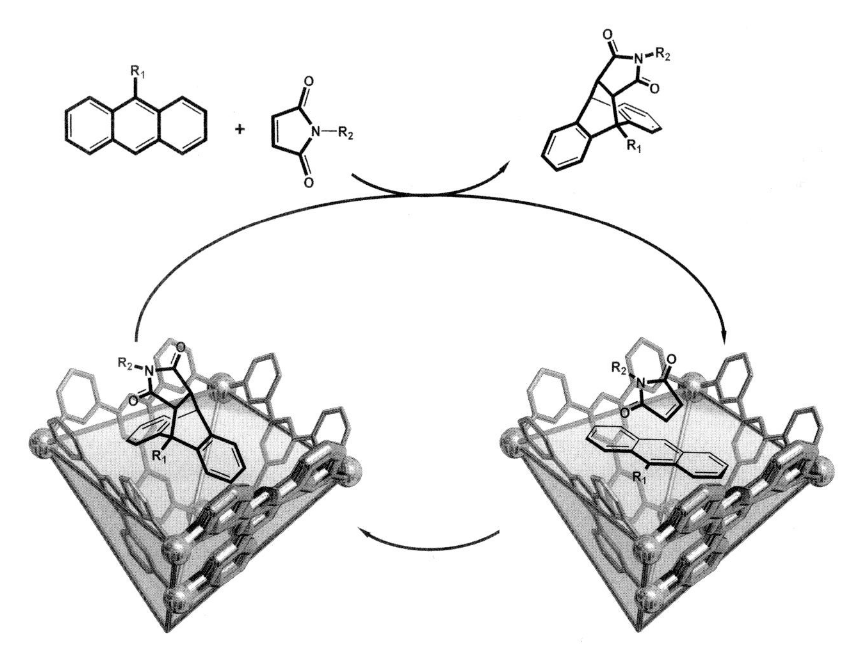

Figure 4.26 A diene and dienophile react efficiently in the hydrophobic cavity of the self-assembled cavitand; the product is a poor fit for the cavity and turnover occurs.

turnover activity [83]. The reaction was complete in 5 or 24 h when using 10 or 1% mol of the cavitand, respectively; in the absence of the catalyst, the reaction did not occur. The acceleration observed was not consequence of the Lewis acidity provided by the presence of several Pd(II) metal centers as confirmed by control experiments performed with (en)Pd$(NO_3)_2$. Instead, it was likely that confinement of both reagents in the cavity amplified the local concentration and accelerated the reaction. The turnover of the system was a consequence of the difficult binding of the bent 9,10-cycloadduct within the cavitand, while the planar substrate anthracene can displace the product thanks to better π–π interactions.

The proper combination of metal–ligand partners based on the hard–soft matching between Lewis acids and bases allows the preparation of precise structures starting from polydentate ligands in the presence of two different metal complexes. Such methodology is called ligand-templated direct assembly [84]. An important example of this strategy for the self-assembly of cavitands was developed by Reek and collaborators that prepared an active organometallic species immersed in a cavity. The key to the success was the employment of tris-*m*-pyridyl phosphine, which bears a hard nitrogen donor that binds preferentially a Zn(II)-porphyrin, while the soft phosphorous atom prefers to coordinate soft Lewis acids like Pd(0) and Rh(I). The reaction between Pd(0) and tris-*m*-pyridyl phosphine with six equivalents of porphyrin led to the formation of mono-phosphine Pd complex due to steric crowding characterized by vase-like trigonal pyramid with the active soft metal center on the bottom. The latter species was catalytically active in the Heck coupling between styrene and iodobenzene since free sites for substrate activation were present on the metal surrounded by the porphyrin walls. Without porphyrins, the system was completely inactive (Figure 4.27) [85].

The same strategy was used with [Rh(acac)(CO)(tris-*m*-pyridyl phosphine$)_2$] as catalytic metal complex. The system was studied in the hydroformylation of 1-octene to give the elongated linear and branched aldehydes with and without addition of six equivalents of Zn-porphyrin [41]. The effect of the porphyrin was to coordinate to the nitrogen atoms of tris-*m*-pyridyl phosphine thus preventing the formation of bis-phosphine Rh complexes, leaving monophosphine active

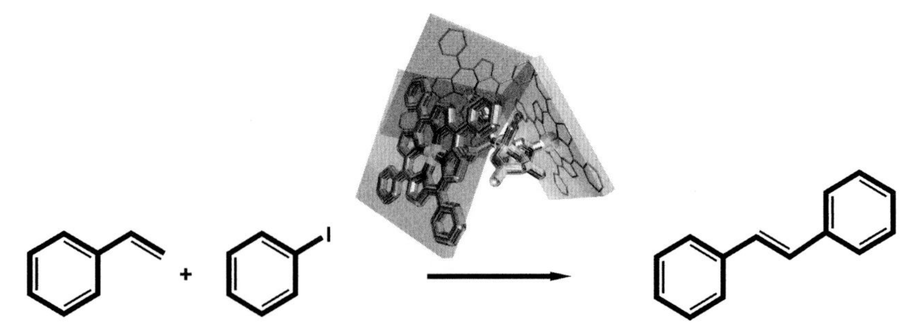

Figure 4.27 Heck C–C coupling between styrene and iodobenzene is mediated by Pd(0) catalysts surrounded by Zn(II)-porphyrin walls that impart unique activity and selectivity.

Rh(I) species. Worthy of note is the effect of such catalyst nesting on the linear/branched selectivity of the aldehyde products. The selectivity ratio decreased from 2.8 to 0.6 upon Zn(II)-porphyrin addition, as a consequence of the more hindered space around the Rh center that favored the formation of elongated products (Figure 4.28a) [85].

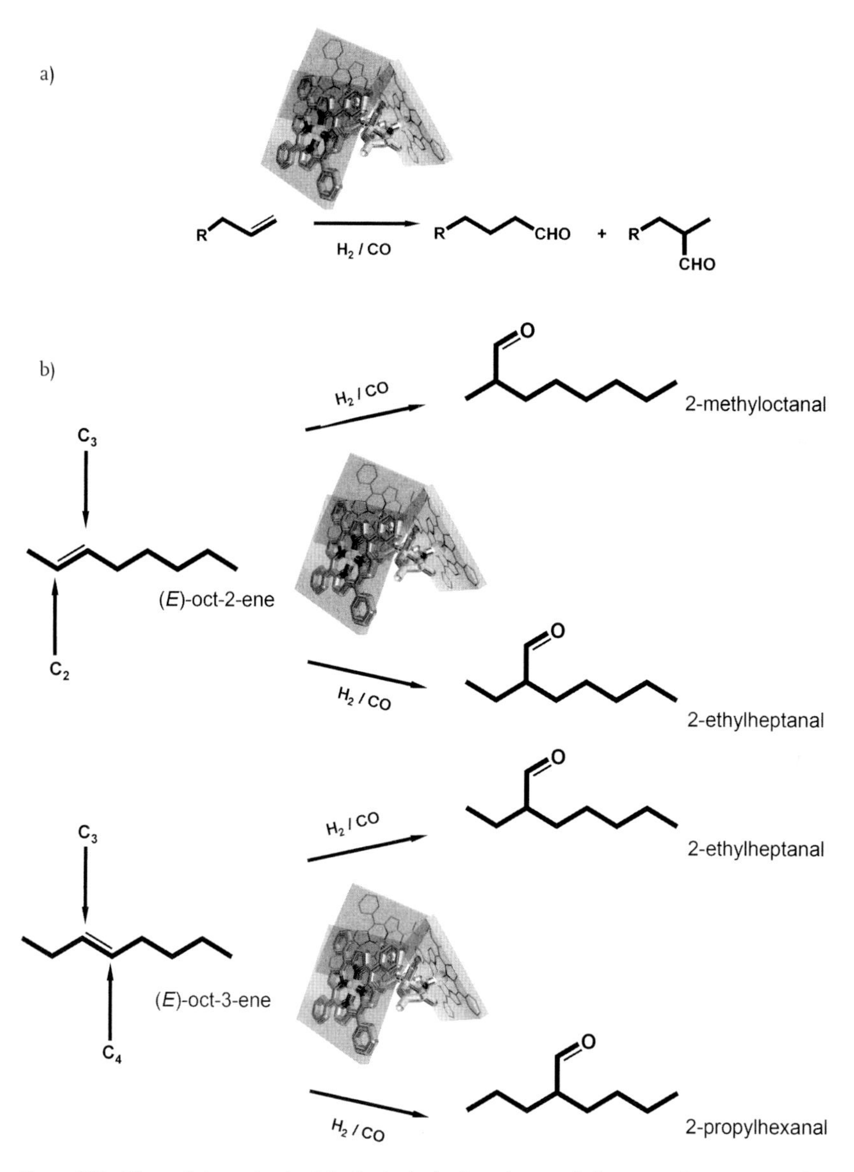

Figure 4.28 The activity and selectivity in the hydroformylation of alkenes in the presence of sheltered Rh(I) catalysts within Zn(II)-porphyrin walls is sensitive to the size and shape of the substrate.

The effect on the product distribution allowed by this supramolecular catalyst was further investigated in the hydroformylation of internal alkenes like *trans*-2-octene and *trans*-3-octene. While the reaction without Zn-porphyrin addition led to equal amounts of the two aldehyde products, the presence of Zn(II)-porphyrin with trans-2-octene produced preferentially 2-ethylheptanal with 88% selectivity and with trans-3-octene as substrate 2-propylhexanal was preferred with 75% selectivity (Figure 4.28b) [86]. No isomerization of the double bond was observed. The selectivity observed was interpreted in terms of hydride migration as the rate-determining step. This requires rotation of the substrate that is hindered by the inner surface of the capsule.

Zn(II)-salen walls could be employed also to surround the Rh(I) tris-pyridyl phosphine complex, and hydroformylation tests showed that the self-assembled catalyst was more active than the metal precursor, providing more branched aldehyde products. This is again in agreement with the tight control over substrate positioning and folding in the cavity of the vase-shaped assembly [87] (Figure 4.29).

4.2.2.2 Self-Assembled Capsule Catalysts

Cavitands and capsules with inner–outer amphiphilic character are also good candidates to behave as phase-transfer agents. Rebek and collaborators developed a water-soluble version of the structure reported in Figure 4.4 by fusing a 2,5-dihydro-1*H*-imidazol-2-yl acetic acid derivative to the walls of the resorcin[4]arene scaffold [88]. The water soluble water soluble cavitand was employed as a phase-transfer agent for hydrophobic guests in the Michael addition of a water-soluble thiol to hydrophobic maleimides (Figure 4.30a) [89]. The role of the cavitand was to extract the maleimide derivative into water by insertion of the apolar part of the molecule within the cavity. This allowed the thiol nucleophile to react with the double bond of the maleimide leading to the Michael adduct which is water soluble and l eaves the cavitand permitting the next catalytic cycle. The system showed turnover, and 2% mol of cavitand allowed conversions of up to 98% in few hours. Noteworthy were both the substrate selectivity and rate enhancements displayed

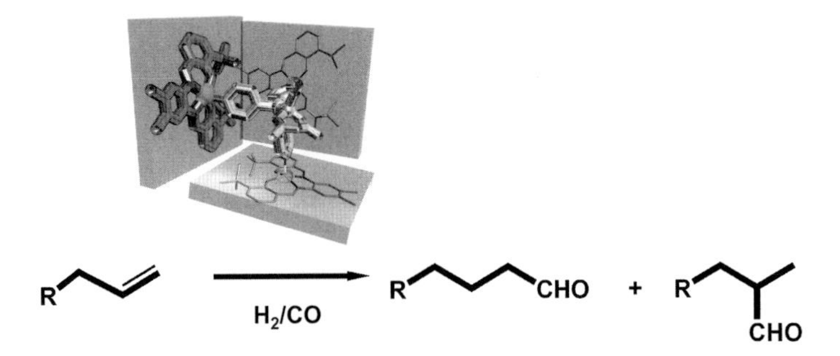

Figure 4.29 The selectivity on the hydroformylation reaction mediated by sheltered Rh(I) catalysts within Zn(II)-salen walls is influenced by the size and shape of the substrate.

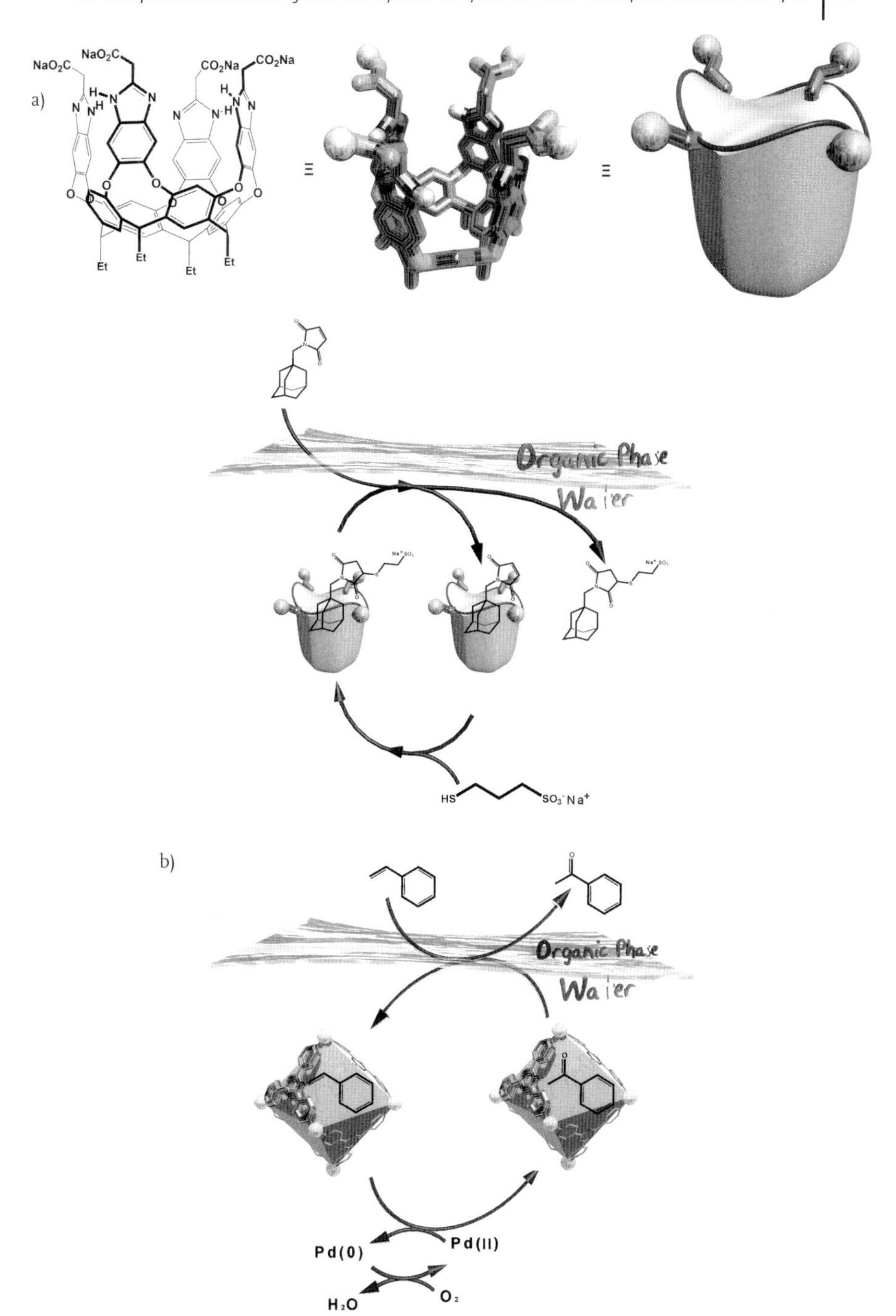

Figure 4.30 Examples of self-assembled hosts successfully employed as phase transfer catalysts: (a) water-soluble cavitand drives Michael addition of water-soluble thiols to hydrophobic maleimide reagents; (b) a water-soluble capsule based on ligand to metal coordination enables solubilization of hydrophobic styrene favoring contact with Pd(II) catalyst and leading to Wacker oxidation to the corresponding methyl ketone.

compared to the background reaction. For more hydrophobic substrates the cavitand enhanced the solubility in water and played a stronger role in the catalysis providing higher activity compared to the reaction in the absence of the cavitand.

Another example of phase transfer supramolecular catalyst was developed by Fujita and coworkers used the self-assembled capsule reported in Figure 30b to favor solubilization of styrene in water followed by Wacker oxidation to acetophenone by molecular oxygen catalyzed by (en)Pd(NO_3)$_2$ species present in solution (Figure 4.30b) [90]. Without the aid of the capsule, no reaction occurred and no products were observed if a better guest like 1,3,5-trimethoxybenzene was added to the system. This competes with styrene and further confirms the "shuttle" character played by the capsule.

4.2.2.2.1 **Stoichiometric Systems**

In the present section, more direct examples of the involvement of capsules in the catalysis of irreversible reactions are presented. Self-assembled molecular capsules are intrinsically more susceptible to suffer from product inhibition, especially when two reagents lead to one product because of entropic difficulty of removing the product by the incoming reagents. Exceptions occur when the product is, for instance, not well-accommodated in the cavity. Nevertheless, the study of stoichiometric capsules in supramolecular catalysis sets further milestones on the road to a better understanding of transition-state stabilization and supramolecular activation. One of the earliest examples of synthetic self-assembled capsules able to promote irreversible chemical transformations was based on the ability of the so-called softball dimeric capsule to encapsulate hetero combinations of guests, thus increasing the local concentration and favoring the chemical reaction. The softball was characterized by a pseudo-spherical cavity that in *p*-xylene-d_{10} could easily host combinations of dienes and dienophiles, as in the case of *p*-quinone and 1,3-cyclohexadiene (Figure 4.31) [91]. While the latter guest can be unequivocally observed in the capsule only at high concentrations because of the competition with the solvent for the cavity, *p*-quinone bound strongly showing two molecules of the dienophile in the cavity as resting state. After few hours, signals corresponding to the Diels–Alder cycloadduct arose blocking the cavity and hampering turno-

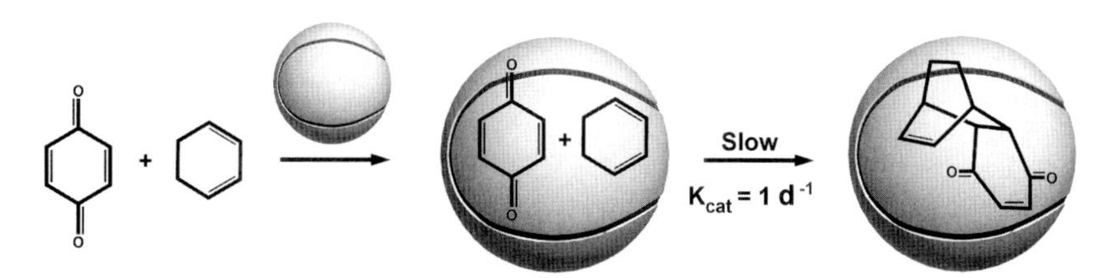

Figure 4.31 A Diels–Alder reaction between *p*-quinone and 1,3-cyclohexadiene was accelerated by the dimeric self-assembled softball capsule. Hetero-coencapsulation of the two reagents is not preferred over homo-coencapsulation of quinone, but when it occurs the lifetime of the complex favors the reaction. The product is the best guest available and no turnover occurs.

ver ability. Nevertheless, the effect of the capsule on the reaction rate was impressive with a 200-fold acceleration compared to the background reaction. It also displayed a typical Michaelis–Menten kinetic profile with saturation profile of the initial rate of the reaction with increasing substrate concentration as a consequence of the saturation of all the molecules of the capsule with the combination of reagents (Figure 4.31) [92]. In agreement with the latter observation, competitive guests like benzene as well as 2,2-paracyclophane inhibited the reaction blocking the cavity, and reagents like *p*-naphthoquinone (which is too big to be encapsulated) did not react as an example of substrate size selectivity. Further control experiments with a capsule monomer with no hydrogen-bonding donors or with a capsule monomer unable to self-assemble into the capsule because of its wrong shape confirmed that the acceleration provided by the capsule arose solely by a concentration effect rather than specific chemical juxtaposition of active functional groups.

Proper size and shape complementarity is crucial for the catalysis of bimolecular reactions with capsules. One example of correct reciprocal orientation of the reagents for a cycloaddition reaction was reported with the cylindrical capsule (Figure 4.32a). The capsule preferred the coencapsulation of phenyl azide with phenylacetylene with both the aromatic residues near the tapered ends of the capsule. This positioned the reactive functions near the center of the cavity and enhanced the formation of the 1,3-dipolar cycloadduct triazole which occurred within days [93]. The capsule accelerated the reaction by enhancing the local concentration of both reagents; larger substrates like 1-naphthylazide and biphenylazide that could not fit into the cavity with the alkyne failed to provide the corresponding products. The cavity also played a major role in the selectivity observed, since the elongated shape of the cavity favored specifically the 1,4-triazole isomeric product, while in solution both isomeric 1,4 and 1,5 products were observed in comparable amounts. From a mechanistic point of view, NMR evidence revealed that the system closely resembled the Michaelis complex in enzymes where the host binds the reagents and brings them close in the proper orientation prior to favor reaction leading to the product which remains in the cavity and hampers turnover. After all, product inhibition was discovered – and is still commonly encountered – in enzymology.

Another example of product selectivity imparted by the cylindrical dimeric capsule appeared in the reaction between coencapsulated carboxylic acids and isonitriles [94]. If performed in the bulk solvent, the reaction provides at 40 °C the formamide derivative together with anhydride. In the supramolecular host, the main species detected in the cavity is a rearrangement *N*-butyl-*N*-formyl-2-(4-methylphenyl)acetamide product as reported in Figure 4.32b. The unique lining of the cavity of the capsule arranges the acid and isonitrile in the appropriate orientation and pushes the reactive centers toward the middle of the capsule where the seam of hydrogen bond donors and acceptors can stabilize polar transition states. No traces of the intermediate species (*E*)-(butylimino)methyl (4-methylphenyl) acetate were observed.

Gibb and Ramamurthy applied the water-soluble octa-carboxylic acid cavitand reported in Figure 4.33 to the photochemical isomerization of substrates with the

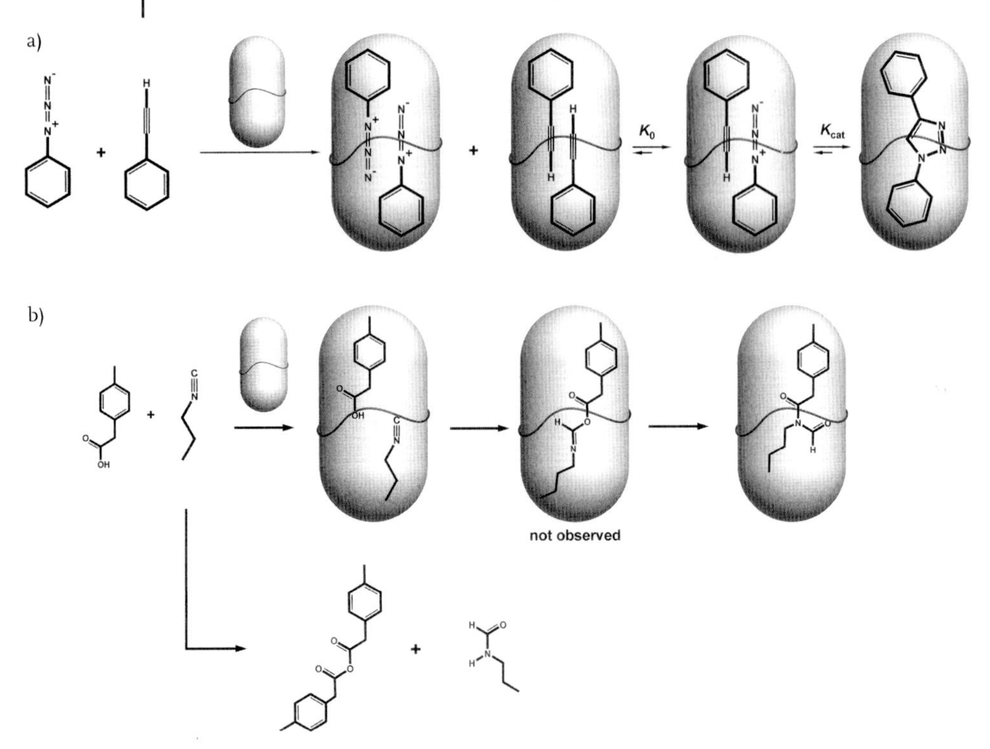

Figure 4.32 Examples of product selectivity driven by the dimeric cylindrical capsule: (a) preferred formation of the sole 1,4-isomer of the diphenyl-1*H*-1,2,3-triazole without formation of the 1,5-isomer; (b) preferred formation of *N*-butyl-*N*-formyl-2-(4-methylphenyl)acetamide in the cavity, while in the bulk solvent only the formamide derivative and the anhydride was observed.

aim of studying the peculiar product selectivities imparted by the host [95]. In borate buffer, the cavitand is negatively charged and in the presence of suitable hydrophobic elongated aromatic guests like *trans*-4,4′-dimethylstilbene and *trans*-stilbene, it self-assembles. The complex is held together by means of hydrophobic effect into a dimeric capsule with a 2:1 host/guest stoichiometry (Figure 4.33a) [96]. Irradiation of *trans*-4,4′-dimethylstilbene in common organic solvents led to a final pseudo-stationary state that consisted in 76% *cis* and 18% *trans* alkenes and phenanthrene as residual species. In contrast, the reaction performed in the presence of the octa-carboxylic acid cavitand in water gave the equilibrium composition of 15% *cis* and 85% *trans*. The impact of the capsule was traced to guest positioning: the *trans* isomer has the methyl groups near the tapered poles of the capsule, while the *cis* isomer methyls were near the equator of the capsule. The same experiments with *trans*-stilbene (lacking the methyl groups) showed similar *cis*/*trans* selectivity regardless of the medium used, either hexane or water with the octa-carboxylic acid cavitand. Further studies with 4-methylstyrene as substrate

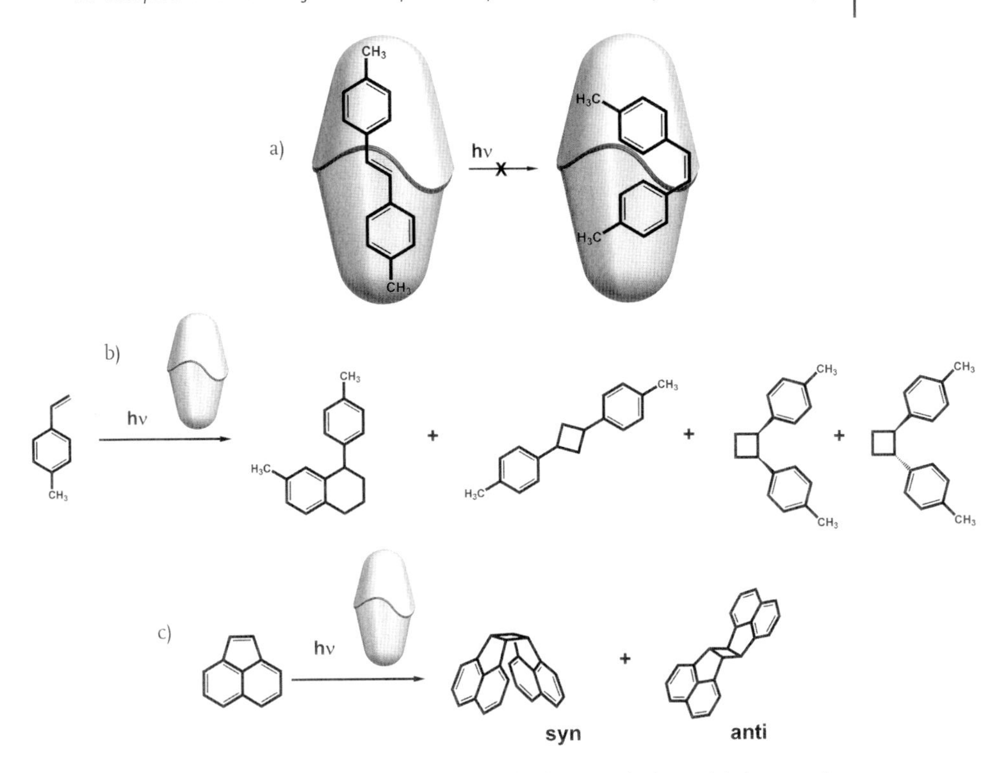

Figure 4.33 Photochemical reactions mediated by water-soluble hosts that provide hydrophobically driven self-assembly into capsules: (a) trans-4,4'-dimethylstilbene cannot isomerize to *cis* when encapsulated; (b) 4-methyl-styrene in the presence of the capsule leads to tetrahydronaphthalenes and 1,3-cyclobutanones rather than 1,2-cyclobutanes; (c) acenaphthylene dimerizes preferentially to the *syn* rather than the *anti* isomer.

gave two 2 : 2 complexes where the methyl groups were close to the poles of the cavity and the C=C double bonds experienced two possible relative orientations. Irradiation of such species led to the preferred formation of dimeric products such as tetrahydronaphthalenes and 1,3-cyclobutanes that are completely different from 1,2-cyclobutane derivatives commonly observed in organic solvents (Figure 4.33b) [95]. The photochemical behavior of acenaphthylene in the hydrophobic dimeric capsule was studied as well, and irradiation of the 2 : 2 assembly led to the formation of >99% *syn* dimer, while in water in the absence of the capsule the *syn* selectivity was 40% at best (Figure 4.33c) [97].

Dibenzyl ketones are suitable guests for this dimeric capsule and exchange slowly on the chemical shift timescale. Photoirradiation of the smaller members of this class of encapsulated guests and subsequent analysis of photolysis products demonstrated the unusual massive formation of rearrangement products derived by recombination of radical residues and minor amounts of decarbonylation

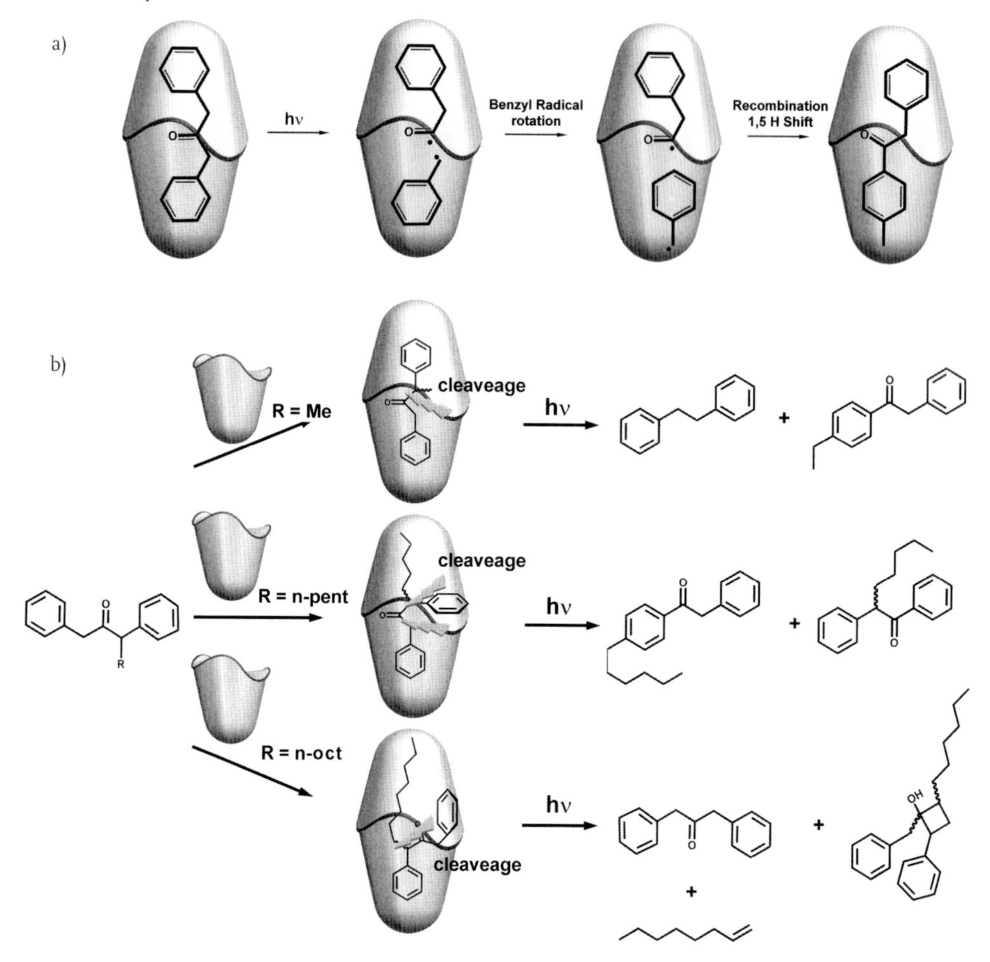

Figure 4.34 (a) Photolysis of methyl substituted dibenzoyl ketone encapsulated in the hydrophobic dimeric capsule leads to selective formation of 1-(4-methylphenyl)-2-phenylethanone with no side products derived by decarbonylation, which is the pathway in the absence of the capsule. (b) α-Alkyl dibenzoyl ketones undergo photocatalytic cleavage with different product distributions as a consequence of their substituents and of the arrangement that they assume within the capsule.

products that were the only observed products when the reaction was performed in the absence of the hydrophobic capsule (Figure 4.34a) [98]. Moreover, as long as decarbonylation products are concerned, photolysis within the capsule provided only mixed products and no homodimers, confirming that the recombination of radicals occurs in the capsule faster than in-out exchange.

α-Alkyl dibenzoyl ketones were also investigated, and the effect of the alkyl chain length on the positions assumed by the guests within the capsule was initially evident. The small methyl group led to preferential accommodation of the aryl residues on the tapered end of the capsule, while with the longer *n*-pentyl residue the alkyl moiety preferred to position toward the end of the cavity. With the even longer *n*-octyl residue, a third preferred orientation in the cavity was observed (Figure 4.34b). This had a direct effect on the outcome of the corresponding photolysis products, while in the bulk solution no effect of the alkyl chain was observed on the product distribution [99]. This behavior is very similar to what experienced in enzymatic catalysis where the catalytic site displays high substrate selectivity as well as extreme regio and stereoselectivity in the reaction due to proper binding and orientation of the substrate.

The same hydrophobic encapsulation strategy was used to explore the oxidation of cyclic alkenes by singlet oxygen produced in water in the presence of water-soluble Rose Bengal (RB) or insoluble dimethylbenzil (DMB) as sensitizers [100]. All cyclic substrates provided quaternary 2:2 capsular self-assembled structures where the methyl alkenes were positioned toward the narrow end of the cavity and, in the presence of singlet oxygen, they led to the preferential formation of allyl hydroperoxides bearing an α-methyl residue. The same reaction in acetonitrile without host led to a complex mixture of hydroperoxides (Figure 4.35). The high product selectivity was the result of the tight binding of the guests within the cavity with a precise orientation, favoring abstraction of the least-hindered allylic hydrogen by singlet oxygen. The overall process is reminiscent of signaling between enzymes [101]. As reported in Figure 4.35b, the capsule mimics an enzyme, where the photosensitizer is followed by opening of the capsule and release of the energy to oxygen. The singlet oxygen produced acts as a signal that moves to a second enzyme mimetic that, if opened, allows the signal to be transmitted to the encapsulated guests with selective formation of one over several possible isomeric products.

Another example of the tight orientation of substrates provided by this hydrophobic host arose from the photochemical Fries rearrangement of 1-naphthyl esters. When the reaction was performed in hexane, up to nine products were detected derived by the cleavage of the acyl residue and subsequent rearrangement to 1-naphthols bearing the acyl moiety in the *ortho* and *para* position, together with other coupling reactions. When the reaction was performed in water with the octa-carboxylic host, complete chemo- and regioselective formation of the *ortho* isomer was observed. This result is a direct consequence of the shielding effect provided by the capsule that did not allow the naphthyl residue to rotate before reaction with the photocleaved acyl residue (Figure 4.36) [102].

In the realm of metal–ligand self-assembled capsules, the tetrahedral assembly developed by Raymond has emerged as an invaluable nanometric reaction chamber to investigate organic as well as organometallic irreversible chemical transformations in mechanistic detail. The capsule was used to host the chiral racemic monomeric Ir(III) complex $Cp^*(PMe_3)Ir(Me)(CH_2CH_2)$, leading to two diastereoisomeric complexes as a consequence of the intrinsic chirality of the capsule with modest diastereoselectivity (Figure 4.37a) [103, 104]. The organometallic species readily reacted with aldehydes R–CO–H leading to C–H activation of the carbonyl compound, followed by decarbonylation of the residue and displacement of ethylene from the complex thus forming chiral $Cp^*(PMe_3)Ir(CO)(R)$ species. For

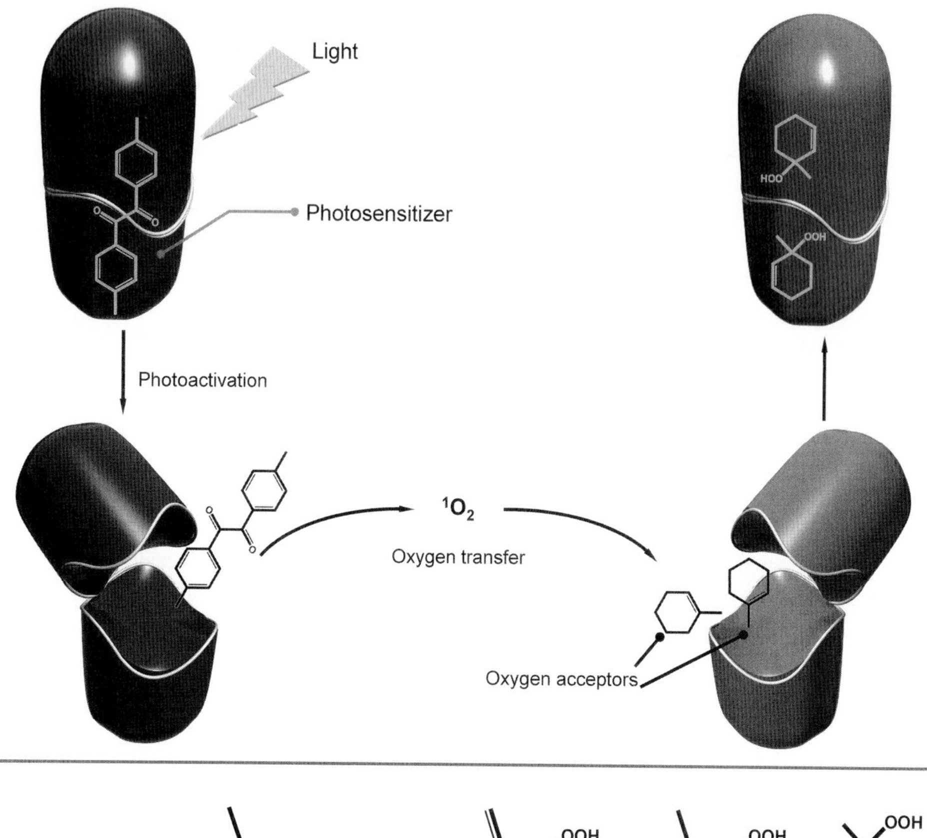

n	Medium/Sensitizer ( )n, n = 1, 2, 3 → $h\nu$ $^1O_2$	OOH ( )n +	OOH ( )n +	OOH ( )n
1	CH_3CN/RB	4	43	53
	H_2O-octaacid/RB	-	5	95
2	CH_3CN/RB	44	20	36
	H_2O-octaacid/RB	10	-	90
3	CH_3CN/RB	4	48	48
	H_2O-octaacid/RB	5	5	90

Figure 4.35 An encapsulated sensitizer is excited with light and it releases the energy to molecular oxygen that oxidizes with great product selectivity a second hydrophobic guest to the corresponding hydroperoxide. The water-soluble hydrophobic capsule mimics in both steps enzymes protecting active species and steering of product selectivity.

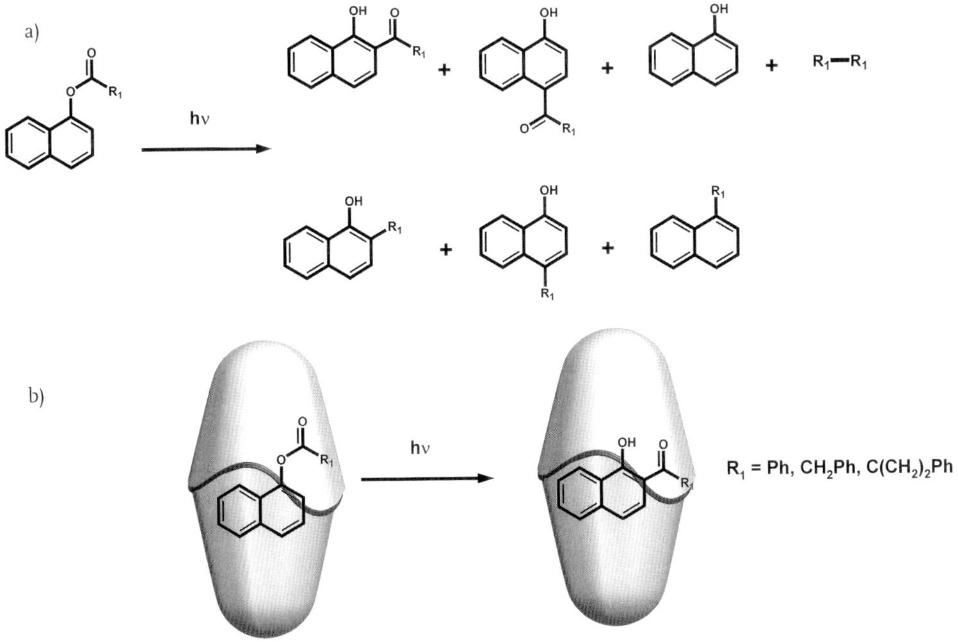

Figure 4.36 Naphthalen-1-yl esters are suitable substrates for the photochemical Fries rearrangement. Product distribution provides (a) several different species in solution or (b) a single 1-(1-hydroxynaphthalen-2-yl) ketone if the reaction is carried within the dimeric hydrophobic capsule.

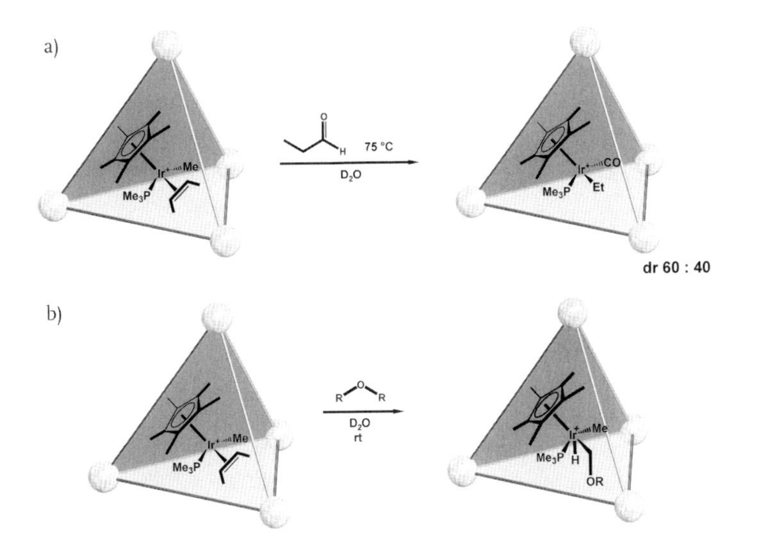

Figure 4.37 Encapsulated organometallic Ir(III) species reacts with (a) aldehydes leading to carbonyl splitting and formation of new entrapped chiral species; (b) methyl alkyl ethers forming carbene iridium species. In both cases, high substrate selectivity imparted by the size and shape of the capsule was observed. n.r. = no reaction observed.

the encapsulated organometallic precursor, the reaction with the aldehyde was highly substrate sensitive, both in terms of size selectivity (*n*-pentanal did not react while *n*-butanal did) as well as shape selectivity (2-methyl-butanal did not react while 3-methyl-butanal did). The chiral products were formed in a chiral cavity; therefore diastereoselective formation of the two stereoisomeric Ir(III) complexes occurred, with selectivity that could be as high as 70 : 30. This is another example of cavity-directed substrate and product selectivity, where the surrounding of the catalytic site influences the pathway of the reactions also from a stereochemical point of view.

The same encapsulated organometallic precursor was active toward methyl alkyl ethers leading to reaction with the oxygen atom and loss of methane with production of carbene iridium species (Figure 4.37b). As in the former reaction, excellent substrate selectivity was observed, with ethers composed of up to three carbon atoms that reacted readily, while bigger ones did not lead to carbene iridium products. Diastereoselective formation of two possible stereoisomeric complexes was observed with diastereoselectivity up to 88 : 12 for dimethyl ether [104], which is a rather remarkable result considering the steric constrain and that the general PC observed is about 55%.

Fujita and collaborators have extensively investigated several unique features of the metal–ligand self-assembled coordination capsule reported in Figure 4.38. One of the earliest examples concerned the polycondensation of phenyltrimethoxysilane that only in the presence of the self-assembled capsule led to the entrapment of trimers that would be otherwise unstable in the bulk solution [77, 105]. The reaction occurred following the "ship in the bottle" principle with initial binding of the apolar substrate in the self-assembled capsule and subsequent reaction that halted at the trimer level. Ninety-two percent yield and no further condensation to higher oligomers were observed, which instead were the principal products when the reaction was performed in the absence of capsule. Once encapsulated in the cavity, the bound trimer was stable for at least 1 month at room temperature but because of the tight binding of the product, no turnover activity could be observed. Further experiments performed with other aryl-substituted trimethoxysilanes

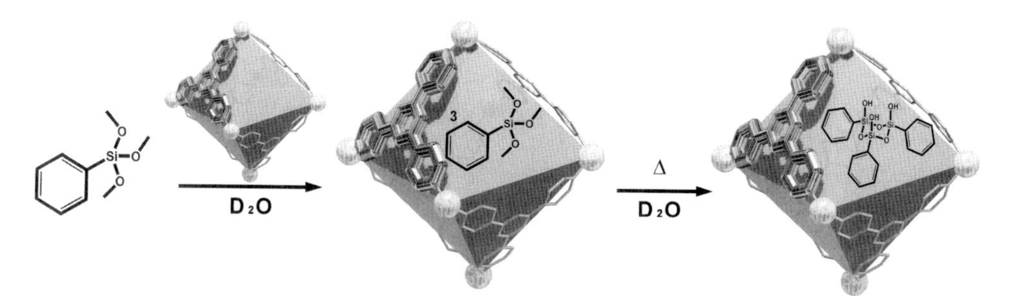

Figure 4.38 Phenyltrimethoxysilane reacts with water providing several condensation products but once encapsulated in a hydrophobic self-assembled capsule it provides only trimers with an all-*cis* configuration.

eventually confirmed the all-*cis* configuration of the cyclic trimeric products that was favored by the geometry imparted by the cavity to the reagents before cyclization [106]. The same capsule was employed to promote Diels–Alder reactions. Proper combination of diene and dienophile characterized by overall shape and size complementarity within capsule's cavity allowed preferable coencapsulation of the two reagents in the self-assembled reaction chamber. Tight binding was essential to freeze reagents' reciprocal motion into particular topochemical arrangements like in the cases of 9-hydroxymethylanthracene and *N*-phenyl maleimide reported in Figure 4.39a [83]. Simple heating of the system allowed the formation of the remarkable 1,4-Diels–Alder adduct, while in the absence of capsule the main product was the 9,10-adduct. The particular regioselectivity observed with a wide range of anthracenes was forced by the size and shape of the cavity and, more importantly, by the fixed reciprocal orientation of the reagents.

Larger polycyclic aromatic substrates could be coencapsulated with the proper dienophile, as in the case of triphenylene with *N*-cyclohexyl maleimide. Subsequent heating of the system (100 °C, 24 h) led to the quantitative conversion of the reagents into a single [4+2] thermal Diels–Alder adduct reported in Figure 4.39b. Analogous reactivity and unique regioselectivity were observed with perylene with the same dienophile leading to quantitative *endo* Diels–Alder adduct. It is worth noting that such unusual products have never been reported before in any organic solvents and were exclusively present when the reaction was mediated by the capsule [107]. Irradiation of the coencapsulation complex bearing perylene and *N*-cyclohexyl maleimide led to quantitative [2+2] photoaddition Diels–Alder reaction with *syn* stereochemistry imparted by the cavity (Figure 4.39c). The size of the alkyl group attached on the nitrogen atom of the maleimide reagent played a crucial role; Me, Et, Ph, or $PhCH_2$ derivatives did not provide the desired products, indicating that tight binding of guests was essential to force the proper reciprocal orientation to promote entropically unfavorable reactions pathways.

Several other photomediated cycloaddition reactions were studied in this capsule, in particular homodimerization reactions of olefins. Reaction occurred after saturation of the capsule with the alkenes and irradiation for 30 min. Acenaphthylene could be easily encapsulated in pairs within the cavity and led to the quantitative formation of the dimer together with unusual *syn* regioselectivity (Figure 4.40a) [78]. The presence of the capsule was crucial, both to accelerate the reaction and to steer the stereoselectivity toward the *syn* isomer. No reaction occurred without capsule in water and a completely different stereochemistry was observed in organic solvents (benzene, *syn* 19%, *anti* 17%). 1-Methylacenaphthylene and 2-methylnaphthoquinone (Figure 4.40b) could be homodimerized and among the four possible isomers, quantitative formation of the *syn* isomers and head-to-tail selectivity of 98% and 96%, respectively, were observed. These astonishing results are further evidence of the impressive stereochemical control imparted by the supramolecular catalyst. Turnover was still a problem since the products were generally better guests than the reactants.

The employment of well defined supramolecular capsules allows the proper choice of reagents to promote the heterodimerization over the homodimerization

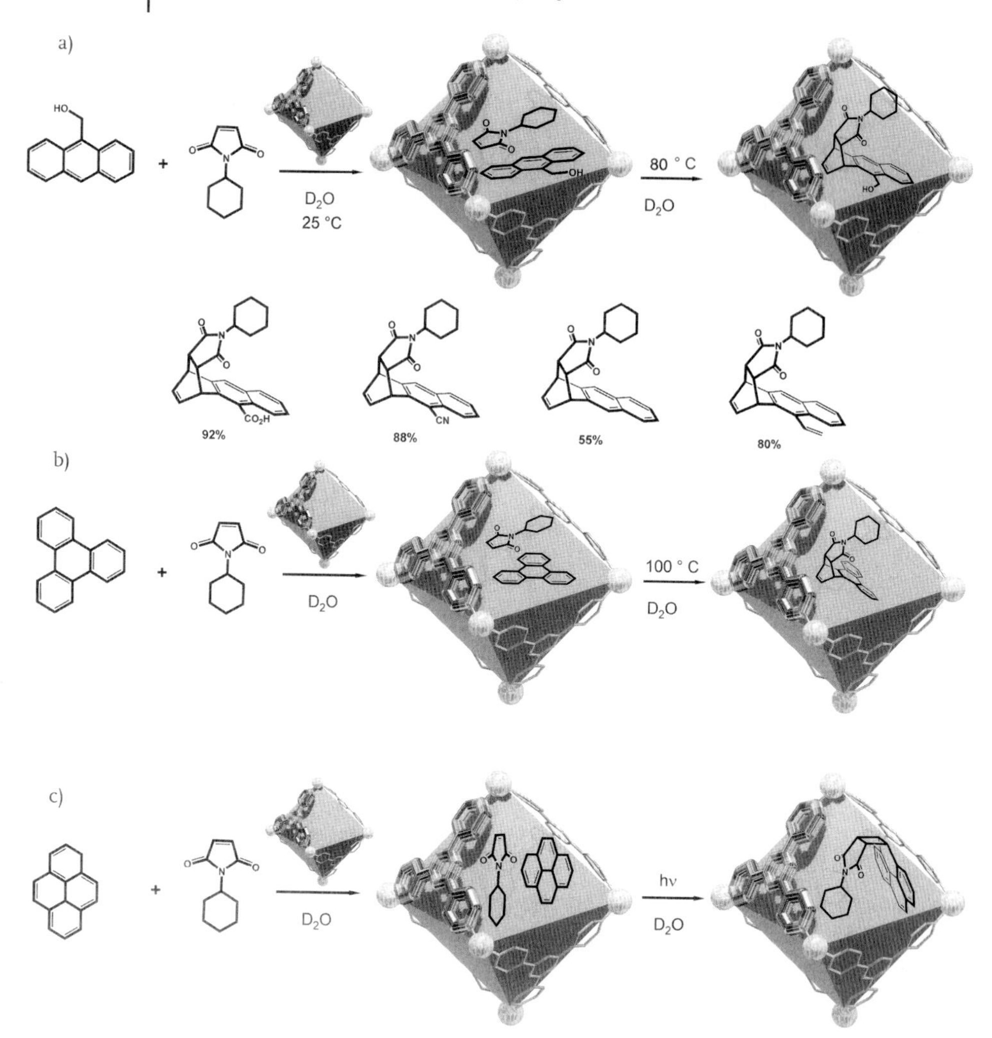

Figure 4.39 Diels–Alder reactions between bulky substrates occur within hydrophobic capsules leading to unconventional regioisomers due to the particular arrangement assumed by the reagents in the cavity: (a) anthracene, (b) triphenylene, and (c) pyrene derivatives reaction with maleimide dienophiles.

process. the heterodimerization to be the favored reaction channel inside the capsule. Acenaphthylene and 5-ethoxynaphthoquinone could be coencapsulated in a pairwise fashion and irradiation led to photo-cross [2+2] dimerization as reported in Figure 4.41a [108]. It is worth noting that, as a consequence of the shape and size of the cavity, only one molecule of acenaphthylene could be accommodated and the residual space available could be occupied only by a different molecule, thus favoring cross-dimerization reaction over homodimerization. The latter effect was strictly a function of the substituents present on the reagents, as

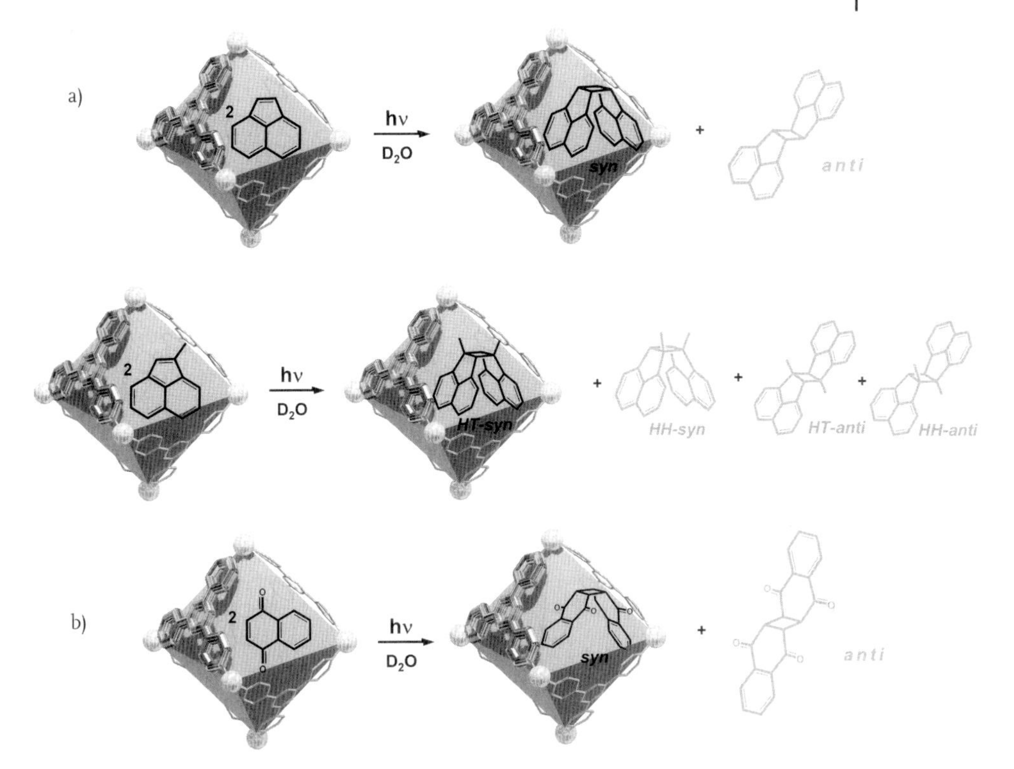

Figure 4.40 High degrees of stereochemical control were achieved with the water-soluble self-assembled capsule in the [2+2] photochemical reaction: (a) acenaphthylenes and (b) naphthoquinones can lead to a series of different stereoisomers but in the presence of the capsule *syn* products are selectively formed.

replacement of the ethoxy with the smaller methoxy group on naphthoquinone caused the formation of only 22% of the homodimeric structures. Generally *syn* selectivity was observed in these dimerization reactions with high yields, although no turnover was possible. The unique pairwise selectivity coupled with tight binding allowed activation of otherwise unreactive olefins like maleimides which was cross-dimerized with acenaphthylene or dibenzosuberenon quite efficiently, while the same reaction did not occur at all in the absence of the capsule (Figure 4.41b). All the reactions were strictly controlled by the capsule which plays different roles in rate acceleration, stereochemical outcome and most importantly pairwise selectivity. Quite remarkable was the filtering effect displayed by the capsule whose UV absorption falls where the [2+2] adducts have their absorptions, while at longer wavelengths the capsule is transparent. This allowed direct dimerization of olefins and simultaneously avoided photodissociation at lower wavelength.

The presence of metal corners on the supramolecular capsule prompted Fujita and collaborators to investigate the possible asymmetric induction from the ligands coordinated to the metal centers into the space of the cavity where

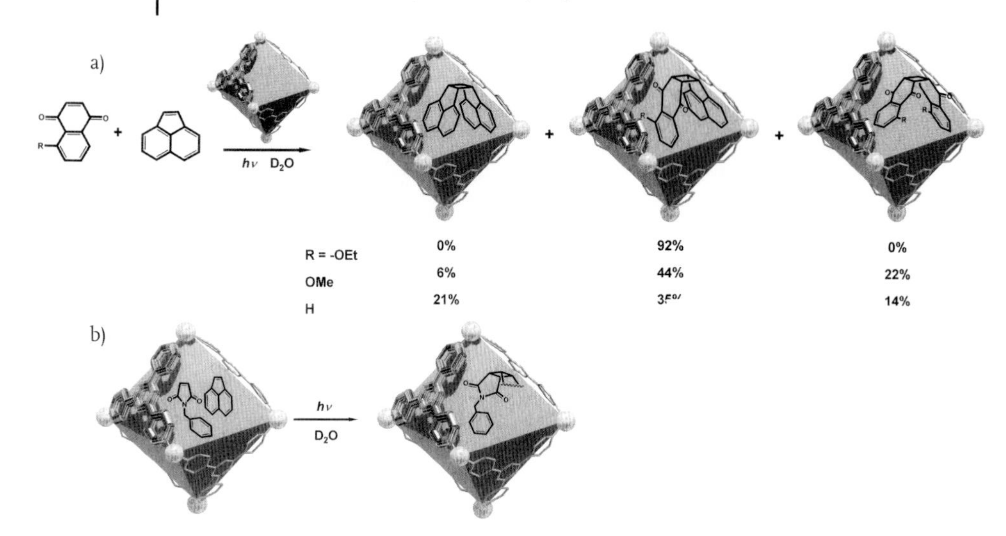

Figure 4.41 (a) Acenaphthylene and 5-alkoxynaphthoquinone are suitable reagents for photodimerization. General *syn* selectivity is observed and product distribution is sensitive to the alkoxy group present on the naphthoquinone moiety. (b) Otherwise unreactive olefins like maleimides reacted with acenaphthylene promoted by the capsule.

the reaction takes place. A series of three chiral diamines were used to prepare the corresponding chiral enantiopure capsules that were employed in the [2+2] photoaddition of fluoranthene with maleimide [109]. The reaction led to formation of the cross-products with moderate enantiomeric excess up to 40%. The asymmetric induction was more efficient with (1*R*, 2*R*)-*N*,*N*′-diethyl-1,2-diaminocyclohexane as chiral ligand for Pd, and with more hindered substrate like 3-methylfluoranthene an ee of 50% was achieved. It is important to emphasize that there is no direct contact between the asymmetric moieties and the reagents, but only the asymmetric space of the inner of the capsule is responsible for the enantioselectivity observed (Figure 4.42).

Photochemical radical reactions were studied using the capsule reported in Figure 4.43 as a nanoreactor, with the aim of limiting the possible products by means of compartmentalization effect. Combinations of *o*-quinones and bulky toluene derivatives could be hosted in the capsule and upon photoirradiation, they were converted into unusual 1,4-adducts derived by addition of benzyl radicals to quinones [110]. The capsule acted as a stoichiometric catalyst favoring heteroencapsulation but also it strictly juxtaposed the two reagents in such a way that the initially formed radical reacted with the partner on the oxygen atom leading only to 1,4-adducts rather than to more common 1,2-adduct or other isomeric species.

The supramolecular cage could act directly as a photochemical catalyst toward encapsulated guests, as demonstrated in the photochemical oxidation of trapped adamantane (Figure 4.44) [111]. Four molecules of adamantane were coencapsulated within the cavity and upon irradiation under aerobic conditions at room temperature, formation of almost quantitative oxidation prod-

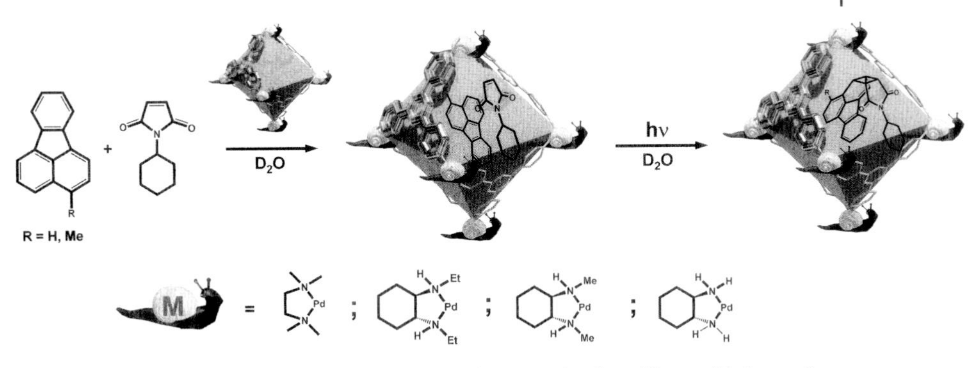

Figure 4.42 Chiral metal complexes on the corners of supramolecular self-assembled capsule impart asymmetry to the cavity and induce stereoselective [2+2] photoaddition reactions between fluoranthenes and maleimides.

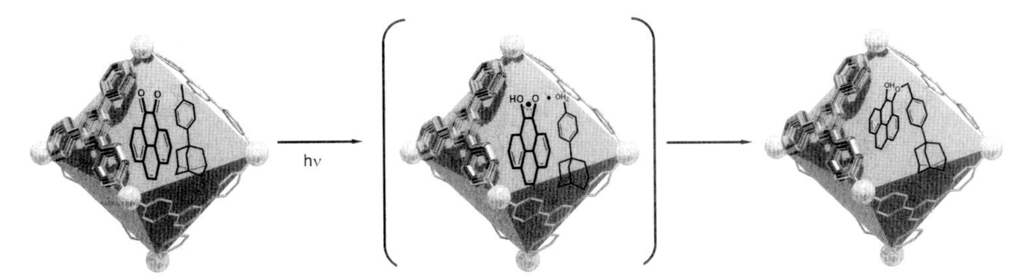

Figure 4.43 Controlled photogeneration of radicals within a self-assembled capsule induces unique regioselective formation of 1,4-adducts and not the more common 1,2-adduct or other isomeric products between *o*-quinones and large substituted toluenes.

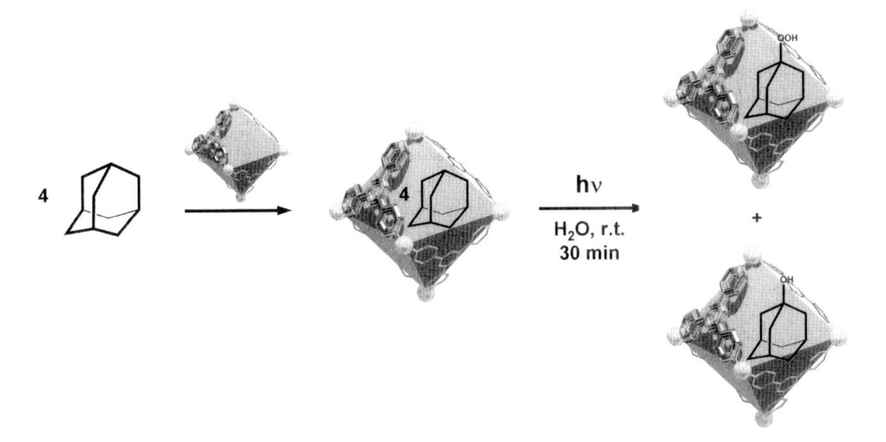

Figure 4.44 Aromatic self-assembled capsule containing triazine core catalyzes the photo-oxidation of adamantane substrates with selective formation of the corresponding alcohol and hydroperoxide.

ucts 1-adamantylhydroperoxide and 1-adamantanol (4 : 1) arising from the almost complete (96%) oxidation of one of the four molecules of tricyclic alkane was observed. No oxidation products were observed in the absence of the capsule or with either the ligand or metal components of the capsule. Analogously, hydroperoxides and ketones were prepared when cyclic alkanes (C_6–C_8) were employed, but with larger ones like decaline or perhydrofluorene that are too large to be encapsulated, no reaction occurred. The key role was played by the triazine core of the ligand, but it was active only on the self-assembled aggregate where the irradiation led to the formation of radical species on the walls of the capsule. These were transferred to a close-contacted encapsulated adamantane giving rise to adamantyl radicals immediately trapped by molecular oxygen under aerobic conditions to give oxygenated products.

4.2.2.2.2 Catalytic Systems

One of the landmark examples of true catalysis with self-assembled capsules is based on the softball applied to Diels–Alder reactions. Substrate choice is crucial truly sub-stoichiometric catalysts and the combination of *p*-quinone and thiophene dioxide turned out to be positively influenced by the presence of the softball catalyst with 55% yield after 2 days at room temperature and 75% after 4 days. Under the same conditions, control experiments showed 10% and 17% yields, respectively (Figure 4.45) [112]. More importantly, the capsule provided true turnover because of the higher steric clashes of the product compared to the coencapsulation of two molecules of *p*-quinone, which was the resting state of the catalyst in the cycle. It is worth noting that the same supramolecular catalyst with *p*-quionone and 1,3-cyclohexadiene as substrates did not show turnover ability (4.2.2.2.1).

The tetrahedral self-assembled structure reported in Figure 4.6a displayed an important catalytic effect on the unimolecular sigmatropic 3-aza-Cope rearrangement of cationic enammonium substrates. The reaction led to iminium cations which were subsequently hydrolyzed to the corresponding γ,δ-unsaturated aldehydes (Figure 4.46a) [113, 114]. The driving force for the encapsulation was a combination of hydrophobic effects and cation–π interactions, with acceleration of the sigmatropic rearrangements up to almost three orders of magnitude compared to the uncatalyzed reactions displayed first-order kinetics for all substrates. The effect of the capsule was strictly related to the selection of the proper conformation of the substrate. This is similar in structure to the highly organized chair-like transition state of the reaction, as evidenced by the negative entropy of activation observed for the reaction. The effect was more pronounced for the more bulky substrates and was evidenced by 2D-NOESY experiments that confirmed the folded arrangements of the substrates within the cavity. The products observed were readily hydrolyzed in solution (but not while still in the cavity) giving the corresponding aldehydes and dimethyl amine. These products did not compete for binding in the cavity and allowed a new molecule of reagent to enter the catalytic cycle (Figure 4.46b). When a competitive guest as tetraethyl ammonium cation was present in solution, inhibition took place confirming the close resemblance of such system with common enzymes. Similarly, propargyl enammonium

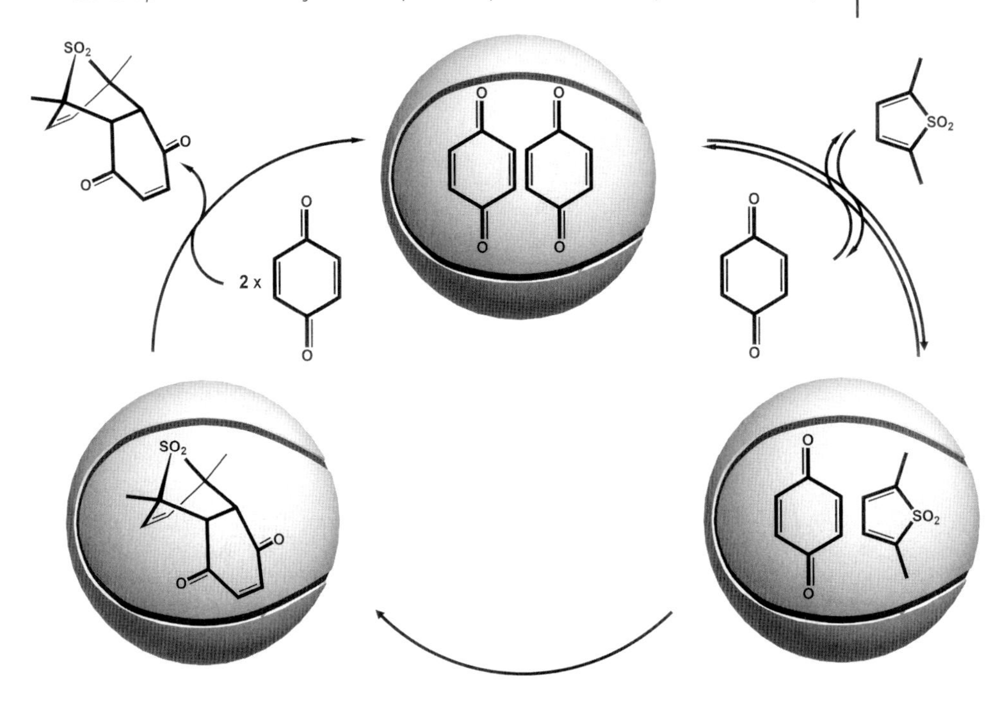

Figure 4.45 Catalytic Diels–Alder reaction mediated by hydrogen-bonded softball.

cationic substrates showed up to a 184-fold acceleration by the capsule [115]. The tetrahedral capsule is chiral because of the octahedral coordination of the ligands to the gallium corners. Enantiopure $\Delta,\Delta,\Delta,\Delta$ and $\Lambda,\Lambda,\Lambda,\Lambda$ capsule was prepared by resolution with (–)-*N*′-methylnicotinium iodide and was tested in the asymmetric version of the reaction. The reaction was extremely sensitive to the substituents of the prochiral enammonium cation leading to moderate to good conversions to the chiral aldehyde with ee up to 78% [116]. It is worth noting that such level of enantioselectivity is the highest so far reported for reactions within self-assembled hosts originated exclusively by the helicity of the metal centers that are not directly involved in the catalysis.

Small, cationic Rh(I) complexes such as $[(PMe_3)_2Rh(COD)]^+$ and $[(dmpe)Rh(COD)]^+$ are suitable guests for the tetrahedral capsule and, once sheltered in the cavity, they showed unique substrate selectivity compared to their behavior in the bulk solvent. Clear cut examples were reported for the double bond isomerization of allylic alcohols to the corresponding carbonyl compounds and of allylic ethers to enol ethers (Figure 4.47) [117]. The reaction in the cavity showed strict shape and size selectivity, as observed for methyl substituted alcohols because of the difficulty for these species to enter the capsule through the channels formed between the ligand walls. Conversely, the same substrates with free complex formed between solution provided the corresponding aldehydes regardless of the substitution pattern of the double bond.

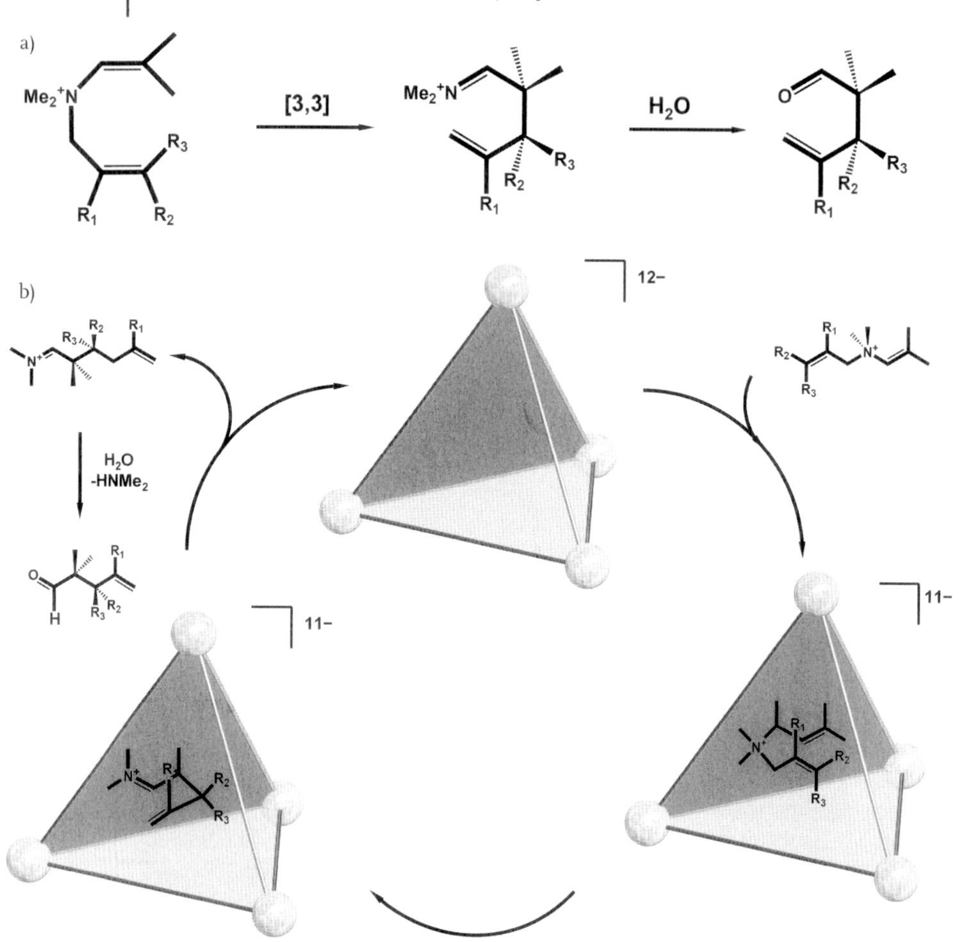

Figure 4.46 (a) The sigmatropic 3-aza-Cope rearrangement of cationic enammonium substrates is promptly catalyzed by the self-assembled capsule leading to γ,δ-unsaturated aldehydes. (b) proposed reaction mechanism.

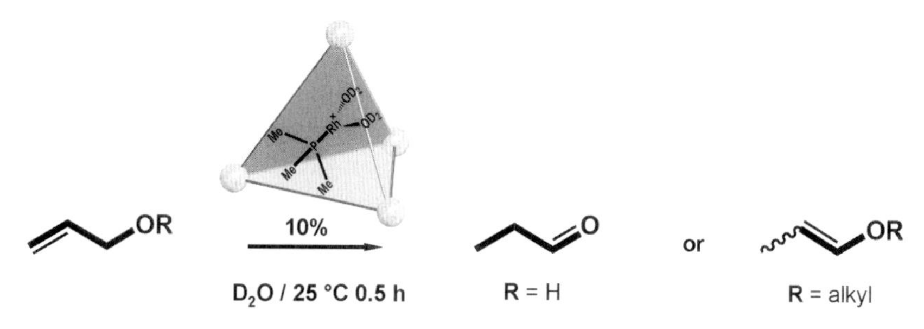

Figure 4.47 Isomerization of terminal allylic substrates catalyzed by $[(PMe_3)_2Rh(OD_2)_2{}^+]$ in water. The reaction in the presence of capsule is sensitive to the length and size of the substrate, while without capsule the reaction is only sensitive to double bond substitution.

The strong affinity of the capsule cavity for cationic species is reinforced by its polyanionic nature. This feature is at the heart of the strong catalytic effect displayed by such supramolecular catalyst in the acceleration of organic reactions that share the formation of cationic transition states derived from protonation of the reagents. Examples of this behavior were reported for the hydrolysis of acetals in water (Figure 4.48) [118]. With 5% mol of capsule, the reaction showed good catalytic activity with features typical of enzymes like inhibition by tetraethylammonium cation (that was a better guest for the capsule), catalytic turnover ensured by the formation of neutral products and strong substrate selectivity imparted by the shape and size of the cavity. Great differences in reactivity were observed in comparing 1,1-dimethoxy-heptane and the longer (unreactive) 1,1-dimethoxy-nonane, or 4,4-dimethoxydecane and the unreactive 2,2-dimethoxy-undecane. It is worth noting that the capsule provided unique nanoenvironment, which was isolated from the bulk solvent. Acetals hydrolysis is generally accomplished under acid conditions, but within the capsule the transformation occurs even under basic pH conditions. In this regard, the tetrahedral capsule mimics those enzymes that can drastically alter the pK_a of acids at their active sites. Detailed kinetic investigation clearly showed that the effect of the capsule is to modify the mechanism of the reaction compared to the same process in the bulk solvent enabling rate acceleration of up to three orders of magnitude [119].

Analogously, the principle was applied to the orthoformate hydrolysis in the tetrahedral capsule, providing an acid-favorable pocket to catalyze reactions while the bulk solvent had a basic pH (Figure 4.49) [120, 121]. In this case, acceleration of the reaction was up to 900-fold, with extra features typical of enzymatic catalysis like substrate selectivity, competitive inhibition, and very interesting Michaelis–Menten kinetic profiles. First-order rate dependence on catalyst concentration was observed, while for the substrate first or zeroth order was present depending if the reaction was performed with a stoichiometric amount of orthoformate or with a large excess under saturation conditions. The formulated mechanistic hypothesis is based on the encapsulation of the substrate within the cavity, followed by its protonation leading to encapsulated stabilized intermediate cationic species and release of two alcohol molecules, with subsequent formation of formate anion and empty capsule, which is available for further catalytic cycles.

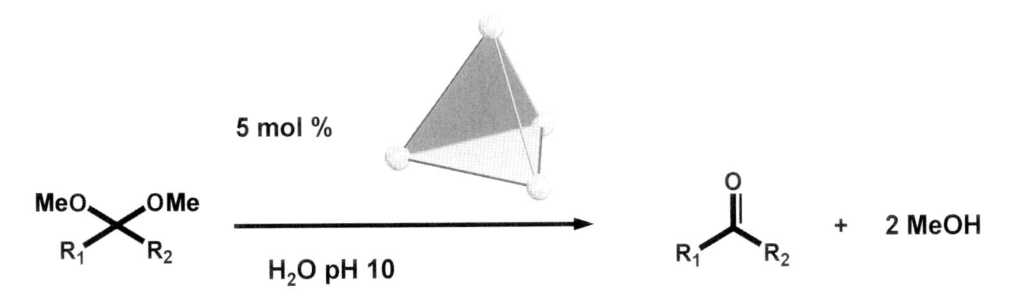

Figure 4.48 Dimethyl acetals are readily hydrolyzed in water in the presence of the capsule only if their size and shape allow encapsulation.

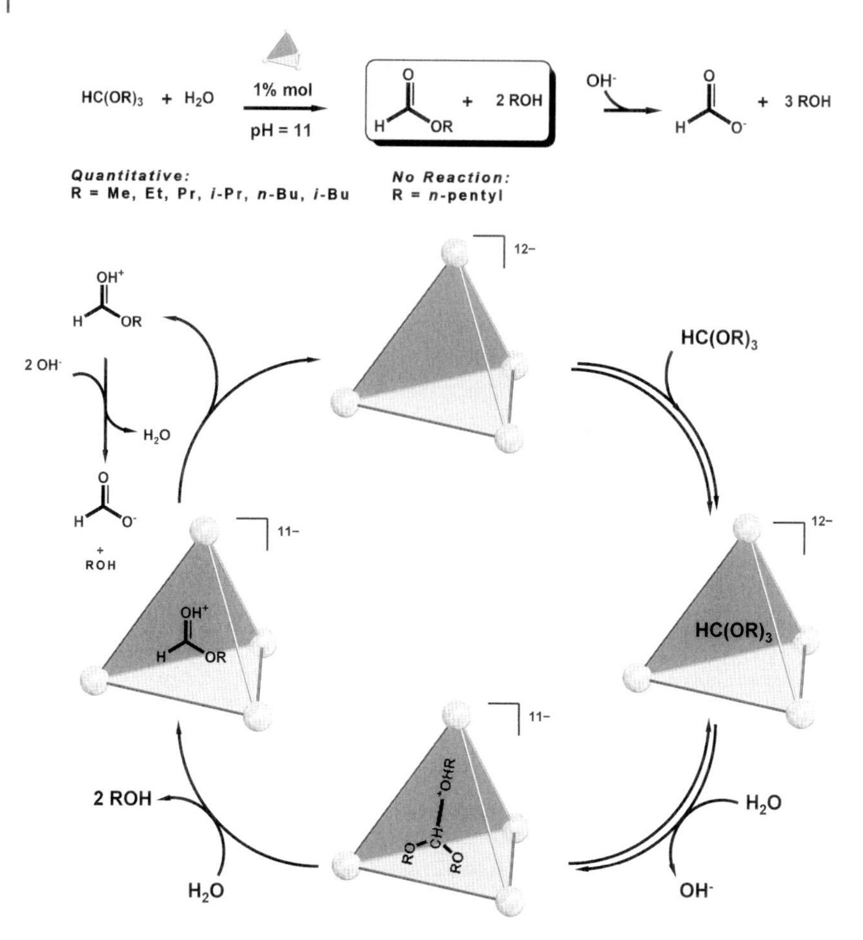

Figure 4.49 Orthoformates are readily hydrolyzed in water in the presence of the tetrahedral capsule showing Michaelis–Menten kinetic profiles.

Very recently, proving that the field of supramolecular catalysis allows further pushing the art of catalysis toward enzyme mimic, Raymond and Bergman beautifully developed an enzyme-like reaction mediated by the tetrahedral capsule. Exploiting the ability of the supramolecule to accelerate common acid catalyzed reactions, the Nazarov cyclization of a series of 1,4-dien-3-ol derivatives was investigated (Figure 4.50) [122]. The reaction provides substituted ciclopentadienes that are competitive guests for the cavity of the capsule; in fact the reaction was accelerated by the presence of the capsule but rather rapidly deactivated because of product inhibition thus limiting turnover ability. In order to overcome such a limit, maleimide was added to the system to react quickly with the product of the cyclization leading to a bulky cycloadduct that does not compete for the host. With this strategy and after optimization of the experimental conditions, the tetrahedral

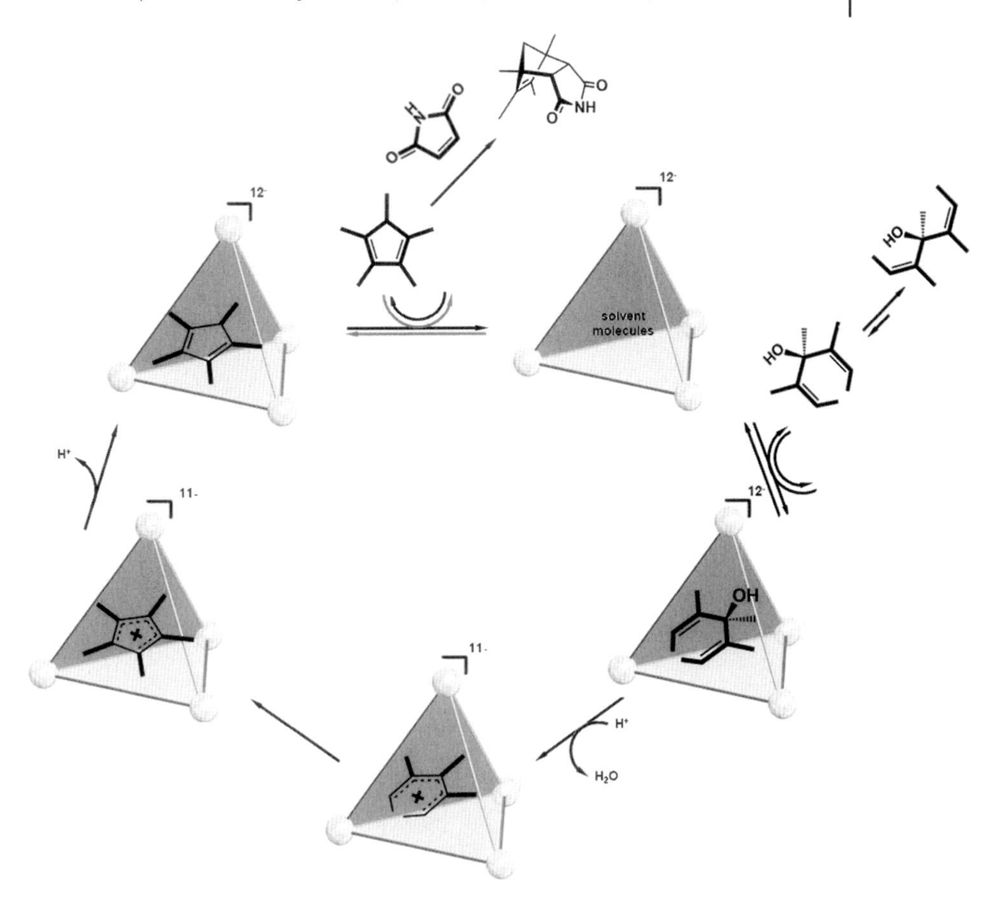

Figure 4.50 Enzyme-like catalytic Nazarov cyclization mediated by tetrahedral capsule in water showing million-fold acceleration compared to uncatalyzed reaction.

capsule ensured typical Michaelis–Menten kinetic profiles with turnover up to 160 and more importantly acceleration up to 2,100,000 times compared to the uncatalyzed reaction, really rivaling with enzymes in terms of ability to activate transition states.

A completely original supramolecular box was developed by Hupp and Nguyen assembling dimeric porphyrins containing Mn or Sn endowed with pyridyl residues with achiral or chiral residues. These were coordinated to tris-Zn-porphyrin structures leading to supramolecular boxes as reported in Figure 4.51 where overall seven building blocks are assembled giving only one possible aggregate with defined structure and geometry [123]. The box containing achiral Mn metal centers surrounded by chiral residues not reciprocally connected was tested in the asymmetric sulfoxidation of thioethers observing low but significant asymmetric induction with ee up to 12% in the oxidation of methyl-*p*-tolyl sulfide. The

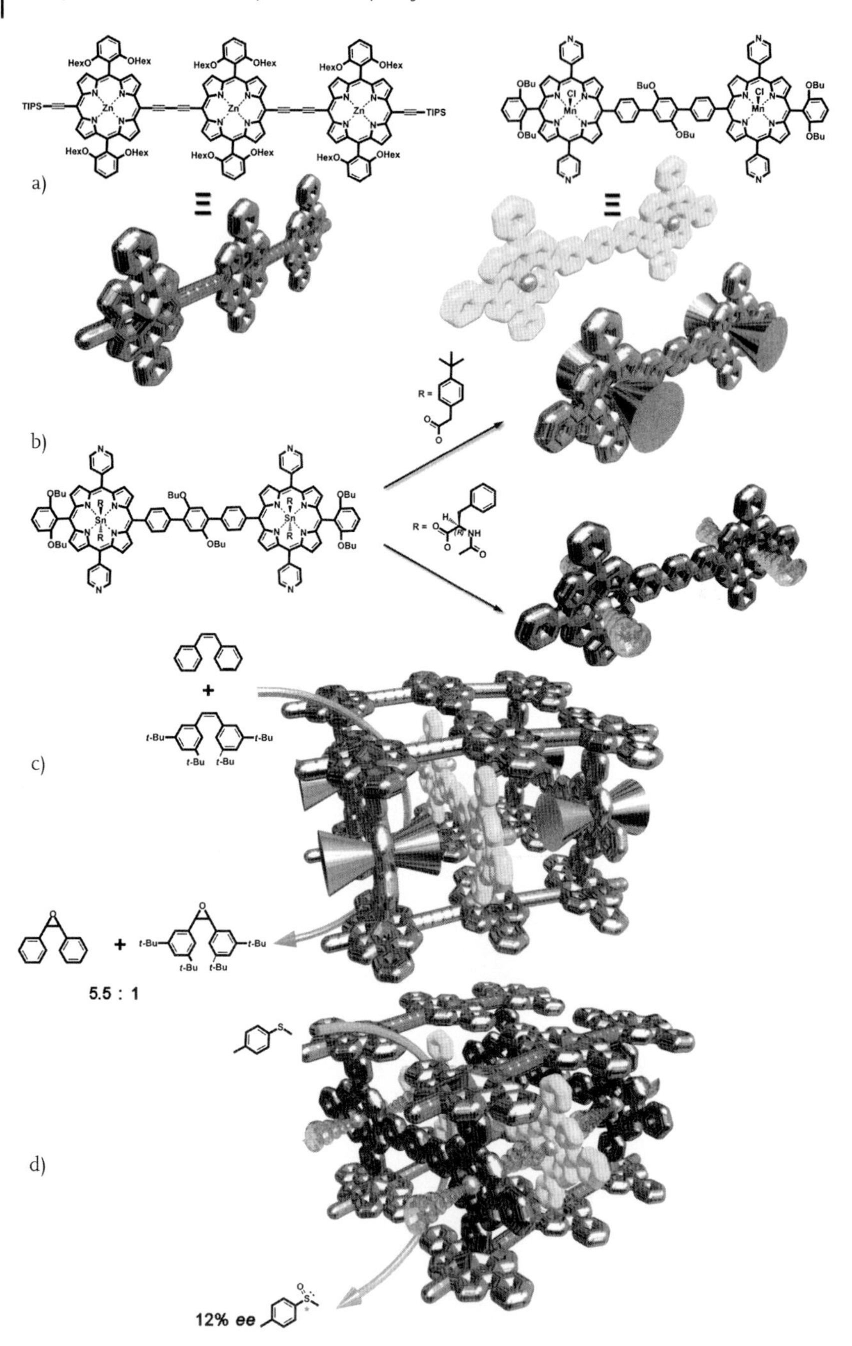

Figure 4.51 Supramolecular box catalyst for oxidation reactions: (a) tris-Zn-porphyrin as scaffold and bis-Mn-porphyrin as catalytic unit; (b) chiral and achiral bis-Sn-porphyrin as lateral walls of the box. (c) Substrate size selective oxidation of alkenes to epoxides with the achiral box and (d) enantioselective conversion of thioethers to sulfoxides mediated by chiral version of the supramolecular box.

supramolecular box showed good substrate selectivity; in fact while alkenes such as *cis*-stilbene were readily oxidized to the corresponding epoxide, *cis*-3,3′-5,5′-tetra(*tert*-butyl)stilbene was more than 5.5 times slower due to access inhibition because of steric requirements.

4.3 Conclusions

After the Nobel Prize was awarded to Cram, Pedersen, and Lehn in 1987 for their development of the field, supramolecular chemistry has witnessed an exponential growth and is booming. Already at that time, the potential of the emerging discipline was extremely high as is evident from the presentation speech of Professor Salo Gronowitz of the Swedish Royal Academy of Sciences: "*Their research has been of enormous importance for the development of coordination chemistry, organic synthesis, analytical chemistry, bioinorganic and bioorganic chemistry, it is no longer science fiction to prepare supermolecules which are better and more versatile catalysts than the highly specialized enzymes. The dream may soon become reality. Through their work Cram, Lehn and Pedersen have shown the way*" [124]. This prediction proved correct; in fact supramolecular chemistry has substantially aided the understanding of enzymes through the development of synthetic enzymes. Therefore a comparable boost in this research area is ongoing.

Nevertheless, the existing gap between synthetic and natural enzymes is still far from being closed and can be compared to sailing across the Atlantic Ocean from Spain to the Caribbean. As Cristoforo Colombo sought India, the scientific community is deeply involved in this challenging endeavor and, as happened to Colombo, the continuous endeavor for understanding chemical transformation mediated by enzymes by studying synthetic analogs could lead to a new unexplored continent. Even though the original theory was wrong (India is still far from America going west), the new discovery would greatly expand our knowledge and is absolutely worth the effort paid.

4.4 Outlook

The recent development of new water-soluble capsules and cavitands [125] represents one of the most recent challenges in the development of synthetic enzymes and toward the understanding of the peculiar role played by water in several natural chemical transformations. Moreover, new combinations of residues able to provide ion pairing and hydrogen bonding in water are emerging as potential candidates to be implemented in new supramolecular catalysts.

As soon as the control over the selectivity of synthetic enzymes reaches high levels, the development of asymmetric supramolecular catalysts is a natural consequence. Even though supramolecular chirality is growing as a new discipline

[126], still the implementation of chiral residues in cavitands and capsules for catalytic features is minimal. Chirality can be present in the cavity by encapsulation of chiral reactive metal complexes, or implemented in the subunits that compose the cavitand or the capsule. It is also possible to obtain chiral capsules from achiral units. It is worth noting that asymmetric discrimination between two pathways leading to enantiomeric products is a challenging issue that can provide good results if strong steric interactions are present or if attractive forces are present between the substrate and the chiral part. Because of the average small size of most supramolecular catalysts, only few examples have been reported but the time is ripe for a rapid growth of this aspect of supramolecular catalysis.

Acknowledgments

A.S. and G.B. are grateful to Università Ca' Foscari di Venezia for support; J.R. Jr. acknowledges The Skaggs Institute. The authors thank G. Strukul for reviewing the manuscript.

Abbreviations

^{1}H-NMR	proton nuclear magnetic resonance
2D-NMR	bidimensional nuclear magnetic resonance
acac	acetyl acetonate
Cp	cyclopentadienyl
Cp*	pentamethyl-cyclopentadienyl
COD	1,5-cyclooctadiene
DABCO	1,4-diazabicyclo[2.2.2]octane
DCC	dicyclohexylcarbodiimide
d.e.	diastereoisomeric excess
DIEA	di-isopropyl-ethyl-amine
DMB	dimethylbenzil
DMF	dimethylformamide
dmpe	1,2-dimethylphosphinoethane
ee	enantiomeric excess
EM	effective molarity
en	1,2-ethylendiamine
NOESY	Nuclear Overhauser Effect SpectroscopY
PC	packing coefficient
PNPCC	*p*-nitrophenyl choline carbonate
RB	Rose Bengal
Salen	*N,N*-ethylenebis(salicylimine)
SN_2	bimolecular nucleophilic substitution
S_NAr	nucleophilic aromatic substitution
TFA	trifluoroacetic acid

References

1 Kirby, A.J. (1996) *Angew. Chem. Int. Ed. Engl.*, **35**, 707.
2 Pauling, L. (1948) *Nature*, **161**, 707.
3 Lehn, J.-M. (1995) *Supramolecular Chemistry: Concepts and Perspectives*, Wiley-VCH Verlag GmbH, Weinheim.
4 Lawrence, D.S., Jiang, T., and Levett, M. (1995) *Chem. Rev.*, **95**, 2229–2260.
5 Anslyn, E.V. (2007) *J. Org. Chem.*, **72**, 687.
6 Motherwell, W.B., Bingham, M.J., and Six, Y. (2001) *Tetrahedron*, **57**, 4663.
7 Steinreiber, J. and Ward, T.R. (2008) *Coord. Chem. Rev.*, **252**, 751.
8 Zecchina, A., Groppo, E., and Bordiga, S. (2007) *Chem. Eur. J.*, **13**, 2440.
9 Sanders, J.K.M. (1998) *Chem. Eur. J.*, **4**, 1378.
10 van Leeuwen, P.W.N.M. (2008) *Supramolecular Catalysis*, Wiley-VCH Verlag GmbH, Weinheim.
11 Wilkinson, M.J., van Leeuwen, P.W.N.M., and Reek, J.N.H. (2005) *Org. Biomol. Chem.*, **3**, 2371.
12 Sandee, A.J. and Reek, J.N.H. (2006) *Dalton Trans.*, 3385.
13 Vriezema, D.M., Comellas Aragonès, M., Elemans, J.A.A.W., Cornelissen, J.J.L.M., Rowan, A.E., and Nolte, R.J.M. (2005) *Chem. Rev.*, **105**, 1445.
14 Blokzijl, W. and Engberts, J.B.F.N. (1993) *Angew. Chem. Int. Ed. Engl.*, **32**, 1545–1579.
15 (a) Dwars, T., Paetzold, E., and Oehme G. (2005) *Angew. Chem. Int. Ed.*, **44**, 7174.
(b) Taşcioğlu, S. (1996) *Tetrahedron*, **52**, 11113.
(c) Engberts, J.B.F.N. (1992) *Pure Appl. Chem.*, **64**, 1653.
(d) Lindstrom, U.M. (2002) *Chem. Rev.*, **102**, 2751.
(e) Otto, S., and Engberts, J.B.F.N. (2003) *Org. Biomol. Chem.*, **1**, 2809.
16 For cyclodextrins in catalysis: Takahashi, K. (1998) *Chem. Rev.*, **98**, 2013.
17 Rebek, J. Jr. (2006) *Nature*, **444**, 557.
18 Scarso, A., Pellizzaro, L., De Lucchi, O., Linden, A., and Fabris, F. (2007) *Angew. Chem. Int. Ed.*, **46**, 4972.
19 Shivanyuk, A., Scarso, A., and Rebek, J. Jr. (2003) *Chem. Commun.*, 1230.
20 Koblenz, T.S., Wassenaar, J., and Reek, J.N.H. (2008) *Chem. Soc. Rev.*, **37**, 247.
21 Purse, B.W. and Rebek, J. Jr. (2005) *Proc. Natl. Acad. Sci. USA*, **102**, 10777.
22 Lützen, A. (2005) *Angew. Chem. Int. Ed.*, **44**, 1000.
23 Pitt, M.A. and Johnson, D.W. (2007) *Chem. Soc. Rev*, **9**, 1441.
24 Hossain, M.A. and Schneider, H.-J. (1999) *Chem. Eur. J.*, **5**, 1284.
25 Prins, L., Reinhoudt, D.N., and Timmerman, P. (2001) *Angew. Chem. Int. Ed.*, **40**, 2382–2426, and references therein.
26 Meyer, E.A., Castellano, R.K., and Diederich, F. (2003) *Angew. Chem. Int. Ed.*, **42**, 1210.
27 Schottel, B.L., Chifotides, H.T., and Dunbar, K.R. (2008) *Chem. Soc. Rev.*, **37**, 68.
28 Nishio, M. (2005) *Tetrahedron*, **61**, 6923.
29 (a) Metrangolo, P., and Resnati, G. (2001) *Chem. Eur. J.*, **7**, 2511–2519.
(b) Auffinger, P., Hays, F.A., Westhof, E., and Ho, O.S. (2004) *Proc. Natl. Acad. Sci. USA*, **101**, 16789.
30 Hamilton, G.L., Kang, E.J., Mba, M., and Toste, F.D. (2007) *Science*, **317**, 496.
31 Wang, X. and List, B. (2008) *Angew. Chem. Int. Ed.*, **47**, 1119.
32 Otto, S. (2006) *Dalton Trans.*, 2861–2864.
33 Mecozzi, S. and Rebek, J. Jr. (1998) *Chem. Eur. J.*, **4**, 1016–1022.
34 Cacciapaglia, R., Stefano, S.D., and Mandolini, L. (2004) *Acc. Chem. Res.*, **37**, 113.
35 Schneider, H.-J. and Yatsimirsky, A.K. (2008) *Chem. Soc. Rev.*, **37**, 263.
36 Mayr, H. and Ofial, A.R. (2006) *Angew. Chem. Int. Ed.*, **45**, 1844.
37 Hof, F. and Rebek, J., Jr. (2002) *Proc. Natl. Acad. Sci. USA*, **99**, 4775.
38 Davis, A.V., Yeh, R.M., and Raymond, K.N. (2002) *Proc. Natl. Acad. Sci. USA*, **99**, 4793.
39 Rudkevich, D.M., Hilmersson, G., and Rebek, J. Jr. (1998) *J. Am. Chem. Soc.*, **120**, 12216.

40 Yu, S.-Y., Kusukawa, T., Biradha, K., and Fujita, M. (2000) *J. Am. Chem. Soc.*, **122**, 2665.
41 Slagt, V.S., Kamer, P.C.J., van Leeuwen, P.W.N.M., and Reek, J.N.H. (2004) *J. Am. Chem. Soc.*, **126**, 1526.
42 Conn, M.M. and Rebek, J., Jr. (1997) *Chem. Rev.*, **97**, 1647.
43 Rebek, J. Jr. (2005) *Angew. Chem. Int. Ed.*, **44**, 2068.
44 Kang, J. and Rebek, J. Jr. (1996) *Nature*, **382**, 239.
45 Fujita, M., Umemoto, K., Yoshizawa, M., Fujita, N., Kusukawa, T., and Biradha, K. (2001) *Chem. Commun.*, 509.
46 Amijs, C.H.M., van Klink, G.P.M., and van Koten, G. (2006) *Dalton Trans.*, 308.
47 Leininger, S., Olenyuk, B., and Stang, P.J. (2000) *Chem. Rev.*, **100**, 853.
48 Maurizot, V., Yoshizawa, M., Kawano, M., and Fujita, M. (2006) *Dalton Trans.*, 2750.
49 Fiedler, D., Leung, D.H., Bergman, R.G., and Raymond, K.N. (2005) *Acc. Chem. Res.*, **38**, 349.
50 Davis, A.V. and Raymond, K.N. (2005) *J. Am. Chem. Soc.*, **127**, 7912.
51 Davis, A.V., Fiedler, D., Seeber, G., Zahl, A., van Eldik, R., and Raymond, K.N. (2006) *J. Am. Chem. Soc.*, **128**, 1324.
52 Gibb, C.L.D. and Gibb, B.C. (2004) *J. Am. Chem. Soc.*, **126**, 11408.
53 Cram, D.J., Tanner, M.E., and Thomas, R. (1991) *Angew. Chem. Int. Ed. Engl.*, **30**, 1024.
54 Warmuth, R. (1997) *Angew. Chem. Int. Ed. Engl.*, **36**, 1347.
55 Warmuth, R. and Marvel, M.A. (2000) *Angew. Chem. Int. Ed.*, **39**, 1117.
56 Kusukawa, T. and Fujita, M. (1999) *J. Am. Chem. Soc.*, **121**, 1397.
57 Brumaghim, J.L., Michelis, M., Pagliero, D., and Raymond, K.N. (2004) *Eur. J. Org. Chem.*, 5115.
58 Fiedler, D., Bergman, R.G., and Raymond, K.N. (2006) *Angew. Chem. Int. Ed.*, **45**, 745.
59 Körner, S.K., Tucci, F.C., Rudkevich, D.M., Heinz, T., and Rebek, J., Jr. (2000) *Chem.Eur. J.*, **6**, 187.
60 Chen, J., Körner, S., Craig, S.L., Rudkevich, D.M., and Rebek, J., Jr. (2002) *Nature*, **415**, 385.
61 Iwasawa, T., Wash, P., Gibson, C., and Rebek, J. Jr. (2007) *Tetrahedron*, **63**, 6506.
62 Iwasawa, T., Hooley, R.J., and Rebek, J. Jr. (2007) *Science*, **317**, 493.
63 Hooley, R.J., Iwasawa, T., and Rebek, J. Jr. (2007) *J. Am. Chem. Soc.*, **129**, 15330.
64 Restorp, P. and Rebek, J. Jr. (2008) *J. Am. Chem. Soc.*, **130**, 11850.
65 (a) Scarso, A., Trembleau, L., and Rebek, J. Jr. (2003) *Angew. Chem. Int. Ed.*, **42**, 5499.(b) Scarso, A., Trembleau, L., and Rebek, J. Jr. (2004) *J. Am. Chem. Soc.*, **126**, 13512.
66 Hooley, R.J., Restorp, P., Iwasawa, T., and Rebek, J. Jr. (2007) *J. Am. Chem. Soc.*, **129**, 15639.
67 Hooley, R.J. and Rebek, J. Jr. (2005) *J. Am. Chem. Soc.*, **127**, 11904.
68 Iwasawa, T., Mann, E., Rebek, J., Jr., and Scarso (2006) *J. Am. Chem. Soc.*, **128**, 9308.
69 A., J. and J. Rebek Jr. (2004) *J. Am. Chem. Soc.*, **126**, 8956.
70 Pluth, M.D., Bergman, R.G., and Raymond, K.N. (2007) *J. Am. Chem. Soc.*, **129**, 11459.
71 Ziegler, M., Brumaghim, J.L., and Raymond, K.N. (2000) *Angew. Chem. Int. Ed.*, **39**, 4119.
72 Dong, V.M., Fiedler, D., Carl, B., Bergman, R.G., and Raymond, K.N. (2006) *J. Am. Chem. Soc.*, **128**, 14464.
73 Purse, B.W., Gissot, A., and Rebek, J. Jr. (2005) *J. Am. Chem. Soc.*, **127**, 11222.
74 Butterfield, S.M. and Rebek, J. Jr. (2007) *Chem. Commun.*, (**16**), 1605.
75 Shenoy, S.R., Pinacho Crisòstomo, F.R., Iwasawa, T., and Rebek, J., Jr. (2008) *J. Am. Chem. Soc.*, **130**, 5658.
76 Gibson, C. and Rebek, J. Jr. (2002) *Org. Lett.*, **4**, 1887.
77 Yoshizawa, M., Kusukawa, T., Fujita, M., Sakamoto, S., and Yamaguchi, K. (2001) *J. Am. Chem. Soc.*, **123**, 10454.
78 Yoshizawa, M., Takeyama, Y., Kusukawa, T., and Fujita, M. (2002) *Angew. Chem. Int. Ed.*, **41**, 1347.
79 Hooley, R.J. and Rebek, J. Jr. (2007) *Org. Biomol. Chem.*, **22**, 3631.
80 Gissot, A. and Rebek, J. Jr. (2004) *J. Am. Chem. Soc.*, **126**, 7424.

81 Richeter, S. and Rebek, J. Jr. (2004) *J. Am. Chem. Soc.*, **126**, 16280.
82 Zelder, F.H., and Rebek, J., Jr. (2006) *Chem. Commun.*, 753.
83 Yoshizawa, M., Tamura, M., and Fujita, M. (2006) *Science*, **312**, 251.
84 Kleij, A.W. and Reek, J.N.H. (2006) *Chem. Eur. J.*, **12**, 4219.
85 Slagt, V.F., Reek, J.N.H., Kamer, P.C.J., and van Leeuwen, P.W.N.M. (2001) *Angew. Chem. Int. Ed.*, **40**, 4271.
86 Kuil, M., Soltner, T., van Leeuwen, P.W.N.M., and Reek, J.N.H. (2006) *J. Am. Chem. Soc.*, **128**, 11344.
87 Kleij, A.W., Lutz, M., Spek, A.L., van Leeuwen, P.W.N.M., and Reek, J.N.H. (2005) *Chem. Commun.*, 3661.
88 Hof, F., Trembleau, L., Ullrich, E.C., and Rebek, J., Jr. (2003) *Angew. Chem. Int. Ed.*, **42**, 3150.
89 Hooley, R.J., Biros, S.M., and Rebek, J., Jr. (2006) *Angew. Chem. Int. Ed.*, **45**, 3517.
90 Ito, H., Kusukawa, T., and Fujita, M. (2000) *Chem. Lett.*, 598.
91 Kang, J. and Rebek, J., Jr. (1997) *Nature*, **385**, 50.
92 Kang, J., Hilmersson, G., Santamaria, J., and Rebek, J., Jr. (1998) *J. Am. Chem. Soc.*, **120**, 3650.
93 Chen, J. and Rebek, J. Jr. (2002) *Org. Lett.*, **4**, 327.
94 Hou, J.-L., Ajami, D., and Rebek, J. Jr. (2008) *J. Am. Chem. Soc.*, **130**, 7810.
95 Liu, S. and Gibb, B.C. (2008) *Chem. Commun.*, 3709.
96 Parthasarathy, A., Kaanumalle, L.S., and Ramamurthy, V. (2007) *Org. Lett.*, **9**, 5059.
97 Kaanumalle, L.S. and Ramamurthy, V. (2007) *Chem. Commun.*, 1062.
98 Kaanumalle, L.S., Gibb, C.L.D., Gibb, B.C., and Ramamurthy, V. (2004) *J. Am. Chem. Soc.*, **126**, 14366.
99 Gibb, C.L.D., Sundaresan, A.K., Ramamurthy, V., and Gibb, B.C. (2008) *J. Am. Chem. Soc.*, **130**, 4069.
100 Natarajan, A., Kaanumalle, L.S., Jockusch, S., Gibb, C.L.D., Gibb, B.C., Turro, N.J., and Ramamurthy, V. (2007) *J. Am. Chem. Soc.*, **129**, 4132.
101 Greer, A. (2007) *Nature*, **447**, 273.
102 Kaanumalle, L.S., Gibb, C.L.D., Gibb, B.C., and Ramamurthy, V. (2007) *Org. Biomol. Chem.*, **5**, 236.
103 Leung, D.H., Fiedler, D., Bergman, R.G., and Raymond, K.N. (2004) *Angew. Chem. Int. Ed.*, **43**, 963.
104 Leung, D.H., Bergman, R.G., and Raymond, K.N. (2006) *J. Am. Chem. Soc.*, **128**, 9781.
105 Yoshizawa, M., Kusukawa, T., Fujita, M., and Yamaguchi, K. (2000) *J. Am. Chem. Soc.*, **122**, 6311.
106 Kusukawa, T., Yoshizawa, M., and Fujita, M. (2001) *Angew. Chem. Int. Ed.*, **40**, 1879.
107 Nishioka, Y., Yamaguchi, T., Yoshizawa, M., and Fujita, M. (2007) *J. Am. Chem. Soc.*, **129**, 7000.
108 Yoshizawa, M., Takeyama, Y., Okano, T., and Fujita, M. (2003) *J. Am. Chem. Soc.*, **125**, 3243.
109 Nishioka, Y., Yamaguchi, T., Kawano, M., and Fujita, M. (2008) *J. Am. Chem. Soc.*, **130**, 8160.
110 Yamaguchi, T. and Fujita, M. (2008) *Angew. Chem. Int. Ed.*, **47**, 2067.
111 Yoshizawa, M., Miyagi, S., Kawano, M., Ishiguro, K., and Fujita, M. (2004) *J. Am. Chem. Soc.*, **126**, 9172.
112 Kang, J., Santamaria, J., Hilmersson, G., and Rebek, J., Jr. (1998) *J. Am. Chem. Soc.*, **120**, 7389.
113 Fiedler, D., Bergman, R.G., and Raymond, K.N. (2004) *Angew. Chem. Int. Ed.*, **43**, 6748.
114 Fiedler, D., van Halbeek, H., Bergman, R.G., and Raymond, K.N. (2006) *J. Am. Chem. Soc.*, **128**, 10240.
115 Hastings, C.J., Fiedler, D., Bergman, R.G., and Raymond, K.N. (2008) *J. Am. Chem. Soc.*, **130**, 10977.
116 Brown, C.J., Bergman, R.G., and Raymond, K.N. (2009) *J. Am. Chem. Soc.*, **131**, 17530–17531.
117 Leung, D.H., Bergman, R.G., and Raymond, K.N. (2007) *J. Am. Chem. Soc.*, **129**, 2746.
118 Pluth, M.D., Bergman, R.G., and Raymond, K.N. (2007) *Angew. Chem. Int. Ed.*, **46**, 8587.
119 Pluth, M.D., Bergman, R.G., and Raymond, K.N. (2009) *J. Org. Chem.*, **74**, 58–63.
120 Pluth, M.D., Bergman, R.G., and Raymond, K.N. (2007) *Science*, **316**, 85.

121 Pluth, M.D., Bergman, R.G., and Raymond, K.N. (2008) *J. Am. Chem. Soc.*, **130**, 11423.

122 Hastings, C.J., Pluth, M.D., Bergman, R.G., and Raymond, K.N. (2010) *J. Am. Chem. Soc.*, **132**, 6938–6940.

123 Lee, S.J., Cho, S.-H., Mulfort, K.L., Tiede, D.M., Hupp, J.T., and Nguyen, S.T. (2008) *J. Am. Chem. Soc.*, **130**, 16828–16829.

124 http://nobelprize.org/nobel_prizes/chemistry/.

125 Biros, S.M., and Rebek, J. Jr. (2007) *Chem. Soc. Rev.*, **36**, 93.

126 Scarso, A., and Borsato, G. (2009) Optically active supramolecules, in *Chirality at the Nanoscale* (ed. D. Amabilino), Wiley-VCH Verlag GmbH, Weinheim, p. 29.

5
Photocatalysts: Nanostructured Photocatalytic Materials for Solar Energy Conversion

Kazunari Domen

5.1
Principles of Overall Water Splitting Using Nanostructured Particulate Photocatalysts

5.1.1
Introduction to Photocatalytic Water Splitting

Large-scale hydrogen production from water using solar energy remains an ultimate goal for the supply of clean and recyclable energy. Of the various approaches employed in attempts to realize this reaction, the use of a heterogeneous particulate photocatalyst is one of the most promising. This approach is derived from the photoelectrochemical (PEC) approach proposed by Honda and Fujishima in 1972 [1], which originally involved the use of a TiO_2 photoanode and Pt cathode. In that reaction, n-type band bending at the interface between the TiO_2 photoanode and aqueous solution induces macroscopic charge separation of photoexcited electrons and holes, thereby driving H_2 evolution at the Pt electrode and O_2 evolution at the TiO_2 electrode. Overall water splitting using particulate photocatalytic systems, with particle sizes of a few micrometers to nanometers, was first demonstrated unequivocally in 1980 [2–4]. Although there were several reports claiming to have achieved water splitting using particulate systems prior to this time [5, 6], the studies lacked clear evidence of simultaneous and catalytic evolution of H_2 and O_2 from water. The three systems reported in 1980, using TiO_2 [4] and $SrTiO_3$ [2, 3], were developed for gaseous water splitting. Subsequent to these works, however, most photocatalytic water splitting reactions have been designed for aqueous solutions, since the reaction rate in aqueous solution is generally faster than that in the gaseous phase [7]. However, research on overall water splitting using heterogeneous photocatalysts diminished rapidly after the early 1980s due to the lack of suitable materials for the reaction under visible light. A small number of research groups continued research on photocatalysts for water splitting under ultraviolet (UV) light, leading to the development of some unique photocatalysts with unusual structures based on ion-exchangeable layered oxides [8, 9]. In such systems, high activity is achieved due to the intercalation of water molecules into the complex

Selective Nanocatalysts and Nanoscience, First Edition. Edited by Adriano Zecchina, Silvia Bordiga, Elena Groppo.

photocatalyst structure. The reaction mechanism in these systems therefore differs from that in the original PEC system.

To develop photocatalysts for overall water splitting that are responsive to visible light, one of the most successful strategies has been to use nonoxide materials, which were previously believed to be unstable in this reaction. After surveying a range of materials, it was found that certain (oxy)nitrides [10–12] and oxysulfides [13] are not only stable in the water oxidation and reduction reactions but also achieve overall water splitting under visible light with appropriate nanoscale modification [14].

In this chapter, some selected examples of particulate systems for overall water splitting are described with an emphasis on the role of the nanostructure.

5.1.2
Energetics and Materials

Thermodynamically, the overall water splitting reaction is an uphill reaction requiring considerable energy conversion:

$$H_2O \xrightarrow{h\nu} H_2 + 1/2O_2;\ \Delta G^\circ = 238\ (\text{kJ/mol})$$

Figure 5.1 shows a schematic illustration of the thermodynamics requirements for the overall water splitting reaction using a heterogeneous particulate photocatalyst. The H^+/H_2 redox potential is located at 0 V, and the O_2/H_2O redox potential is located at +1.23 V. Under irradiation with photons of energy greater than the band gap of the photocatalyst, electrons in the valence band are excited into the conduction band, leaving (positive) holes in the valence band. The subsequent

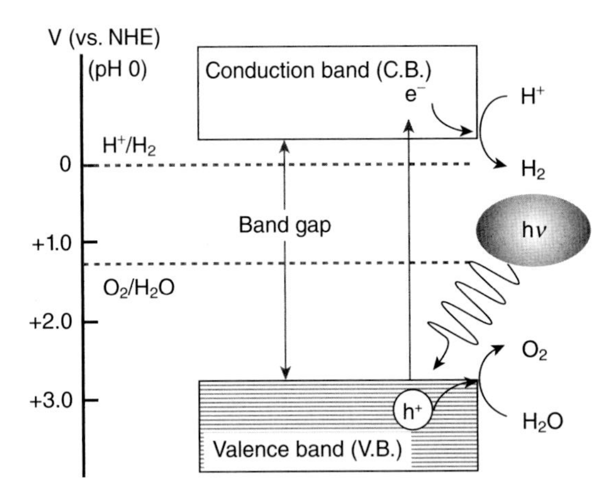

Figure 5.1 Basic principle of overall water splitting using a heterogeneous photocatalyst. Vertical axis indicates potential electron energy versus NHE at pH = 0. Maeda, K.,Domen, K., "New non-oxide photocatalysts designed for overall water splitting under visible light", *J. Phys. Chem. C*, 2007, *111*, 7851. Copyright American Chemical Society, reproduced with permission.

relaxation of electrons to the bottom of the conduction band and holes to the top of the valence band has a picosecond (10^{-12} s) timescale, whereas the electrons and holes at the band edges have lifetimes typically on the order of microseconds (10^{-6} s), which is sufficiently long to participate in photocatalytic reactions. To accomplish the overall water splitting reaction, the bottom of the conduction band must be more negative than the H^+/H_2 redox potential, and the top of the valence band must be more positive than the O_2/H_2O redox potential. Therefore, the minimum energy required for the band gap of the photocatalyst, and for the photons, is 1.23 eV, which corresponds to a wavelength of 1000 nm. This is valid in the case of overall water splitting using a photocatalyst assuming two photons for the production of one H_2 molecule. This wavelength (1000 nm) is in the near-infrared region, suggesting the possibility that the entire spectral range of visible light could be exploited for the water splitting reaction. However, activation barriers exist for the electron transfer processes between reactants and photocatalysts (or overpotentials are necessary), and therefore larger energy is required in order to drive the reaction at a reasonable reaction rate. In addition to these conditions, the reverse reaction from the products, H_2 and O_2, must be effectively suppressed. A more serious problem for many photocatalysts is the stability of the particulate material during photocatalytic water splitting, in which the photocatalysts often undergo photodegradation.

The quantum efficiency (QE) for the overall water splitting reaction is defined as

$$\mathrm{QE}\ (\%) = \frac{\text{number of produced } H_2 \text{ molecules}}{\text{number of absorbed photons}} \times 2 \times 100$$

As it is usually difficult to determine the number of absorbed photons precisely, the number of incident photons is used in this equation instead. The QE thus calculated is regarded as an apparent QE or photonic efficiency [15].

5.1.3 Hydrogen and Oxygen Evolution Sites

Figure 5.2 shows a schematic illustration of H_2 or O_2 evolution using TiO_2 from various aqueous solutions. TiO_2 alone displays little photocatalytic activity for overall water splitting. As it is generally difficult to achieve overall water splitting due to the uphill nature of the reaction, sacrificial reagents such as methanol or silver nitrate, which act as a hole or electron scavenger, are often used to evaluate the photocatalytic activity of a material. In the presence of such sacrificial reagents, TiO_2 becomes an active photocatalyst for H_2 or O_2 evolution, depending on the sacrificial reagent employed. The rate of H_2 evolution is much lower than that of O_2 evolution, indicating that TiO_2 hosts suitable surficial active sites for O_2 evolution but not for H_2 evolution.

It has been demonstrated that the modification of a semiconductor photocatalyst with noble metal (e.g., Pt, Rh, Au, or Ag) [16–19] or transition-metal oxide (e.g., NiO_x or RuO_2) [2, 7, 8] in nanoparticulate form improves the efficiency of electron

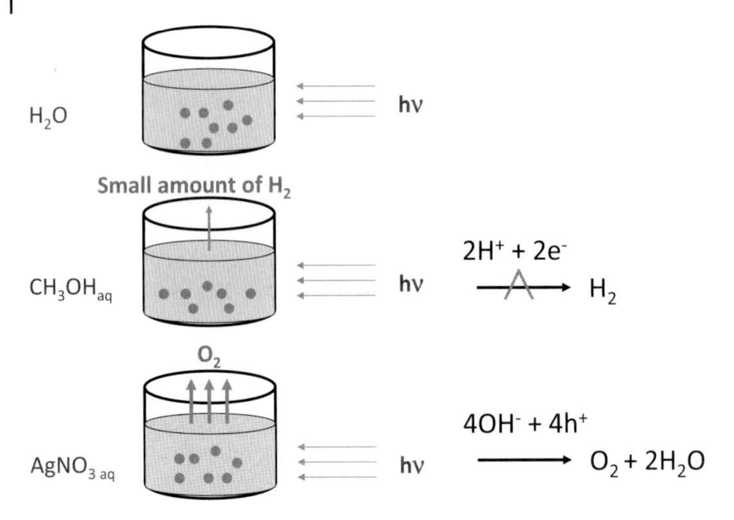

Figure 5.2 Schematic illustration of H_2 and O_2 evolution using TiO_2 in aqueous solutions containing sacrificial reagents.

transfer in the catalytic reaction. For water splitting, photogenerated electrons in the photocatalyst migrate to the nanoparticulate cocatalyst and reduce adsorbed H^+ to H_2. The loaded cocatalyst therefore provides active sites for H_2 formation with lower activation energy, thereby increasing the activity for H_2 evolution. Modification of TiO_2 with nanoparticulate Pt also results in appreciable H_2 evolution, demonstrating the importance of appropriate modification to promote H_2 evolution in the development of a highly efficient photocatalyst. Modification with an appropriate O_2 evolution cocatalyst also improves the activity for that side of the reaction [12, 20]. In the case of nonoxide photocatalysts such as (oxy)nitrides, modification with a cocatalyst to promote O_2 evolution is important not only from the viewpoint of improving O_2 evolution activity but also for suppressing oxidative decomposition of the material [12].

5.2 Oxide Photocatalysts for Overall Water Splitting

5.2.1 Nanostructures of Particulate Photocatalysts

5.2.1.1 NiO/Ni/$SrTiO_3$

$SrTiO_3$ was known to be an active photoanode for overall water splitting without external bias when it was first used in a PEC cell with a Pt counter electrode [21]. The rutile-TiO_2 photoanode requires external bias to allow overall water splitting to proceed because the bottom of the conduction band is located slightly more positive than the redox potential of H^+/H_2 [1]. $SrTiO_3$ powder with various

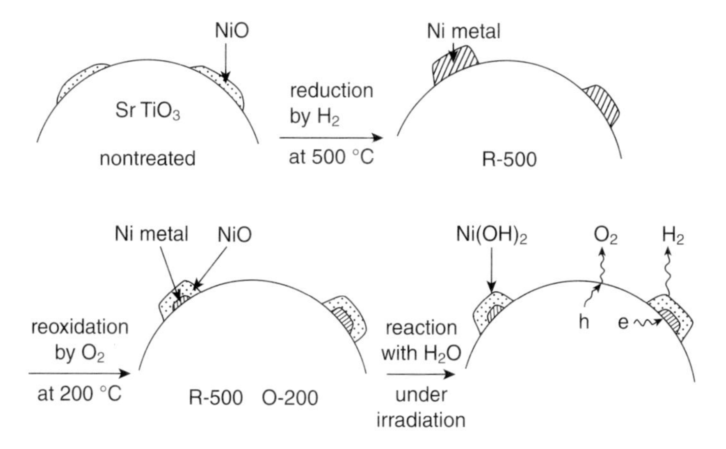

Figure 5.3 Schematic view of the structures of $NiO/Ni/SrTiO_3$ after various treatments [22].

modifications has been examined as a photocatalyst for overall water splitting. Without modification, $SrTiO_3$ does not catalyze the reaction. In the case of Pt/ $SrTiO_3$, a small amount of H_2 was detected in the gas phase, but no O_2 was evolved. Although modification with various transition-metal oxides has been attempted, including Cr, Mn, Fe, Co, Ni, and Cu, only NiO modification was found to result in an active photocatalyst for overall water splitting [22]. This system is prepared by loading the particulate $SrTiO_3$ (ca. 1 μm) with particulate NiO (ca. 20 nm) by impregnation from aqueous $Ni(NO_3)_2$ solution to achieve a final loading rate of 5 wt% NiO. The catalyst is then activated by reduction under H_2 at 773 K for 2 h, followed by oxidization under O_2 at 473 K. During treatment, the nanoparticulate Ni is reduced to Ni metal, and the surface is then reoxidized to NiO. This procedure results in a core/shell structure, as shown in Figure 5.3. In an aqueous solution, the outermost layer of the NiO/Ni structure becomes hydrated to Ni–OH. Reoxidation at high temperature (e.g., 773 K) oxidizes Ni metal to NiO, causing destruction of the core/shell structure and loss of activity. These structures have been confirmed by XAFS, XPS, SEM, and TEM. Typical time courses of overall water splitting using the $NiO/Ni/SrTiO_3$ photocatalyst are shown in Figure 5.4. It is obvious that the nanostructure of the modified $SrTiO_3$ is essential in order to achieve appreciable activity for the overall water splitting reaction. It is considered that the Ni metal between NiO and $SrTiO_3$ facilitates electron transfer between the two oxide materials, although the reason why only NiO modification has been successful remains clear. NiO is known to exhibit some activity for the H_2/D_2 exchange reaction, which indicates that the recombination of H atoms proceeds on NiO. In contrast, however, Co_3O_4 is also known to possess high activity for the H_2/D_2 exchange reaction, yet has been found to be ineffective for overall water splitting even after various kinds of pretreatment. The activity of the $NiO/Ni/SrTiO_3$ system was enhanced in the presence of concentrated NaOH solution [7].

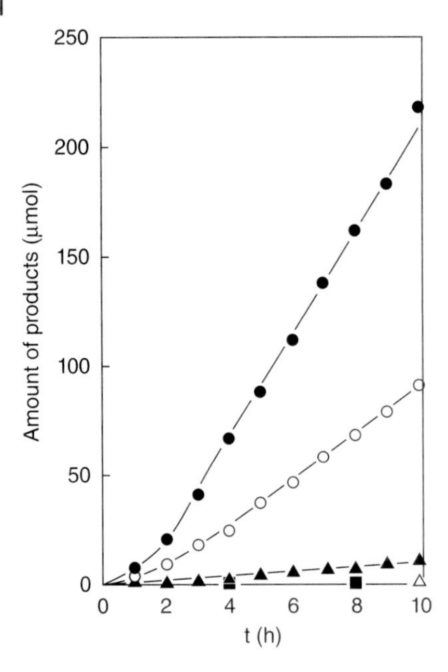

Figure 5.4 Time course of overall water splitting reaction using $NiO/Ni/SrTiO_3$ after various pretreatments. (■) H_2, nontreated catalyst: (•) H_2, (○) O_2, reoxidized at 473 K: (▲) H_2, (△) O_2, reoxidized at 773 K. Catalyst, 0.5 g; 450 W high-pressure Hg lamp; Inner irradiation type Pyrex cell. Domen, K.; Kudo, A.; Onishi, T., "Mechanism of photocatalytic decomposition of water into H_2 and O_2 over $NiO–SrTiO_3$", *J. Catal.* 1986, *102*, 92.

5.2.1.2 $NiO/NaTaO_3$:La

It has been reported that many tantalates are active photocatalysts for overall water splitting even without modification with a cocatalyst such as NiO [23–29]. La-doped $NaTaO_3$ modified with NiO as a cocatalyst ($NiO/NaTaO_3$: La) has been found to display the highest activity for overall water splitting among the tantalate systems examined, achieving a QE of 56% under excitation at 270 nm [29]. Gas evolution using this catalyst is clearly observed as a continuous bubbling, evolving at rates of 19.8 mmol/h H_2 and 9.7 mmol/h O_2 under a 400 W high-pressure Hg lamp. The activity of this system also shows no sign of deactivation even over continuous periods of irradiation of more than 400 h. The reaction scheme for water splitting using the $NiO/NaTaO_3$: La photocatalyst has been clarified by nanoscale characterization, as shown in Figure 5.5. Electron microscopy has revealed that the particle size of the $NaTaO_3$: La crystal (0.1–0.7 μm) is smaller than that of the undoped $NaTaO_3$ crystal (2–3 μm), and that the ordered surficial nanosteps are created by La doping. The small particle size and high crystallinity are advantageous for increasing the probability of reaction between photogenerated electrons–holes and water molecules and suppressing the rate of electron–hole recombination. TEM and EXAFS analyses indicate that the NiO cocatalyst is loaded on the edges of the nanostep structures in the form of ultrafine NiO particles. The H_2 evolution site at the edge is therefore effectively separated from the O_2 evolution site in the groove of the nanostep structure. This physical separation is advantageous for water splitting by suppressing the backward reaction.

5.2.2 Photocatalysts with Ion-Exchangeable Layered Structures

5.2.2.1 $NiK_4Nb_6O_{17}$

$A_4Nb_6O_{17}$ (A = K, Rb) is a layered compound with ion-exchange functionality. The structure and appearance of $K_4Nb_6O_{17}$ are shown in Figure 5.6. $Nb_6O_{17}^{4-}$ niobate sheets are stacked along the *c* axis, and K^+ cations are located in the interlayer space. The anisotropy and the stacking of niobate sheets result in the formation of two different types of alternating interlayer spaces. One of these interlayer spaces (I) is readily hydrated, and the alkaline metal cations in the space can be replaced by divalent metal cations such as Ni^{2+} [30]. The alternate interlayer space (II) is more resistant to hydration, and the alkaline metal cations cannot be substituted. Under ambient conditions, interlayer space I is hydrated whereas II is not, resulting in the formation of $K_4Nb_6O_{17} \cdot 3H_2O$ and/or $K_4Nb_6O_{17} \cdot 4.5H_2O$. In aqueous solution, interlayer II is also hydrated due to the excess of hydrated water.

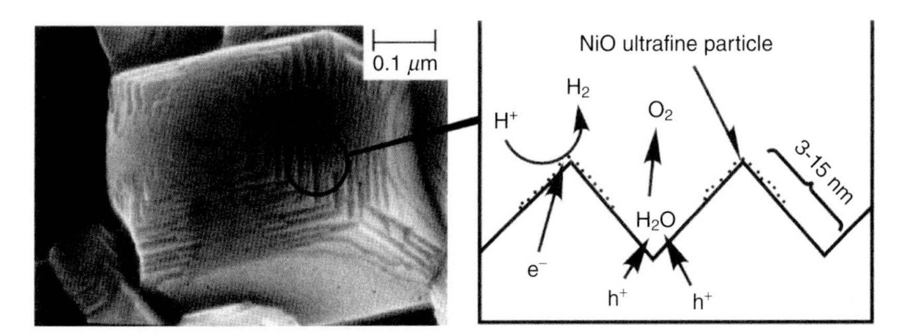

Figure 5.5 Mechanism of water splitting on a surface nanostep of the $NiO/NaTaO_3$: La. Kudo, A.; Kato, H.; Tsuji, I., "Strategies for the development of visible-light-driven photocatalysts for water splitting", *Chem. Lett.* 2004, *12*, 1534.

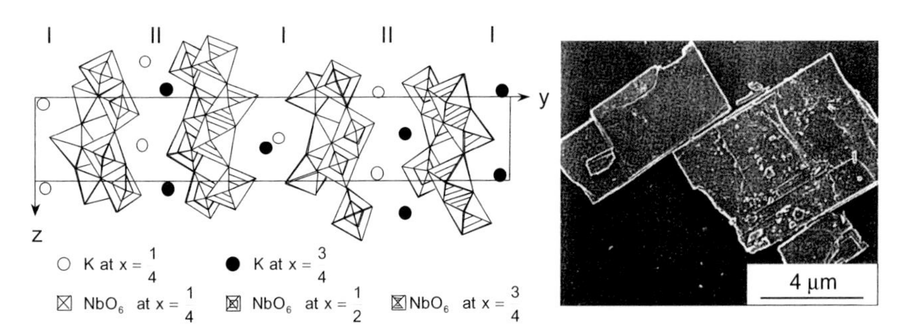

Figure 5.6 SEM photograph and schematic view of the layer structure of $K_4Nb_6O_{17}$.

Table 5.1 Photodecomposition of water over $K_4Nb_6O_{17}$ loaded with metal oxides [31].

Metal oxide[a] (1 wt%)	Pretreatment[b]	Amount of gas evolved (μmol)	
		H_2	O_2
None	R773-O473	63	7
Cr_2O_3	R773-O473	27	0
Mn_3O_4	R773-O473	6	0
Fe_2O_3	R873-O473	8	0
Co_3O_4	R823-O473	32	0
NiO	R773-O473	630	310
CuO	R773-O473	12	1
PtO[c]	Untreated	19	0
	R773-O473	12	0
RuO_2[c]	Untreated	77	16
	R773-O473	9	0
Rh_2O_3[c]	Untreated	18	0
	R773-O473	3	0

Note. Catalyst, 1 g; H_2O, 300 mL; light source, high-pressure Hg lamp (450 W); reaction cell, inner irradiation reaction cell; reaction time, 10 h.

a) The metal oxides represented are the probable forms under the reaction conditions but were not confirmed.

b) R773-O473 means reduction by H_2 at 773 K, then oxidation by O_2 at 473 K.

c) 0.5 wt% loading.

Unmodified particulate $K_4Nb_6O_{17}$ displays a small degree of activity for water splitting under UV light, although at low and nonstoichiometric rates of H_2 and O_2 evolution. Modification of $K_4Nb_6O_{17}$ with various transition metals has been attempted, as shown in Table 5.1. Among these modifications, loading with Ni and Ru results in better catalytic performance than the unmodified $K_4Nb_6O_{17}$. Ni-modified $K_4Nb_6O_{17}$ is prepared as shown in Scheme 5.1 [31, 32]. By immersion of $K_4Nb_6O_{17}$ powder in aqueous $Ni(NO_3)_2$ solution, Ni^{2+} cations are substituted for K^+ cations in interlayer space I at a rate equivalent to 0.1 wt%. After reduction under H_2 at 773 K, intercalated Ni^{2+} is reduced to nanoparticles of Ni metal within interlayer space I. This structure has been confirmed by XPS, XAFS, ESR, and TEM. The active structure of $Ni/K_4Nb_6O_{17}$ is depicted in Figure 5.7a, and a typical time course of H_2 and O_2 evolution is shown in Figure 5.8. The activity decays gradually as the reaction proceeds. Based on the structure, the proposed reaction mechanism for the overall water splitting is shown in Figure 5.7b [32]. According to this mechanism, both interlayer spaces are hydrated, and photoexcited electrons migrate to the Ni metal nanoparticles in interlayer space I, where H^+ (or H_3O^+) cations are reduced to H_2. Photoexcited holes react with water molecules in interlayer space II to form O_2. The sites of H_2 and O_2 evolution are thus physically separated by niobate sheets, which results in relatively high QE as shown below. The degradation of activity, as shown in Figure 5.8a is mainly attributable to

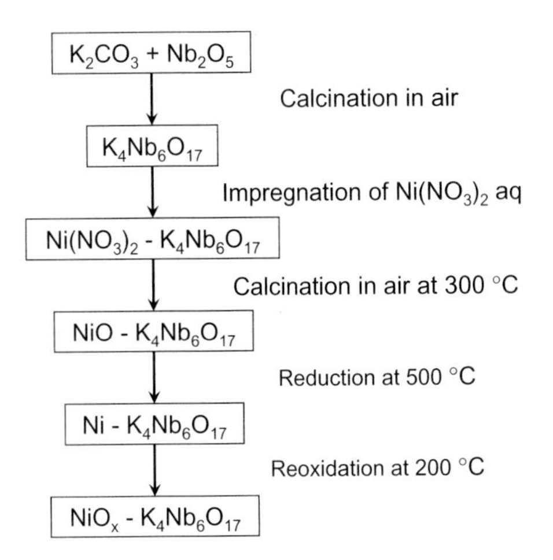

Scheme 5.1 Preparation procedure of $NiO_x/K_4Nb_6O_{17}$.

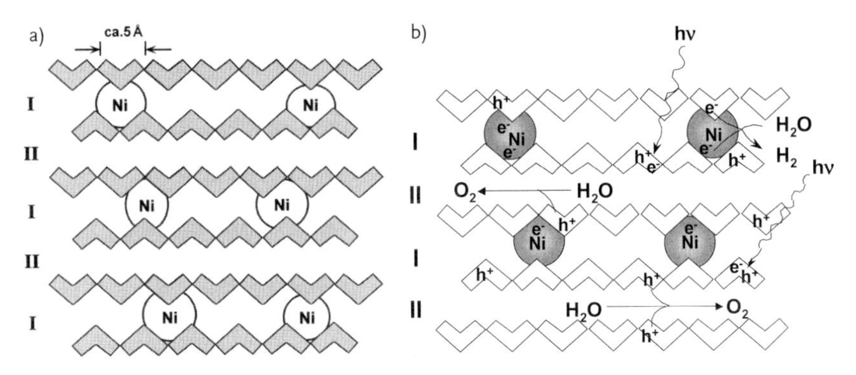

Figure 5.7 (a) Active structure and (b) plausible reaction mechanism for overall water splitting on $Ni/K_4Nb_6O_{17}$.

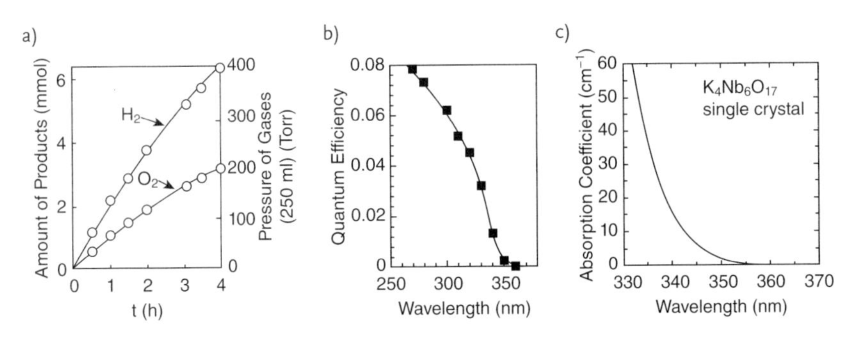

Figure 5.8 (a) Time course of overall water splitting reaction using $Ni/K_4Nb_6O_{17}$ (0.1 wt% Ni). Catalyst, 1 g; light source, 500 W high-pressure Hg lamp; quartz reaction vessel. Domen, K., *Petrotech* 1992, *15*, 940. (b) Wavelength dependence of quantum efficiency of water splitting reaction and (c) absorption spectrum of $K_4Nb_6O_{17}$.

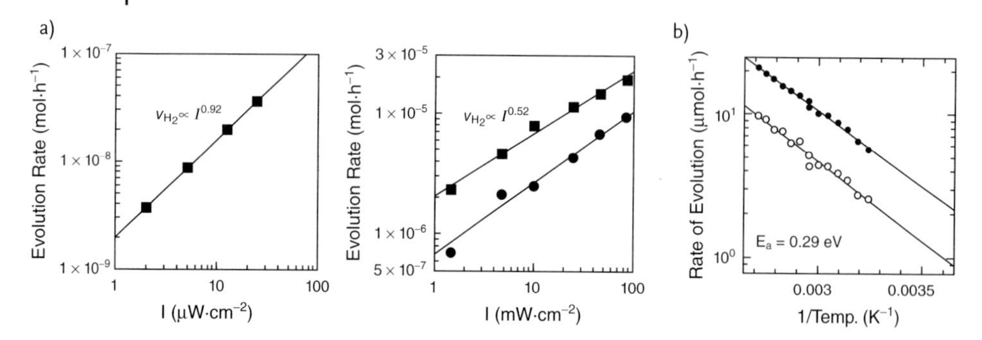

Figure 5.9 (a) The light-intensity dependence of the water decomposition reaction on $K_4Nb_6O_{17}$ at lower and higher intensity region. (▪) H_2, (•) O_2 [33]. (b) Arrhenius plot of water decomposition using $K_4Nb_6O_{17}$.

Table 5.2 Activation energy of water decomposition.

Photocatalyst	Activation energy (kJ mol^{-1}) (eV)	
	H_2O	D_2O
$K_4Nb_6O_{17}$	28 (0.29)	26 (0.17)
$Ni/K_4Nb_6O_{17}$	37 (0.38)	22 (0.23)

oxidation of the Ni metal nanoparticles, which indicates that photoexcited holes also react with Ni particles in interlayer space I. The quantum efficiencies of overall water splitting using $K_4Nb_6O_{17}$ are plotted for various excitation wavelengths in Figure 5.8b. A distinct dependence on wavelength can be seen, and the onset wavelength of photocatalytic activity is almost coincident with the absorption edge of $K_4Nb_6O_{17}$ (Figure 5.8c). The dependence of the H_2 and O_2 evolution rates on light intensity are shown in Figure 5.9a (S. Tabata and K. Domen, unpublished results), and an Arrhenius plot of H_2O decomposition on $K_4Nb_6O_{17}$ is shown in Figure 5.9b. The activation energy for photocatalytic overall water splitting can be readily identified from these data. Table 5.2 summarizes the activation energies for H_2O and D_2O decomposition on $K_4Nb_6O_{17}$ and $Ni/K_4Nb_6O_{17}$ [33]. The activation energy on $Ni/K_4Nb_6O_{17}$ is lower than that on $K_4Nb_6O_{17}$, and that for H_2O is lower than that for D_2O. As the Ni nanoparticles function as H_2 evolution sites, it seems reasonable to conclude that hydrogen (H_2 or D_2) evolution is the rate-determining step of overall water splitting using $K_4Nb_6O_{17}$-based photocatalysts. $Rb_4Nb_6O_{17}$ exhibits very similar behavior for the water splitting reaction [34], and with appropriate treatments, $Ni/Rb_4Nb_6O_{17}$ displays even higher activity than $Ni/K_4Nb_6O_{17}$.

5.2.2.2 $NiO/Ni/Rb_2La_2Ti_3O_{10}$

As mentioned above, $A_4Nb_6O_{17}$ is a unique layered oxide that possesses two different alternating interlayer spaces. Most ion-exchangeable layered compounds, however, have only one type of interlayer space. The Ruddlesden–Popper layered perovskite oxides are examples of such compounds [35]. The structure of the

Ruddlesden–Popper $Rb_2La_2Ti_3O_{10}$ layered perovskite oxide is shown in Figure 5.10. This material consists of $[La_2Ti_3O_{10}]^{2-}$ perovskite layers with Rb^+ cations hosted in the interlayer space, which can be hydrated to form $Rb_2La_2Ti_3O_{10} \cdot H_2O$ under ambient conditions. It has been found that modification of the external surface of the $Rb_2La_2Ti_3O_{10} \cdot H_2O$ particles (ca. 1 µm in size) with nanoparticulate NiO/Ni (core/shell) in a similar manner to NiO/Ni/$SrTiO_3$ results in appreciable enhancement of activity for overall water splitting, as shown in Figure 5.11 [9]. The

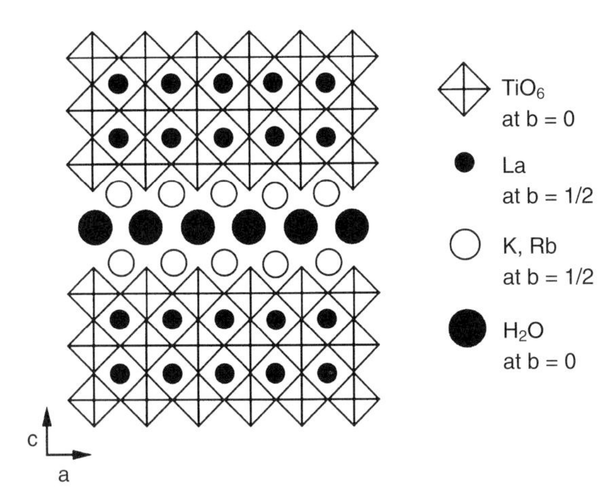

Figure 5.10 Schematic structure of $A_2La_2Ti_3O_{10}$ (A = K, Rb). Takata, T.; Tanaka, A.; Hara, M.; Kondo, J. N.; Domen, K., "Recent progress of photocatalysts for overall water splitting", *Catal. Today*, 1998, *44*, 17.

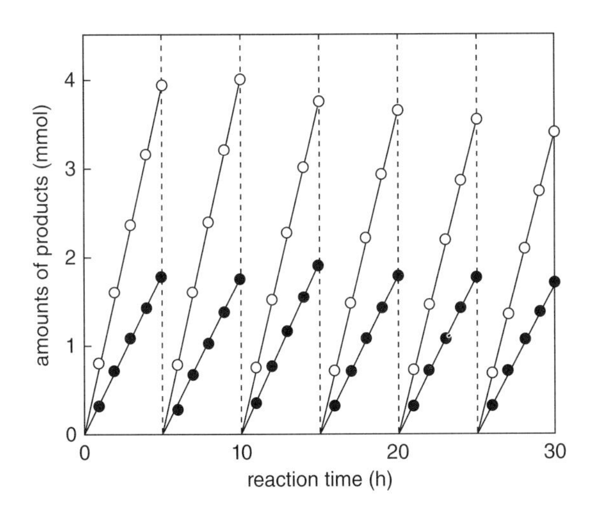

Figure 5.11 Time course of H_2 and O_2 evolution from an aqueous RbOH solution (0.1 M) on NiO/Ni/$Rb_2La_2Ti_3O_{10}$. The gas phase was evacuated at every for 5 h of reaction and the amount of evolved H_2 and O_2 are plotted by open and filled circles, respectively. Catalyst, 1.0 g; H_2O, 330 mL; light source; 450 W high-pressure Hg lamp; inner irradiation type quartz cell [9].

proposed reaction mechanism is shown in Figure 5.12. Photogenerated electrons in the perovskite layer migrate to the NiO/Ni nanoparticles at the external surface, where H^+ cations are reduced to H_2 molecules. Photogenerated holes, which are considered to have low mobility, react with intercalated water molecules to form O_2 molecules at the interlayer space. Evidence for this reaction mechanism, particularly the inclusion of intercalated water molecules as reactants, has been obtained through tests of interlayer hydration. It is possible to reduce the layer charge density by replacing Ti^{4+} with Nb^{5+}, and at a certain layer charge density, hydration of the interlayer space no longer occurs, as shown in Table 5.3. The

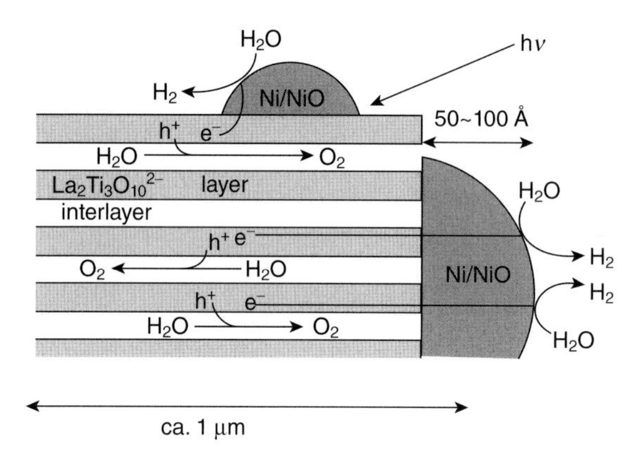

Figure 5.12 Schematic view of the reaction mechanism of H_2O decomposition on layered perovskite photocatalyst [9].

Table 5.3 Photocatalytic activities of various layered perovskites and hydration numbers [9].

Catalyst	Rate of gas evolution (μmol h^{-1})		Optimum condition			Hydration no.[c]
	H_2	O_2	Ni-loading (wt%)[a]	AOH (mol L^{-1})[b]	pH	
$K_2La_2Ti_3O_{10}$	444	221	3	0.1	12.8	1.0
$Rb_2La_2Ti_3O_{10}$	869	430	4	0.1	12.8	1.1
$Rb_{1.5}La_2Ti_{2.5}Nb_{0.5}O_{10}$	725	358	5	0.1	12.6	0.9
$RbLa_2Ti_2NbO_{10}$	79	30	0.3	0.1	12.8	0.0
$Cs_2La_2Ti_3O_{10}$	700	340	3	0.1	10.5	3.5
$Cs_{1.5}La_2Ti_{2.5}Nb_{0.5}O_{10}$	540	265	4	0.1	10.4	2.0
$CsLa_2Ti_2NbO_{10}$	115	50	0.3	0.1	8.5	0.0

Note. Reaction condition: catalyst, 1.0 g; H_2O, 320 mL; high-pressure Hg lamp (450 W); an inner irradiation type quartz reaction cell.

a) Amount of loaded Ni.
b) Concentration of AOH (A = K, Rb, Cs) solution.
c) The number of hydrogen in the formula of $A_{2-x}La_2Ti_{3-x}Nb_xO_{10} \cdot nH_2O$.

activity of the nonhydrated material is considerably lower than that for the hydrated photocatalyst, demonstrating the importance of intercalated water molecules in this reaction.

The $NiO/Ni/SrTiO_3$ photocatalyst could be regarded as a microscale PEC cell. However, the reaction mechanisms of the $Ni/K_4Nb_6O_{17}$ and $NiO/Ni/Rb_2La_2Ti_3O_{10}$ photocatalysts for overall water splitting obviously differ from that of the original PEC cell, and should be regarded as a type of artificial photosynthetic system based on nanostructured inorganic materials.

5.3 Visible Light-Responsive Photocatalysts for Overall Water Splitting

5.3.1 (Oxy)nitrides and Oxysulfides as Photocatalysts

Many oxide photocatalysts have been developed for overall water splitting, yet most are active only under UV light. Certain nonoxide materials, such as CdS and CdSe, are known to have suitable band-gap energies for the absorption of visible light, and also to have suitable band positions for overall water splitting. However, most nonoxide materials are considerably unstable in the water splitting reaction, particularly in the oxidation of water to O_2. Recent surveys of new nonoxide materials such as oxynitrides, nitrides, and oxysulfides have revealed some new stable photocatalysts for the water splitting reaction. Early transition metals such as Ti^{4+}, Ta^{5+}, and Nb^{5+} were initially examined. These transition-metal elements are employed in the highest respective oxidation states, with no remaining *d*-electrons (i.e., d^0 electronic configuration). (Oxy)nitrides containing typical elements such as Ga^{3+} and Ge^{4+} with filled d orbitals have also been examined (d^{10} electronic configuration).

Ta_2O_5, a typical d^0 oxide, has a band gap of 3.9 eV, as shown in Figure 5.13. The top of the valence band mainly consists of O 2p orbitals, and is located at ca. +3.4 V with respect to the normal hydrogen electrode (NHE). The bottom of the conduction band mainly consists of Ta 5d orbitals, and is located at ca. –0.5 V versus NHE. The oxynitride TaON and nitride Ta_3N_5 can be synthesized by nitridation of Ta_2O_5 at 1123–1173 K under flowing NH_3. As the potential energy of the N 2p orbital is higher than that of the O 2p orbital, the tops of the valence bands of TaON and Ta_3N_5 are dominated by N 2p orbitals and are located at ca. +2 V and +1.5 V versus NHE, respectively [36]. The bottoms of the conduction bands mainly consist of empty Ta 5d orbitals and are located at –0.3 to –0.5 V versus NHE, comparable to that of Ta_2O_5. Therefore, the absorption edges of TaON and Ta_3N_5 are located at wavelengths of ca. 500 and 600 nm, respectively, and both materials have suitable band-gap positions for the overall water splitting reaction. Water oxidation using Ta_3N_5 powder in the presence of Ag^+ cations as an oxidizing reagent has been tested under irradiation at visible wavelengths [11]. As shown in Figure 5.14, O_2 evolved smoothly in this reaction, with a QE close to 10%. A small

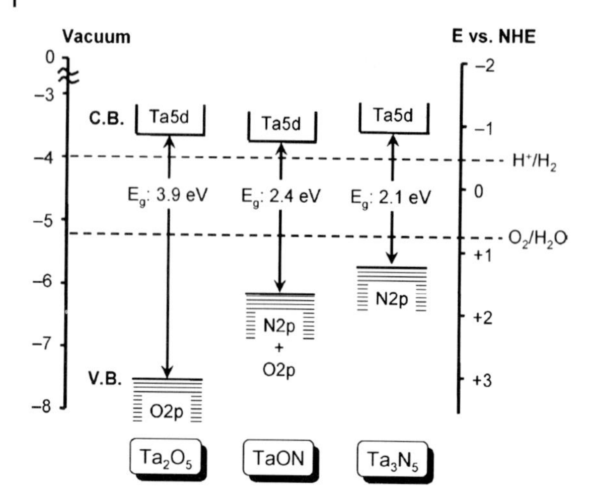

Figure 5.13 Schematic illustration of band structures of Ta_2O_5, TaON, and Ta_3N_5. Maeda, K; Domen, K., "New non-oxide photocatalysts designed for overall water splitting under visible light", *J. Phys. Chem. C* 2007, *111*, 7851.

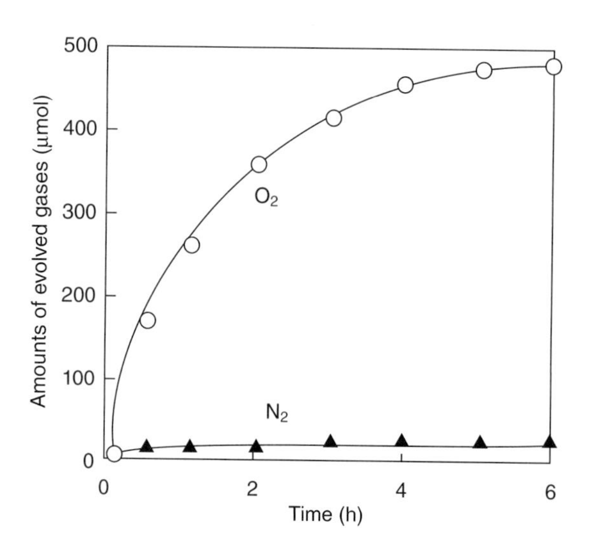

Figure 5.14 O_2 evolution from an aqueous silver nitrate solution using Ta_3N_5. Catalyst, 0.2 g; La_2O_3, 0.2 g; 0.01 M $AgNO_3$ solution, 200 mL; light source, 300 W Xe lamp ($\lambda > 420$ nm). Hara, M.; Hitoki, G.; Takata, T.; Kondo, J. N.; Kobayashi, H.; Domen, K., "TaON and Ta_3N_5 as new visible light driven photocatalysts", *Catal. Today* 2003, *78*, 555.

amount of N_2 formation was observed at the beginning of the reaction, but was almost entirely suppressed under prolonged irradiation, indicating that the nitride is essentially stable during the water oxidation reaction. The small amount of N_2 evolution has been attributed to the oxidation of surface nitrogen. Pt-loaded Ta_3N_5 in aqueous methanol solution clearly evolved H_2, as shown in Figure 5.15, although at a QE of just 0.1%. This behavior has been found to be a common feature of (oxy)nitrides and nitrides with the d^0 electronic configuration. One of the possible reasons for the low H_2 formation activity is the relatively high density of defects in the material bulk and surface. These materials are prepared at temperatures above 1073 K under NH_3 flow, exceeding the decomposition temperature of the (oxy)nitrides (ca. 923 K). It is therefore difficult to avoid the formation of a high density of defects by this preparation method. The wavelength dependence of H_2 and O_2 evolution for these materials is shown in Figure 5.16. These results confirm

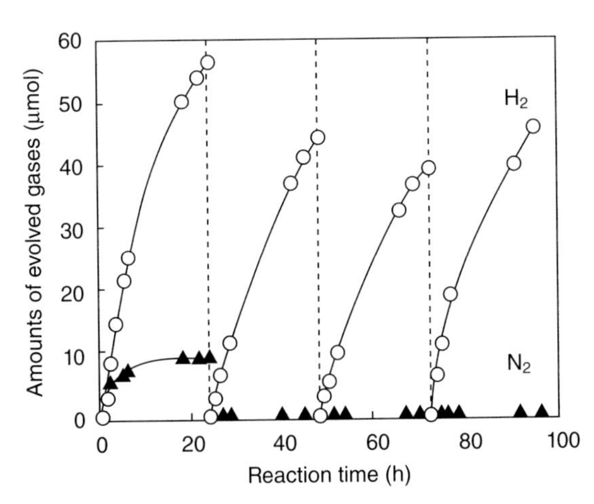

Figure 5.15 H_2 and N_2 evolution from an aqueous methanol solution using Pt/Ta_3N_5 (3.0 wt% Pt). Catalyst, 0.2 g; 10 vol% aqueous methanol solution, 200 mL; light source, 300 W Xe lamp ($\lambda > 420$ nm) [11].

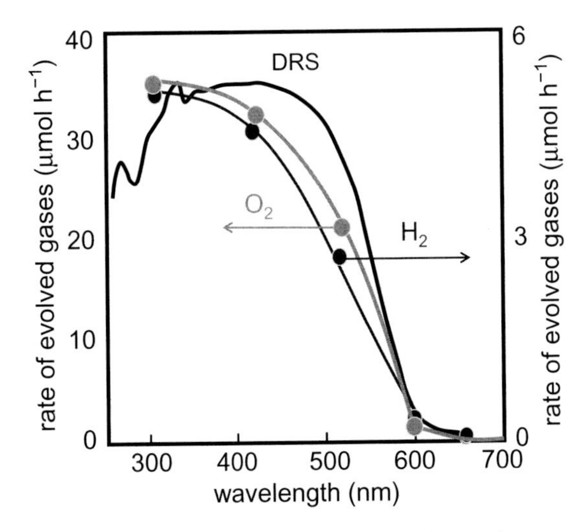

Figure 5.16 Wavelength dependence of H_2 and O_2 evolution using Ta_3N_5.

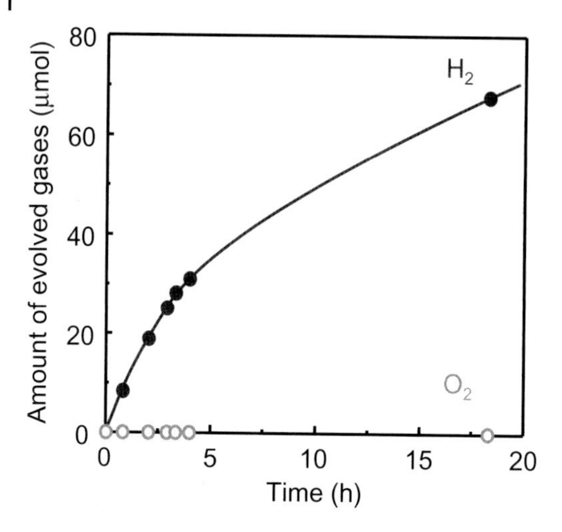

Figure 5.17 Time course of overall water splitting using TaON.

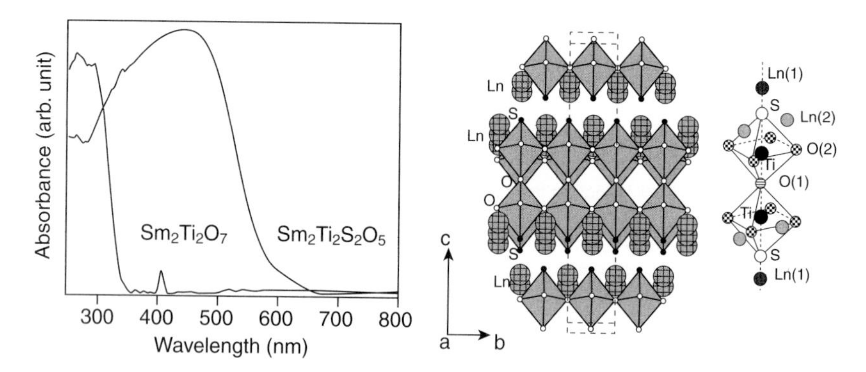

Figure 5.18 UV-visible diffuse reflectance spectra for $Sm_2Ti_2O_7$ and $Sm_2Ti_2S_2O_5$ [13]. The schematic view of the $Ln_2Ti_2S_2O_5$ structure is also shown. Ishikawa, A.; Takata, T.; Matsumura T.; Kondo, J. N.; Hara, M.; Kobayashi, H.; Domen, K. *J. Phys. Chem. B* 2004, *108*, 2637.

that both reactions take place via band-gap transition. A typical time course for the overall water splitting reaction is shown in Figure 5.17. A small amount of H_2 evolution was achieved, but no O_2 evolution was observed. It is considered that the rate of H_2 evolution is too slow to induce the water oxidation reaction, preventing the overall reaction from proceeding under these conditions.

Figure 5.18 shows the UV-visible absorption spectrum and structure for the oxysulfide $Sm_2Ti_2S_2O_5$ [13], and Figure 5.19 shows the evolution of H_2 and O_2 under visible light. $Sm_2Ti_2S_2O_5$ is another stable nonoxide material with potential for overall water splitting, although the reaction has yet to be successfully demonstrated probably due to the high defect density and correspondingly low efficiencies of H_2 and O_2 evolution.

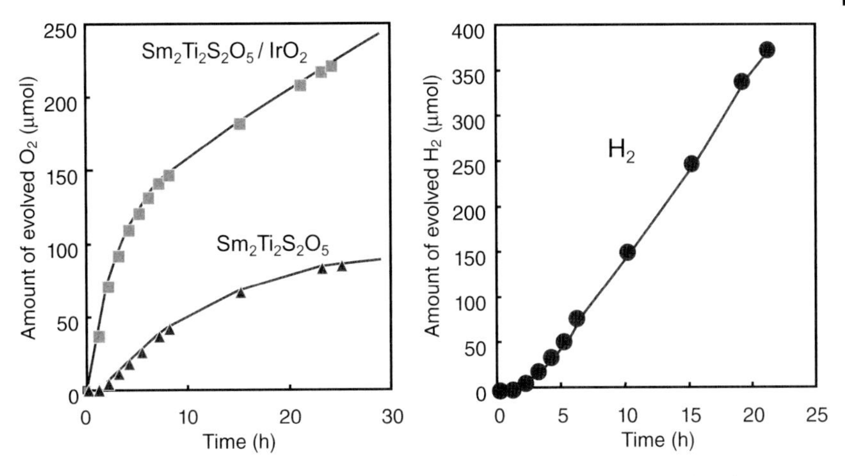

Figure 5.19 Time course of H_2 and O_2 evolution using $Sm_2Ti_2S_2O_5$ under visible light ($\lambda \geq 440$ nm). H_2 evolution reaction: Pt/$Sm_2Ti_2S_2O_5$, 0.2 g; 10 vol% aqueous methanol solution, 200 mL; O_2 evolution reaction: catalyst, 0.2 g; 0.01 M $AgNO_3$ solution, 200 mL; La_2O_3, 0.2 g.

The d^0 nonoxide materials described above are essentially stable during water oxidation and reduction, and have suitable band-gap positions for overall water splitting under visible light. However, overall water splitting reaction has not been achieved using these materials due to presumably the high defect densities in these materials. More sophisticated preparation and modification methods will therefore be required in order to enhance the activity of these materials for both H_2 and O_2 evolution.

5.3.2
Overall Water Splitting on Oxynitride Photocatalysts under Visible Light

Recently, a range of mixed oxides containing typical elements such as Ga^{3+}, Ge^{4+}, and Sb^{5+} were found to be effective for overall water splitting. Nitrides and (oxy) nitrides of these d^{10} systems have been examined as photocatalysts for water splitting. The d^{10} nitride β-Ge_3N_4 exhibits a band gap of ca. 3.8–3.9 eV, yet is pale brown in color due to the high density of defects in the bulk [37]. Alone, it does not photocatalyze the water splitting reaction. When modified with RuO_2, however, this material achieves stoichiometric H_2 and O_2 evolution under UV light in acidic solution, as shown in Figure 5.20. RuO_2/β-Ge_3N_4 was the first example of a non-oxide photocatalyst achieving overall water splitting. The activity of this material, however, decreases gradually with reaction time, but can be recovered by recalcination in air. The decrease in activity is therefore not due to the degradation of the β-Ge_3N_4 itself but due to presumably deterioration of the contact between β-Ge_3N_4 and RuO_2.

Mixtures of GaN and ZnO have also been examined as photocatalyst for overall water splitting [14, 38, 39]. It is well known that both of these materials have a

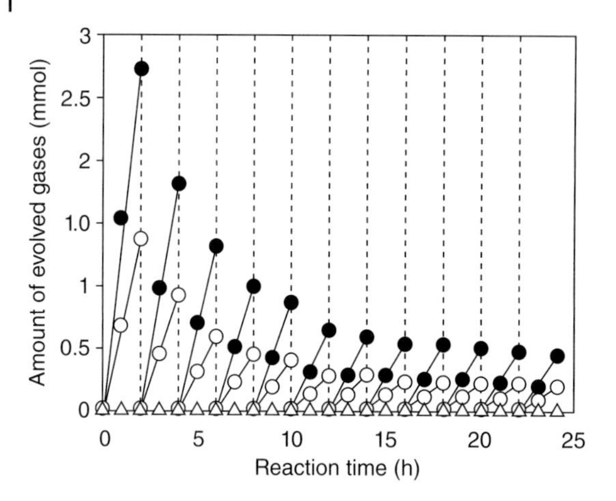

Figure 5.20 Time course of overall water splitting using RuO_2-loaded β-Ge_3N_4. The system (0.5 g of catalysts in 390 mL aqueous solution adjusted to pH 0 by H_2SO_4) was evacuated every 2 h, and the reaction was continued to observe the change in activity. The reaction was performed in a quartz reaction vessel with internal irradiation from a 450 W high-pressure Hg lamp: (•) H_2, (○) O_2, (△) N_2 [37].

wurtzite structure and similar lattice constants (mismatch of 0.8%). The band gaps of GaN and ZnO are 3.4 and 3.2 eV, respectively, and thus neither material is responsive to visible wavelengths. A solid solution of these two compounds can be prepared by heating of Ga_2O_3 and ZnO powders under 1 atm of ammonia at 1123–1173 K. Although the resultant solid solution retains the wurtzite structure, as confirmed by XRD and neutron diffraction measurements [14, 38, 40], the absorption edge of GaN:ZnO is shifted into the visible region. As the ZnO content is increased to ca. 20 at%, the absorption edge is shifted to ca. 480 nm, well within the visible region.

After modification with RuO_2, GaN:ZnO evolves H_2 and O_2 at a stoichiometric ratio under visible light, immediately following the onset of irradiation [14]. This behavior is very different from that of the d^0-type (oxy)nitride photocatalysts. One of the reasons for the success of overall water splitting using GaN:ZnO is the wide conduction band of this system, which consists of Ga 4s4p orbitals. The mobility of photoexcited electrons in this system is thus considered to be higher than that for the d^0-type photocatalysts. This solid solution was the first reproducible example of the system achieving overall water splitting using a photocatalyst with a band gap in the visible region. The QE of this prototype system was 0.23% at 420–440 nm.

To improve the QE, various methods of modification were examined. Instead of the RuO_2 cocatalyst, a mixed oxide of Rh and Cr was investigated, and much higher activity for overall water splitting was achieved [41]. It should be also noted that the optimal loading ratio of Rh to Cr is different with respect to semiconductor photocatalyst employed, suggesting metal–support interaction. A typical time

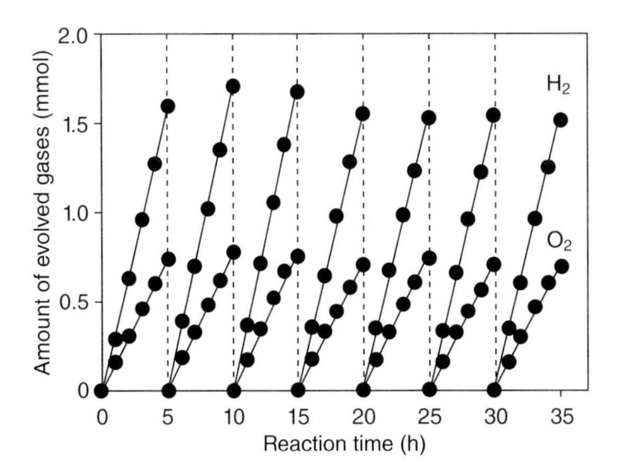

Figure 5.21 Time course of overall water splitting under visible light ($\lambda > 400$ nm) using a GaN:ZnO loaded with Rh and Cr oxide. The reaction was continued for 35 h, with evacuation every 5 h [41].

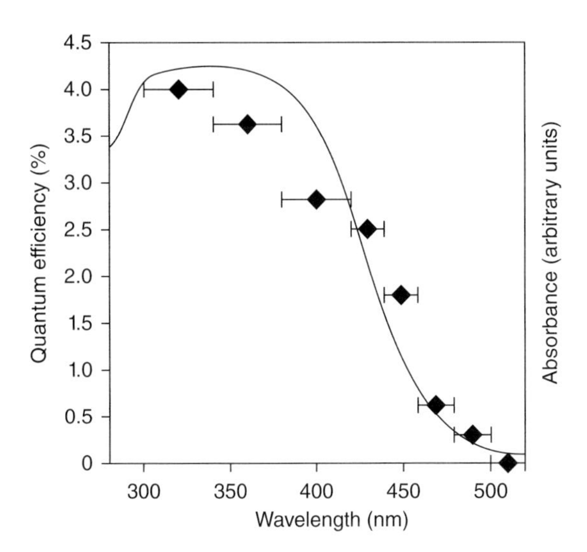

Figure 5.22 Quantum efficiency of the photocatalyst plotted as a function of wavelength of the incident light [41].

course of the reaction is shown in Figure 5.21, and the wavelength dependence of the QE is presented in Figure 5.22. The QE of this system at 420–440 nm is 2.5–3.0%, which is an order of magnitude higher than that achieved by the RuO_2-loaded photocatalyst. SEM and TEM observations have revealed that the Rh and Cr thus loaded form mixed oxide particles of 20–30 nm in size. Through experiments conducted using methanol and Ag^+ as sacrificial reagents, it was found that the Rh–Cr mixed oxide particles function as H_2 evolution sites. It was found previously that RuO_2 particles also function as H_2 evolution sites [42]. These results therefore suggest that the rate-determining step of overall water splitting reaction on modified GaN:ZnO photocatalysts is again the hydrogen evolution reaction. The low rate of this side of the reaction therefore appears to be a common problem for overall water splitting using d^0 and d^{10} nonoxide photocatalysts, necessitating the improvement of the activity of H_2 evolution sites.

5.3.3
Nanostructured Hydrogen Evolution Sites

(Oxy)nitrides exhibit relatively high photocatalytic activity for water oxidation in the presence of an appropriate electron acceptor [10–12, 41]. However, the activity for water reduction is approximately one order of magnitude lower than that for water oxidation. The overall efficiency of these (oxy)nitride-based catalysts can thus be improved by modification to promote water reduction. The conventional modification applied to improve the water reduction activity involves loading the photocatalyst with a suitable cocatalyst by impregnation. This approach produces a random dispersion of active species on the photocatalyst surface, requiring subsequent activation treatment such as reduction or oxidation to obtain high activity. Activation treatment is inherently unsuitable for (oxy)nitrides, which are less thermally stable than the corresponding metal oxides [43]. In contrast, *in-situ* photodeposition allows the cocatalyst to be loaded selectively at reaction sites and does not require subsequent activation treatment [16]. However, most of the cocatalysts suitable for introduction by this method are noble metals (e.g., Rh, Pd, and Pt), which act as a catalyst not only for water reduction but also for water formation from H_2 and O_2, an undesirable backward reaction [4, 17]. A new modification method that achieves the introduction of a water-reducing cocatalyst without the need for activation treatment is therefore desired.

A new type of H_2-evolution cocatalyst was thus developed: noble-metal/Cr_2O_3 (core/shell) nanoparticulate cocatalysts that can be prepared by *in-situ* photodeposition [44]. Figure 5.23 shows HR-TEM images of Rh-loaded GaN:ZnO before and after photodeposition of the Cr_2O_3 shell. In this system, the Rh nanoparticles forming the core induce the migration of photogenerated electrons from the GaN:ZnO bulk to the surface, whereas the Cr_2O_3 shell inhibits water formation from H_2 and O_2 on Rh. According to our preliminary study on the particle size of Rh, decreasing the particle size and narrowing the size distribution resulted in enhancing water splitting rate [45]. GaN:ZnO loaded with smaller Rh core nanoparticles having 1.9 ± 0.6 nm size and Cr_2O_3 shell exhibited 3–4 times higher activity than an analogues sample consisting of large (aggregated) Rh core. In photocatalytic overall water splitting, cocatalysts such as NiO_x, RuO_2, and $Rh_{2-y}Cr_yO_3$ play at least two roles simultaneously, that is, extraction of photogenerated electrons from the photocatalyst bulk and reduction of H^+ to H_2 on the cocatalyst surface. The reverse reaction proceeds very slowly on these oxide cocatalysts, especially on $Rh_{2-y}Cr_yO_3$.

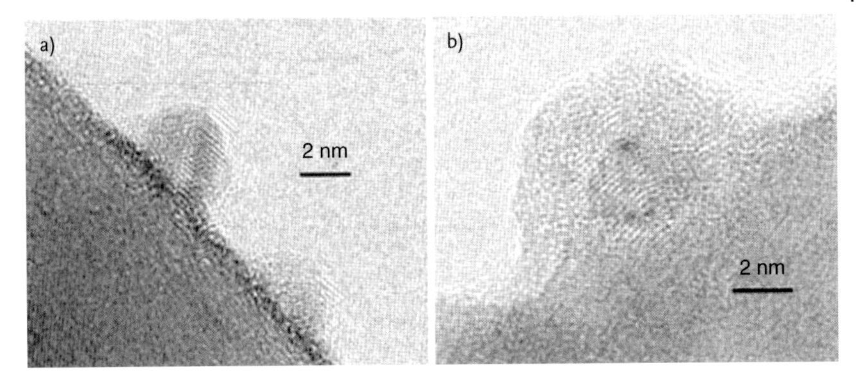

Figure 5.23 HR-TEM images of Rh-loaded GaN:ZnO before (a) and after (b) photodeposition of the Cr shell [44].

Compared with Ni/NiO as another core/shell cocatalyst presented earlier, the noble-metal/Cr_2O_3 core/shell cocatalyst has several advantages, including the possibility of selective introduction of active species for overall water splitting at reduction sites, the possibility of using various noble and transition metals as a core for extraction of photogenerated electrons from the bulk, and elimination of the need for activation treatment by oxidation or reduction. The noble-metal/Cr_2O_3 modification method therefore provides a new strategy for the construction of water-reducing cocatalysts for photocatalytic overall water splitting.

5.4 Conclusions

Overall water splitting is a typical example of so-called uphill reactions. To achieve the reaction on a particulate photocatalyst, effective suppression of the reverse reaction, that is, water formation from H_2 and O_2, is indispensable. In this chapter, several examples of overall water splitting systems were discussed, in which the reverse reaction was effectively avoided in each case by means of nanostructure of photocatalyst. To develop effective solar energy conversion systems to chemical energy, more sophisticated designs of nanostructures of the photocatalysts will be required.

References

1 Honda, K. and Fujishima, A. (1972) *Nature*, **238**, 37.

2 Domen, K., Naito, S., Soma, M., Onishi, T., and Tamaru, K. (1980) *J. Chem. Soc. Chem. Commun.*, **12**, 543.

3 Lehn, J.M., Sauvage, J.P., and Ziessel, R. (1980) *Nouv. J. Chim.*, **4**, 623.

4 Sato, S. and White, J.M. (1980) *Chem. Phys. Lett.*, **72**, 83.

5 Schrauzer, G.N. and Guth, T.O. (1977) *J. Am. Chem. Soc.*, **99**, 7189.
6 Damme, H.V. and Hall, W.K. (1979) *J. Am. Chem. Soc.*, **101**, 4373.
7 Domen, K., Naito, S., Soma, M., Onishi, T., and Tamaru, K. (1982) *Chem. Phys. Lett.*, **92**, 433.
8 Domen, K., Kudo, A., Shinozaki, A., Tanaka, A., Maruya, K., and Onishi, T. (1986) *J. Chem. Soc. Chem. Commun.*, 356.
9 Takata, T., Furumi, Y., Shinohara, K., Tanaka, A., Hara, M., Kondo, J.N., and Domen, K. (1997) *Chem. Mater.*, **9**, 1063.
10 Hitoki, G., Takata, T., Kondo, J.N., Hara, M., Kobayashi, H., and Domen, K. (2002) *Chem. Commun.*, 1698.
11 Hitoki, G., Ishikawa, A., Takata, T., Kondo, J.N., Hara, M., and Domen, K. (2002) *Chem. Lett.*, 736.
12 Kasahara, A., Nukumizu, K., Hitoki, G., Takata, T., Kondo, J.N., Hara, M., Kobayashi, H., and Domen, K. (2002) *J. Phys. Chem. A*, **106**, 6750.
13 Ishikawa, A., Takata, T., Kondo, J.N., Hara, M., Kobayashi, H., and Domen, K. (2002) *J. Am. Chem. Soc.*, **124**, 13547.
14 Maeda, K., Takata, T., Hara, M., Saito, N., Inoue, Y., Kobayashi, H., and Domen, K. (2005) *J. Am. Chem. Soc.*, **127**, 8286.
15 Serpone, N. (1997) *J. Photochem. Photobio. A*, **104**, 1.
16 Kraeutler, B. and Bard, A.J. (1978) *J. Am. Chem. Soc.*, **100**, 4317.
17 Yamaguti, K. and Sato, S. (1985) *J. Chem. Soc. Faraday Trans. 1*, **81**, 1237.
18 Subramanian, V., Wolf, E., and Kamat, P.V. (2001) *J. Phys. Chem. B*, **105**, 11439.
19 Tada, H., Ishida, T., Takao, A., and Ito, S. (2004) *Langmuir*, **20**, 7898.
20 Hosogi, Y., Shimodaira, Y., Kato, H., Kobayashi, H., and Kudo, A. (2008) *Chem. Mater.*, **20**, 1299.
21 Wrighton, M.S., Ellis, A.B., Wolczanski, P.T., Morse, D.L., Abrahamson, H.B., and Ginley, D.S. (1976) *J. Am. Chem. Soc.*, **98**, 2774.
22 Domen, K., Kudo, A., Onishi, T., Kosugi, N., and Kuroda, H. (1986) *J. Phys. Chem.*, **90**, 292.
23 Kudo, A. and Kato, H. (1997) *Chem. Lett.*, **26**, 867.
24 Ishihara, T., Nishiguchi, H., Fukamachi, K., and Takita, Y. (1997) *J. Phys. Chem. B*, **103**, 1.
25 Kato, H. and Kudo, A. (1998) *Chem. Phys. Lett.*, **295**, 487.
26 Shimizu, K., Tsuji, Y., Kawakami, M., Toda, K., Kodama, T., Sato, M., and Kitayama, Y. (2002) *Chem. Lett.*, (**31**), 1158.
27 Machida, M., Yabunaka, J., and Kijima, T. (1999) *Chem. Commun.*, 1939.
28 Otsuka, H., Kim, K., Kouzu, A., Takimoto, I., Fujimori, H., Sakata, Y., Imamura, H., Matsumoto, T., and Toda, K. (2005) *Chem. Lett.*, **34**, 822.
29 Kato, H., Asakura, K., and Kudo, A. (2003) *J. Am. Chem. Soc.*, **125**, 3082.
30 Kinomura, N., Kumada, N., and Muto, F. (1985) *J. Chem. Soc. Dalton Trans.*, 2349.
31 Kudo, A., Tanaka, A., Domen, K., Maruya, K., Aika, K., and Onishi, T. (1988) *J. Catal.*, **111**, 67.
32 Kudo, A., Sayama, K., Tanaka, A., Asakura, K., Domen, K., Maruya, K., and Onishi, T. (1989) *J. Catal.*, **120**, 337.
33 Tabata, S., Ohnishi, H., Yagasaki, E., Ippommatsu, M., and Domen, K. (1994) *Catal. Lett.*, **28**, 417.
34 Sayama, K., Tanaka, A., Domen, K., Maruya, K., and Onishi, T. (1990) *J. Catal.*, **124**, 541.
35 Ruddlesden, S.N. and Popper, P. (1957) *Acta Crystallogr.*, **10**, 538; *Acta Crystallogr.* (1958), **11**, 54.
36 Chun, W.J., Ishikawa, A., Fujisawa, H., Takata, T., Kondo, J.N., Hara, M., Kawai, M., Matsumoto, Y., and Domen, K. (2003) *J. Phys. Chem. B*, **107**, 1798.
37 Sato, J., Saito, N., Yamada, Y., Maeda, K., Takata, T., Kondo, J.N., Hara, M., Kobayashi, H., Domen, K., and Inoue, Y. (2005) *J. Am. Chem. Soc.*, **127**, 4150.
38 Maeda, K., Teramura, K., Takata, T., Hara, M., Saito, N., Toda, K., Inoue, Y., Kobayashi, H., and Domen, K. (2005) *J. Phys. Chem. B*, **109**, 20504.
39 Teramura, K., Maeda, K., Saito, T., Takata, T., Saito, N., Inoue, Y., and Domen, K. (2005) *J. Phys. Chem. B*, **109**, 21915.
40 Yashima, M., Maeda, K., Teramura, K., Takata, T., and Domen, K. (2005) *Chem. Phys. Lett.*, **416**, 225.

41 Maeda, K., Teramura, K., Lu, D., Takata, T., Saito, N., Inoue, Y., and Domen, K. (2006) *Nature*, **440**, 295.

42 Kohno, M., Kaneko, T., Ogura, S., Sato, K., and Inoue, Y. (1998) *J. Chem. Soc. Faraday Trans.*, **94**, 89.

43 Le Gendre, L., Marchand, L., and Laurent, R. (1997) *J. Eur. Ceram. Soc.*, **17**, 1813.

44 Maeda, K., Teramura K., Lu D., Saito N., Inoue Y., and Domen K. (2006) *Angew. Chem. Int. Ed.*, **45**, 7806.

45 Sakamoto, N., Ohtsuka, H., Ikeda, T., Maeda, K., Lu, D., Kanehara, M., Teramura, K., Teranishi, T., and Domen, K. (2009) *Nanoscale*, **1**, 106.

6
Chiral Catalysts

José M. Fraile, José I. García, and José A. Mayoral

6.1
The Origin of Enantioselectivity in Catalytic Processes: the Nanoscale of Enantioselective Catalysis

Enantiomerically pure compounds are extremely important in fields such as medicine and pharmacy, nutrition, or materials with optical properties. Among the different methods to obtain enantiomerically pure compounds, asymmetric catalysis [1] is probably the most interesting and challenging, in fact one single molecule of chiral catalyst can transfer its chiral information to thousands or even millions of new chiral molecules.

Enantioselective reactions are the result of the competition between different possible diastereomeric reaction pathways, through diastereomeric transition state (TS), when the prochiral substrate complexed to the chiral catalyst reacts with the corresponding reagent. The efficiency of the chirality transfer, measured as enantiomeric excess ($\% \text{ ee} = (R - S)/(R + S) \times 100$), depends on electronic and steric factors in a very subtle form. A simple calculation shows that differences in energy of only 2 kcal/mol between these TS are enough to obtain more than 90% ee, and small changes in any of the participants in the catalytic process can modify significantly this difference in energy. Those modifications may occur in the near environment of the catalytic center, at less than 1 nm scale, but also at longer distances in the catalyst, substrate, reagent, solvent, or support in the case of immobilized catalysts. This is the reason because asymmetric catalysis can be considered a nanometric phenomenon that requires a careful control of different variables.

Selective Nanocatalysts and Nanoscience, First Edition. Edited by Adriano Zecchina, Silvia Bordiga, Elena Groppo.

6.2 Parameters Affecting the Geometry of the Metal Environment

6.2.1 The Modification of the Chiral Pocket

Chiral catalysis can be represented in a general picture as a process that takes place in a so-called chiral pocket (Figure 6.1) formed by the catalytic center (in many cases a metal) and the bulky groups in the near environment that restrict the mobility of molecules around the coordinated substrate, provoking the enantioselection. The most obvious method to modify enantioselectivity is the modification of this chiral pocket, either by changing the shape and size of the bulky groups or by changing the coordination of metal, using a different metal or the same one with different oxidation state. When searching the optimal chiral pocket for a given reaction, bulky groups and metal must be considered as a whole, given that the accommodation of the substrate in the chiral pocket and the efficient shielding of one of its prochiral faces are conditioned by the global geometry of this environment.

6.2.2 Distal Modifications and Conformational Consequences

Chiral ligands are usually complicated molecules with ample possibilities of variability not only in the bulky groups forming the chiral pocket but also in positions relatively far from the catalytic center, represented as R_1 and R_2 in Figure 6.1. However, the variations in such distal positions may produce important differences in the conformational preferences of the chiral ligand and hence of the chiral complex, the corresponding reaction intermediate, and the diastereomeric TS. Those conformational variations have also consequences in the relative energy of the TS and hence in the enantioselectivity. This is one of the reasons because sometimes simplified models are not able to explain the enantioselective process, as those conformational effects of *a priori* nonrelevant groups are not considered.

6.2.3 Additional Ligands: Anions, Solvents, and Additives

Other possibility of variation in a chiral catalytic system is the presence of additional ligands on the metal center. The origin of those ligands can be multiple. If the metal is not in zero oxidation state, it will require the presence of an anion that would act as a ligand in the case of coordinating anions (e.g., chloride) or not in the case of noncoordinating anions (e.g., perchlorate). If chiral ligand and substrate are not able to saturate the coordination sphere of the metal, solvent molecules can enter to play this role. Donor ability and bulkiness of solvent molecules will condition the geometry of the chiral pocket, but other parameters such as dielectric constant may also modify the conformational preferences of the whole

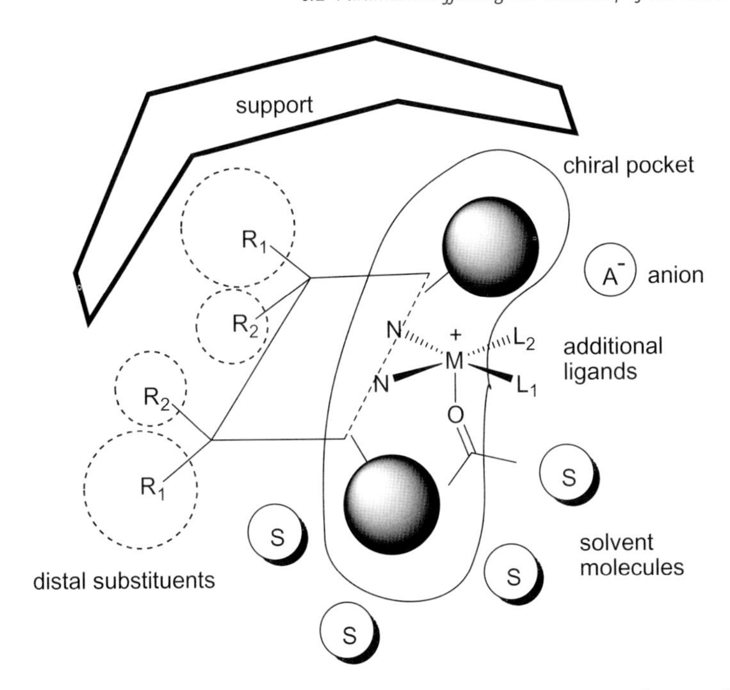

Figure 6.1 Parameters in the nanoenvironment of the catalytic center that can affect enantioselectivity.

complex, with the same effects commented above. Finally, the saturation of the coordination sphere of metal can be produced by additives, whose properties can be tailored to optimize the chirality transfer.

6.2.4
Parameters Beyond the Molecular Scale: Aggregates and Supported Catalysts

All the above considerations assume the existence of ideal catalytic monomeric species in solution that are attacked by a perfectly dissolved reagent. However, this is not the case in many catalytic processes. Depending on the reaction solvent, catalyst molecules may aggregate provoking steric and electronic interactions between catalyst molecules with consequences on enantioselectivity difficult to predict. Finally, in order to facilitate the recovery and reuse of the catalyst, the complex can be supported in a phase different from that of the substrate and reagent, either another liquid phase or a solid phase. In the case of immiscible liquid phases, the reaction may take place either in one liquid phase, due to partial solubility of the components, or in the interface, with possible consequences on enantioselectivity. In the case of catalysts immobilized on solid supports, the existence of possible catalyst-support interactions of different nature (coordinating, steric, diffusion limitations) cannot be discarded and effects on enantioselectivity are expected.

Along the rest of the chapter, different effects of all those parameters will be presented, using as cases of study some well-known reactions from a mechanistic point of view.

6.3
Case of Study (1): Bis(oxazoline)–Cu Catalysts for Cyclopropanation

Cyclopropanation reactions promoted by bis(oxazoline)–copper (Box–Cu)* complexes constitute a good case of study. The mechanistic aspects of the catalysis have been thoroughly studied both from the experimental and theoretical viewpoints and a good model for the stereoselection has been developed in the case of enantioselective homogeneous catalysis. This catalytic system has been shown to be very sensitive to multiple "surrounding" effects, such as solvent, counteranion, remote substituents, and support, in the case of immobilized catalysts (Figure 6.2). In the next sections, these effects will be analyzed, and put in the context of the nanoenvironment effects.

6.3.1
The Mechanism of Chiral Induction

The mechanism of cyclopropanation reactions by copper complexes has been experimentally investigated by several groups [2], demonstrating by kinetic experi-

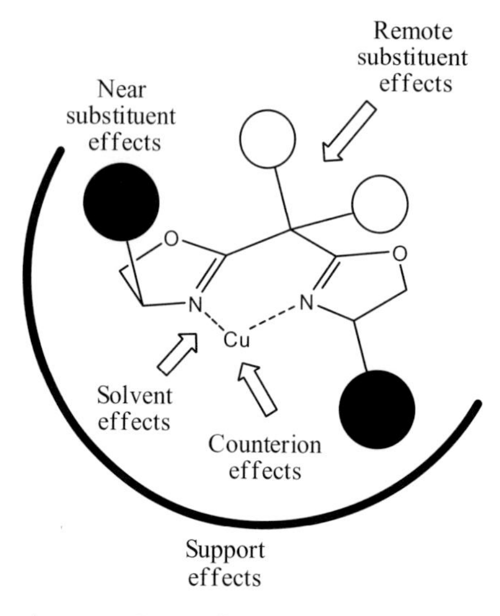

Figure 6.2 Survey of nanoenvironment effects in Box–copper catalytic systems.

* C_2-symmetric Box ligands will be represented as RBox, with R being the same substituent in position 4 of both oxazoline rings. In case of Box ligands with different substitution in each oxazoline ring, both substituents will be presented as RR′Box.

ments that the rate-determining step of the cyclopropanation mechanism is the dinitrogen extrusion from the diazocompound, to form (supposedly) a copper–carbene intermediate, a very reactive species, and therefore very elusive to experimental detection [3]. Subsequent addition of the carbenoid moiety of this intermediate to the olefin double bond results in the cyclopropane products.

As the rate-determining step of the mechanism turns to be the formation of the copper–carbene complex, the mechanistic issues posterior to this step, including the addition of the carbene to the olefin C=C double bond, which is the stereochemistry-determining step, are not accessible for experimental kinetic studies.

Happily, computational mechanistic studies do not suffer from this drawback (they have their own, however!), and hence, the mechanism of the chiral induction in these catalytic systems has been investigated using these techniques. In pioneering works, both Mayoral and coworkers [4] and Norrby and coworkers [5] studied the mechanism of model enantioselective cyclopropanation reactions, catalyzed by Box–Cu(I) complexes. Both studies agreed in the final stereoinduction mechanism proposed. Thus, the *trans*/*cis* selectivity in the cyclopropane products is governed by the steric interaction between the olefin substituent and the ester group linked to the carbene carbon in the addition TS (Figure 6.3). On the other hand, the enantioselectivity is governed by the steric interactions between the ester group linked to the carbene carbon and the substituents on the position 4 of the ligand oxazoline ring, induced by the olefin approach in the different addition TS (Figure 6.3).

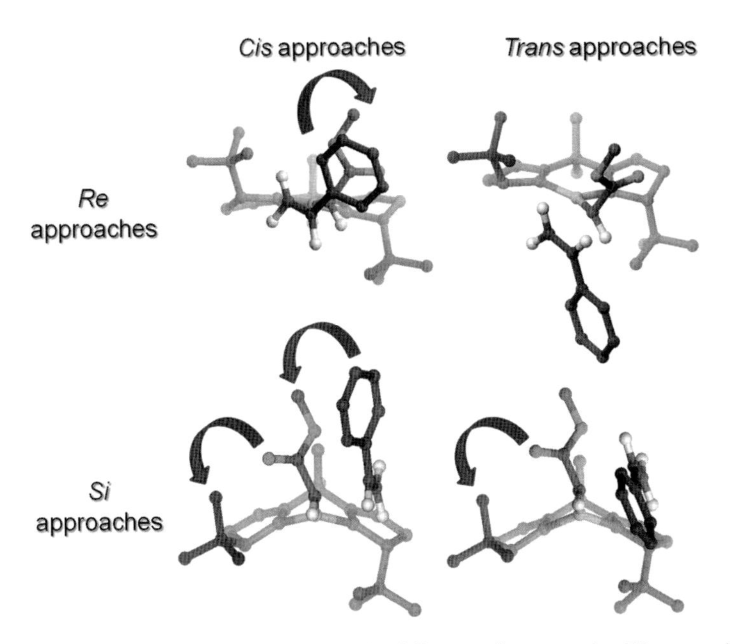

Figure 6.3 Main steric interactions in the different cyclopropanation TS, responsible for the reaction stereoselectivities.

In subsequent works of Mayoral and coworkers, the model system used in mechanistic computational studies using the hybrid QM/MM approach is almost identical to the experimental one [6]. The results obtained with this sophisticated model support the mechanism of the stereoselection schematized in Figure 6.3. In particular, the enantioselectivity is determined by the presence or absence of an intramolecular ester-oxazoline substituent in the addition TS, induced by the approach of the olefin through the different reaction channels.

6.3.2
The Importance of Symmetry: C_1 versus C_2

C_2-symmetric ligands are usually preferred over C_1-symmetric (asymmetric, in the sense of lack of any symmetry element) ligands for catalytic enantioselective transformations. Those C_1-symmetric ligands that are successful for catalytic applications are generally both electronically and sterically asymmetric, as for instance salicylaldimines [7] and phosphinooxazolines (Figure 6.4) [8]. There are evident advantages in using C_2-symmetric ligands: Arguably most important, less reaction channels are possible for the reaction, simplifying the prediction of chiral induction.

Box family ligands display C_2 symmetry in the vast majority of cases. Given that all these ligands have two electronically and sterically equivalent coordinating centers, the possibility exists of modifying the steric surroundings in the proximity of one of these centers, thus leading to electronically equivalent, but sterically different coordinating points. These ligands would be "halfway" between the abovementioned C_1-symmetric ligands and the usual C_2-symmetric ligands. Some illustrative examples based on the oxazoline motif are shown in Figure 6.4.

In general, in those cases in which C_2-symmetric ligands lead to good enantioselectivities, the use of electronically equivalent (e.g., in the sense of a close similarity of the coordinating groups) but sterically nonequivalent analogs results in a dramatic worsening of the results. Analogously, the use of an asymmetric pyridineoxazoline (pybox) in the copper-catalyzed cyclopropanation reaction of styrene with ethyl diazoacetate leads to virtually racemic products [9]. Similar observations have also been described for chiral unsymmetrical 2,2′-bipyridyl ligands in the

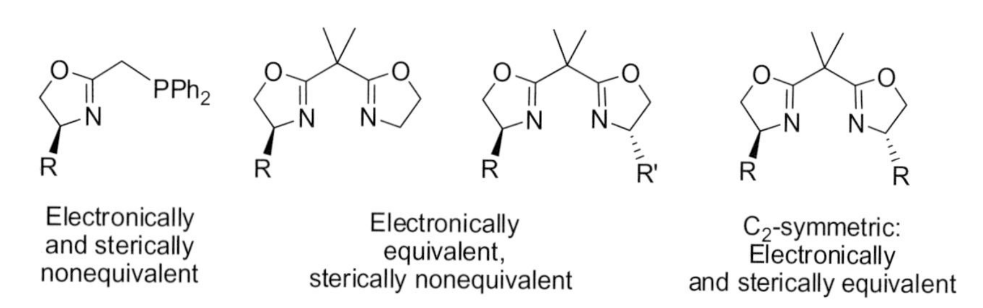

Figure 6.4 Some examples of chiral ligands, based on the oxazoline motif, with different degree of electronic and steric equivalency.

same reaction [10]. However, there is at least one case in which the use of sterically nonequivalent ligands results in enantioselectivities comparable to those obtained with the corresponding C_2-symmetric analogs, namely the so-called single-chiral pybox ligands, described by Nishiyama and coworkers (Figure 6.5) [11].

When these ligands are used in the ruthenium-catalyzed cyclopropanation reaction of styrene with alkyl diazoacetates (Figure 6.6), very good enantioselectivities are obtained for the *trans*-cyclopropanes (up to 94% ee) [12]. A mechanistic explanation for this unusual result has been offered, based on computational studies [12].

Recently, C_1-symmetric Box and azabis(oxazolines) have been described (Figure 6.7) [13] and tested in the homogeneous catalysis of the benchmark cyclopropanation reaction [14].

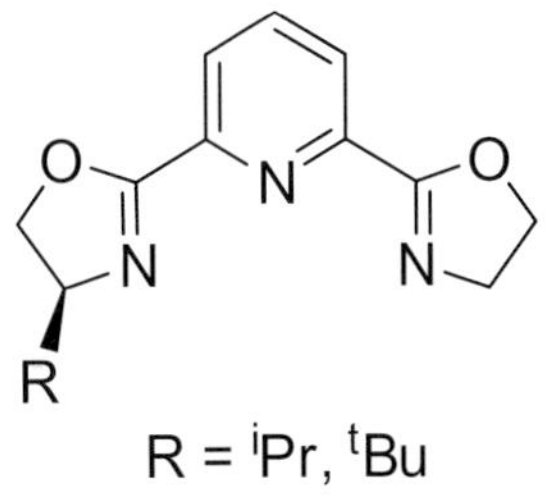
R = ⁱPr, ᵗBu

Figure 6.5 Structure of the Nishiyama asymmetric pybox ligands.

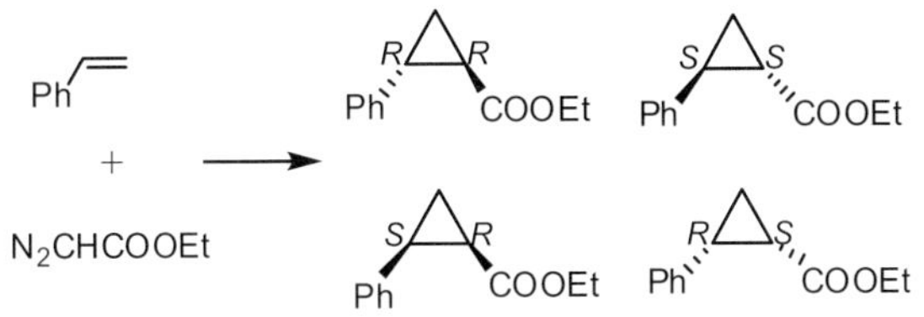
Ph
+
N2CHCOOEt
COOEt
COOEt
COOEt
COOEt

Figure 6.6 Cyclopropanation reaction between styrene and ethyl diazoacetate.

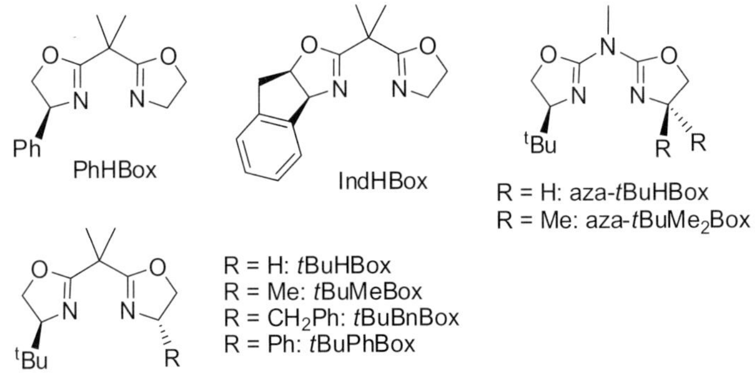
PhHBox
IndHBox
R = H: aza-tBuHBox
R = Me: aza-tBuMe2Box
R = H: tBuHBox
R = Me: tBuMeBox
R = CH2Ph: tBuBnBox
R = Ph: tBuPhBox

Figure 6.7 Structures of the asymmetric Box and azabis(oxazoline) ligands used in the catalytic experiments.

Table 6.1 Results of the cyclopropanation reaction of styrene with ethyl diazoacetate, catalyzed by chiral Box–CuOTf complexes.

Ligand	*trans/cis*	% ee *trans*	% ee *cis*
PhBox	68/32	60	51
PhHBox	71/29	20	8
IndBox	60/40	85	81
IndHBox	69/31	33	25
*t*BuBox	71/29	94	91
*t*BuHBox	68/32	29	8
*t*BuMeBox	67/33	84	79
*t*BuBnBox	64/36	83	75
*t*BuPhBox	72/28	82	69
aza-*t*BuBox	73/27	92	84
aza-*t*BuHBox	73/27	23	9
aza-*t*BuMe$_2$Box	71/29	85	68

Table 6.1 gathers the results obtained with these ligands, and compares them with those obtained with the related C_2-symmetric ligands, when applicable. These results conclude that C_2-symmetry is not mandatory to obtain reasonable levels of enantioselection. Ligands bearing one "big" and one "small" group on the oxazoline rings, like *t*BuMeBox, allow to obtain good levels of enantioselectivity. Even ligands bearing only one stereogenic center, like aza-*t*BuMe$_2$Box, are able to induce stereoselectivity levels close to the best obtained with the classical C_2-symmetric ligands.

The origin of this behavior has been studied through computational mechanistic studies, which show a very good agreement with experimental observations. In particular, the enantioselection mechanism comes from the differently favored reaction channels, leading to one or another cyclopropane enantiomer, as a function of the different steric interactions between the ester group and the bisoxazoline substituents (Figure 6.8). The calculated ee for ligands *t*BuMeBox and aza-*t*BuMe$_2$Box are 88% and 90% ee, respectively, which compare very well with the experimental values. This experimental–theoretical agreement should allow to theoretically investigating the behavior of new ligands before their synthesis and testing, facilitating the design of tailored catalytic systems for this reaction. It is clear from these studies that once the stereoselection mechanism is well understood, and the definition of the chiral pocket is clearly established, C_2 symmetry is no longer required, adding more versatility to the ligand design with specific purposes (for instance, supporting).

6.3.3
Distal Modifications: Substitution in the Methylene Bridge

It is generally assumed that, for a good catalyst enantiodiscrimination, bulky groups defining the chiral pocket must be near the catalytic center. However, distal

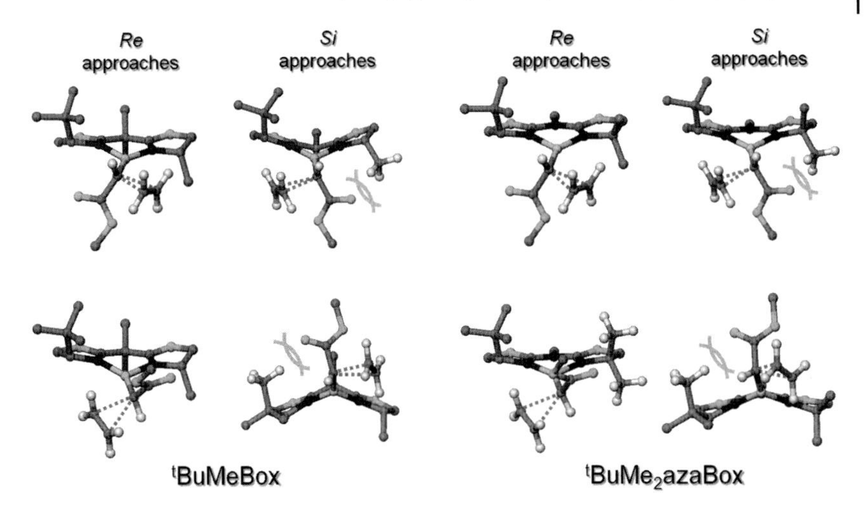

Figure 6.8 Some selected calculated (at the B3LYP/6-31G(d) theoretical level) geometries of transition structures of the reaction of ethylene with methyl diazoacetate, catalyzed by the *t*BuMeBox–Cu(I) and the aza-*t*BuMe$_2$Box–Cu(I) complexes.

substituents may also have a significant role in determining the stereoselectivity of the catalytic reaction.

In the case of Box–Cu complexes, such distal effects have been reported for the cyclopropanation reaction, in connection with the support of these kinds of complexes [15]. Thus, when a homopolymer obtained from a modified Box precursor (Figure 6.9) bearing 4-vinylbenzyl groups in the central methylene bridge is charged with copper and used as catalyst of the benchmark cyclopropanation reaction (Figure 6.6), an unexpected *cis* selectivity is obtained (37:63 *trans*/*cis* ratio, when the usual value with this family of ligands is ca. 70:30). Furthermore, the enantioselectivities in *trans*- and *cis*-cyclopropanes are also lower, when compared with the traditional ligand used in homogeneous phase, bearing an isopropylidene bridge (78% vs. 94% ee in *trans*-cyclopropanes and 72% vs. 90% ee in *cis*-cyclopropanes).

These effects are not due to the presence of the polymeric backbone because when the corresponding Box ligand, dibenzylated in the central methylene bridge, is used in homogeneous catalysis experiments, virtually identical results are obtained, which indicates that the effect is due to the substitution pattern of the methylene bridge of the Box ligand, constituting a genuine case of distal effect on the stereodiscrimination of the catalyst.

It is worth noting that this effect is only clearly observed when the Box ligand bears *tert*-butyl groups in position 4. When these positions are occupied by phenyl groups, the cyclopropanation is less *cis*-selective (52:48 *trans*/*cis* ratio), and enantioselectivities obtained are nearly identical to those obtained in the homogeneous catalysis with the analogous ligand with isopropylidene bridge. It seems that there is an interplay between the substituents in the methylene bridge and in position

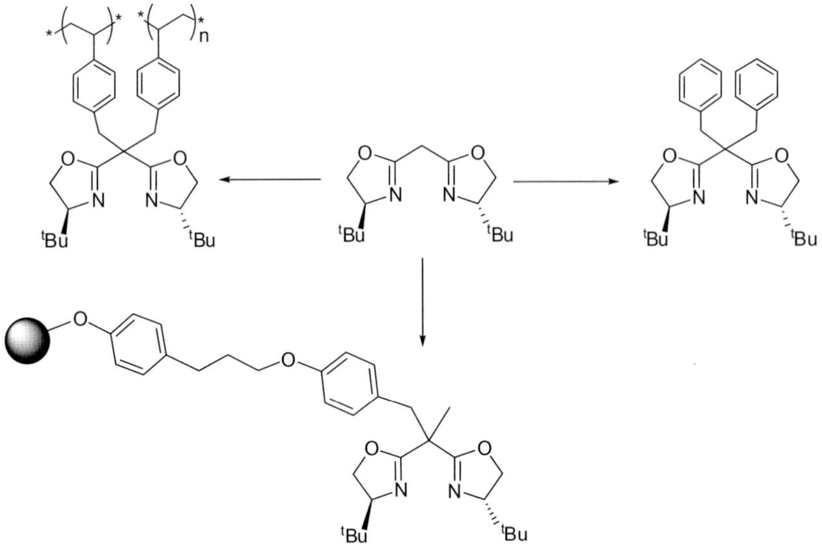

Figure 6.9 Some Box ligands benzylated in the central bridge.

4 of the oxazoline ring to configure the shape of the chiral pocket leading to this unexpected stereoselectivity change. No analogous homogeneous experiments have been carried out with a monobenzylated Box ligand, but a similar system has been described by Annunziata *et al.*, in which the ligand is linked to a polyethylene glycol chain through a spacer containing a single benzyl group bonded to the Box methylene bridge (Figure 6.9).

When the corresponding copper complex is used in the homogeneous catalysis of the benchmark cyclopropanation reaction, up to 77 : 33 *trans*/*cis* selectivity and 91% ee in *trans*-cyclopropanes is obtained, which seems to indicate that the presence of two benzyl groups is necessary to observe their remote effect in the stereoselectivity, probably due to a decrease in the mobility of the catalytic intermediates, and in the number of possible reaction channels (due to the C_2 symmetry). The ultimate reason for this particular behavior remains, however, unveiled.

6.3.4
Effect of Anion

Cationic Box–copper complexes require the presence of anions to keep electroneutrality. These anions usually come from the copper salt, and their nature has an enormous influence on the activity and enantioselectivity of the Box–copper catalysts in homogeneous phase [16]. For instance, when the counteranion is changed from triflate to chloride, the enantioselectivity of the cyclopropanation reaction of styrene with ethyl diazoacetate, catalyzed by the *t*BuBox–Cu(I) complex in dichloromethane, drops from 94% to 3% ee for the *trans*-cyclopropanes, and from 92% to 8% ee for *cis*-cyclopropanes.

When these cationic complexes are immobilized by electrostatic interactions onto anionic supports (through a cation-exchange procedure), the enantioselectivity pattern also follows a similar scheme. Thus, when the anionic moiety has a fluorosulfonate structure (Nafion, Nafion–silica nanocomposites), copper complexes of PhBox lead to results (59% ee in the *trans*-cyclopropanes) almost identical to those obtained in homogeneous phase with copper triflate salts. On the other hand, when other anionic supports, such as clays or sulfonic acid resins, are used, a marked decrease in enantioselectivity is observed, up to only 17% ee in the *trans*-cyclopropanes [17].

This dramatic influence of the counteranion on the enantioselectivity has been ascribed in homogeneous phase (and, in part, also in heterogeneous phase) to its higher or lower coordinating character. The correctness of this hypothesis has been verified through computational studies [18]. Thus, a computational study at the DFT theoretical level of a model cyclopropanation reaction, catalyzed by Box–Cu(I) complexes bearing or not a chloride anion coordinated to the metal, has shown that the geometrical changes induced in the key TS by the presence of the anion are responsible for the decrease in enantiodiscrimination of the catalyst. Figure 6.10 illustrates these differences and the steric interactions responsible for the enantioselectivity.

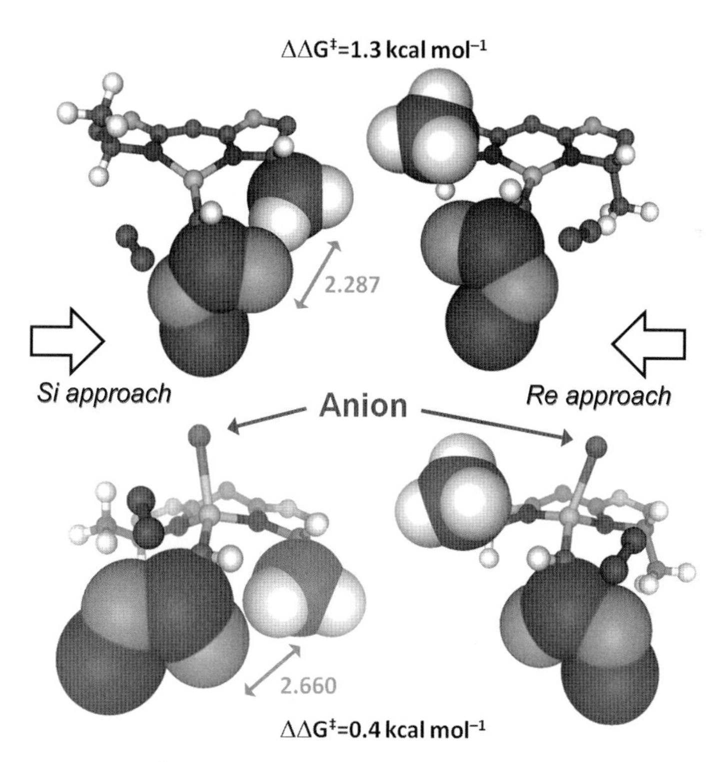

Figure 6.10 Differences in steric enantiodiscriminating interactions induced by the presence of the counteranion.

As stated in Section 6.3.1, the main steric interaction responsible for the enantiodiscrimination lies in the steric repulsion between the ester group and one of the substituents in position 4 of the oxazoline ring, which appears only when the alkene approaches the carbene carbon by its *Si* face. In the case of the cationic complexes with weakly coordinating counteranions, the ester group and the oxazoline substituent become closer in the less-favored *Si* TS (Figure 6.10), increasing the energy difference with regard to the corresponding *Re* TS ($1.3\,kcal\,mol^{-1}$ in the model shown in Figure 6.10). However, when a chloride anion is coordinated to the copper center, the deformation induced in the neighborhood of the metal results in a longer distance between the ester group and the oxazoline substituent, and hence in la lower steric repulsion, giving rise to closer TS energies ($0.4\,kcal\,mol^{-1}$ in the model shown in Figure 6.10), and hence to lower enantioselectivities. Of course, greater or lesser coordinating abilities of the counteranion may lead to different degrees of geometry changes in the nanoenvironment of the metal, giving rise to stereoselectivity changes that may vary from modest to dramatic ones.

6.3.5
Beyond the Coordination Sphere: Supports That Change the Dimensionality

In general, it is considered that immobilized catalysts should be designed to minimize the possible interactions between the catalytic sites and the support, to avoid unpredictable effects of the latter on the stereochemistry of the reaction. However, this interaction can be used to improve and even to change the stereochemical results; in this way, the solid catalyst leads to products difficult to obtain in solution and its use is clearly justified. It must be recalled that the support may block very efficiently some of the reaction channels. Usually, this blocking is at random, due to the amorphous character of the support and/or the lack of a rigid disposition of the catalyst with respect to the support, resulting in the absence of any support-induced stereoselectivity. However, in the case of the electrostatic support of cationic Box–Cu complexes on lamellar anionic solids (clays), through an ion-exchange process, a marked support effect has been reported.

Mayoral and coworkers reported a complete change in the stereoselectivity when cyclopropanation between styrene and ethyl diazoacetate was carried out in styrene as the reaction media using laponite immobilized PhBox–Cu complex as catalyst [19]. Complete reversal of the *trans/cis* diastereoselectivity (31 : 69) was observed and, even more interestingly, the major *cis*-cyclopropane obtained has the opposite absolute configuration, with regard to homogeneous phase results. Furthermore, the effect is not permanent, when the solid used in styrene is recovered and reused in dichloromethane, the "normal" stereochemical results are obtained again. This effect is not due to a particular behavior as it is also observed with other solvents with a low dielectric constant. Depending on the reaction conditions, with the same complex one can pass from 70 : 30 *trans/cis* selectivity and 60% ee in *trans*-1*R* cyclopropane in homogeneous conditions to 20 : 80 *trans/cis* selectivity and 72% ee in *cis*-1*S* cyclopropane in heterogeneous catalysis [20]. Note that enantioselectivity is even better in heterogeneous phase that clearly illustrates the great effect of the nanoenvironment of the catalyst in the case of supported complexes.

An explanation to the reversal of selectivity has been offered, based on the key insertion step of carbene to styrene, responsible for the stereoselectivity. As shown in Figure 6.11, the presence of the support surface disfavors most of TS, and particularly those leading to the major products in homogeneous phase, due to the new steric interactions between styrene and the surface. The only TS lacking these interactions is precisely that leading to the major product obtained in heterogeneous phase. A reaction medium with low dielectric permittivity favors the close proximity of the cationic complex to the anionic support, enhancing the confinement effect. In this case, the planarity of the surface of the support is a key point, because it effectively blocks half of reactive trajectories, resulting in a genuine confinement effect.

It is clear that a closer proximity of the complex to the support is desirable to maximize the effect. Following the model depicted in Figure 6.11, this should be feasible if C_1-symmetric ligands were used instead of the traditional C_2-symmetric ones (Figure 6.12).

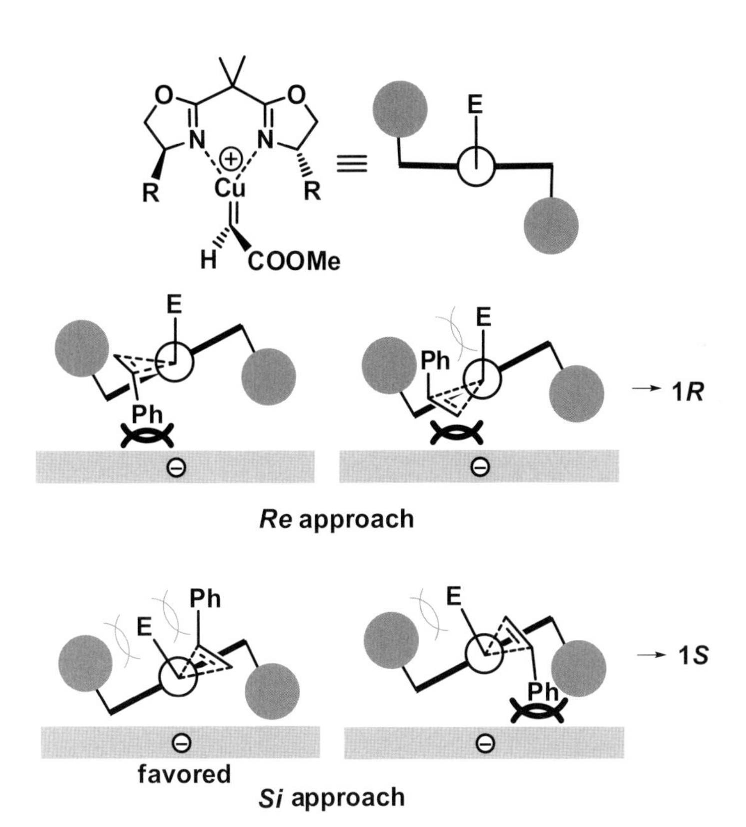

Figure 6.11 Styrene approaches in cyclopropanation reactions catalyzed by Box–Cu complexes immobilized onto laponite.

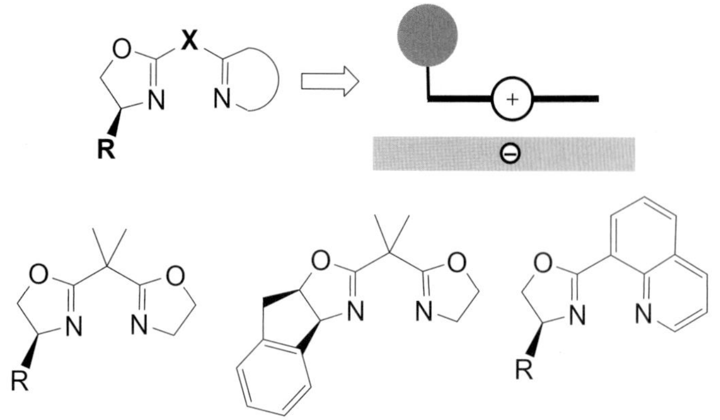

R = Ph : PhHBox
R = ^{t}Bu : *t*BuHBox

IndHBox

R = Ph : PhQox
R = ^{t}Bu : *t*BuQox

Figure 6.12 Structure of C_1-symmetric ligands and its consequences on the catalyst supporting.

Table 6.2 Results of the cyclopropanation reaction of styrene with ethyl diazoacetate catalyzed by chiral Box– and Qox–Cu complexes.

Ligand	Homogeneous (CH_2Cl_2 as solvent)			Heterogeneous (styrene as solvent)		
	trans/cis	**% ee *trans***	**% ee *cis***	***trans/cis***	**% ee *trans***	**% ee *cis***
*t*BuHBox	71/29	20	8	9/91	15	−41[a]
PhHBox	68/32	29	8	15/85	13	−48
IndHBox	69/31	33	25	16/84	30	−32
PhQox	71/29	24	25	14/86	39	30
*t*BuQox	68/32	48	28	23/77	24	33

a) Negative sign indicates that 1*S*-cyclopropanes are the major enantiomers.

Two families of this kind of ligand have been tested in the clay-supported catalysis of the benchmark reaction [20, 21], and some of the most relevant results described are shown in Table 6.2.

Concerning the *trans/cis* selectivity, it is clear that the use of C_1-symmetric ligands supports a clear improvement, given that up to 91% *cis*-cyclopropane can be obtained with *t*BuHBox ligand, that is, that in which the steric asymmetry is the highest (see Figure 6.4). It must be noted that *cis*-cyclopropanes are usually more difficult to obtain, and relatively few catalytic methods have been described that show this preference, most of them based on the use of rather special ligands (see Section 6.5.3 for some examples). Support confinement effects are therefore very useful in this context, since the same ligand can lead preferentially to *trans*-

or *cis*-cyclopropanes depending on it is used in homogeneous or heterogeneous catalysis.

Concerning enantioselectivity, results are much less clear. Homogeneous phase enantioselectivities are consistently low (except maybe the 48% ee obtained in *trans*-cyclopropanes with ligand *t*BuQox). With Box ligands, a reversal in the absolute configuration of the major *cis*-cyclopropanes is obtained, but with low enantioselectivities. Surprisingly, when the quinolinoxazolines are used in heterogeneous catalysis, no such reversal is observed, and the enantioselectivities are similar to those obtained in homogeneous catalysis. These results point to a surface confinement model more complicated than that previously proposed, with more geometrical possibilities of the key carbene intermediate with regard to the support surface. On the one hand, multiple dispositions of the ester group of the carbene intermediate with regard to the oxazoline substituent and to the support surface are possible. On the other hand, intermediates and transition structures have some degree of flexibility, so they can adopt conformations in which the steric repulsion with the support is minimized (for instance, the substituent on the oxazoline ring can adopt a pseudoequatorial disposition, more parallel to the support surface, as can it do the ester group of the carbene moiety). These circumstances lead to an increase in the number of possible reaction channels, and hence to a decrease in the final enantioselectivity.

6.4 Case of Study (2): Catalysts for Diels–Alder Reactions

6.4.1 Enantioselectivity in Diels–Alder Reactions

Enantioselective Diels–Alder reactions promoted by chiral Lewis acids constitute a powerful tool for the preparation of enantiomerically pure cyclic compounds. Therefore, this kind of reactions has been extensively investigated [22], allowing the identification of factors influencing the extension and the sense of the asymmetric induction.

One of the key points for enantiodiscrimination is the control of the conformation of the dienophile. It can adopt an s-*cis* or s-*trans* conformation and each conformer shows reversal topicity of the upper and lower faces, thus leading to different enantiomers (Figure 6.13). In this regard, both theoretical and experimental studies have shown a preference for s-*cis* conformation.

6.4.2 Chiral Pocket in Box–Metal Complexes: Ligand, Metal, and Additives

Several groups have made important contributions related to the use of metal complexes of C_2-symmetric Box as catalysts in Diels–Alder reactions, mainly with oxazolidinone derivatives. In the short range, dienophile is included in the chiral

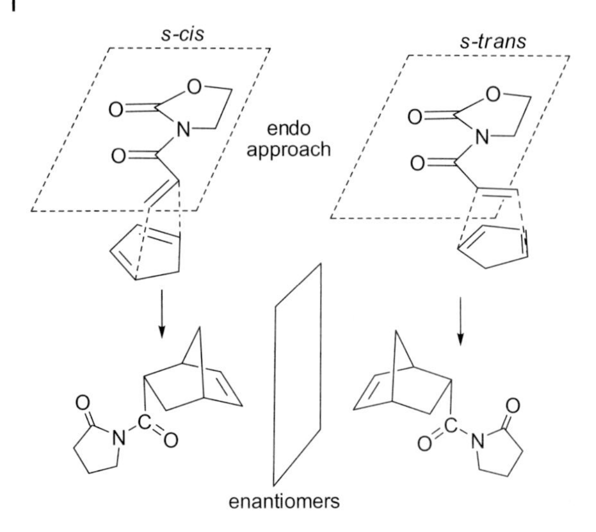

Figure 6.13 Reversal of enantioselectivity with dienophile conformation.

pocket of the complex, which is primarily defined by the bulky groups of the chiral ligand shielding some spatial zones, and the coordinating sphere of the metal that controls the geometry of the dienophile–catalyst complex. In fact bis(oxazolines) are able to form catalytic complexes with a large variety of metals, for example, Fe^{3+}, Mg^{2+}, and Cu^{2+}, all of them efficient catalysts for Diels–Alder reaction. Corey [23], Evans [24], and Gosh [25] showed that both the extension and the sense of the enantioselection (Table 6.3) depend on the geometry of this intermediate, octahedral, tetrahedral, or square-planar (Figure 6.14), imposed by the metal. This coordination geometry modifies the dihedral angle between the Box ligand and the dienophile planes, from perpendicular in the case of Mg to coplanar in the case of Cu.

In this way, the position of the C=C double bond with respect to the bulky groups changes and the unshielded face that suffers the diene attack is different, *Si* face for octahedral and tetrahedral complexes, *Re* face for square-planar complex (Figure 6.15), leading to the observed change in the sense of the asymmetric induction. Another consequence of this change in the orientation of the dienophile in the chiral pocket is the different optimum substituents of the Box ligand depending on the metal, phenyl for Fe and Mg, *tert*-butyl for Cu. Even more subtle effects, such as the increase of bulkiness of R_2 from H to methyl, have different consequences depending on the geometry of the metal–dienophile complex.

Another method to modify the geometry of the chiral pocket is the use of different counterion and/or Lewis bases [26]. In the case of Mg complex of the same Box, the use of $Mg(ClO_4)_2$ leads to a tetrahedral intermediate that favors the attack of the diene on the *Re* face of the dienophile (Figure 6.16). The addition of two equivalents of a monodentate additional ligand, such as water or tetramethylurea (TMU), or one equivalent of a bidentate ligand, such as ethylene glycol, modifies

Table 6.3 Variation of enantioselectivity in Diels–Alder reactions with the shape of the chiral pocket in Box–metal complexes.

Metal	R_1	R_2	% ee (major isomer)
Fe	Ph	H	82 (*R*)
	Ph	Me	86 (*R*)
	^{t}Bu	H	24 (*R*)
Mg	Ph	H	76 (*R*)
	Ph	Me	91 (*R*)
	^{t}Bu	H	0
Cu	Ph	H	30 (*S*)
	Ph	Me	10 (*S*)
	^{t}Bu	H	98 (*S*)
Mg			61 (*S*)
Cu			99 (*R*)

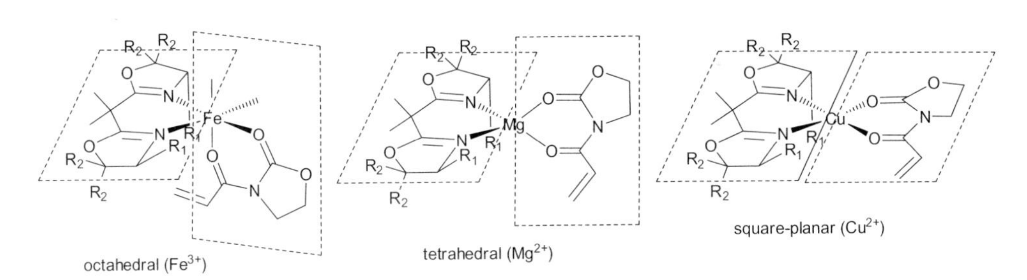

Figure 6.14 Modification of the geometry of the chiral pocket with different Box–metal complexes.

the geometry of the intermediate, leading to a preference for octahedral coordination, with the additional ligands in *cis* relative position (Figure 6.16) that favors the attack of the diene on the *Si* face of the alkene, and consequently, the sense of the asymmetric induction changes (Table 6.4). In the case of the more coordinating triflate anion, the intermediate is always octahedral, but both triflates are placed in *trans* relative position (Figure 6.16), favoring an attack on the *Si* face of

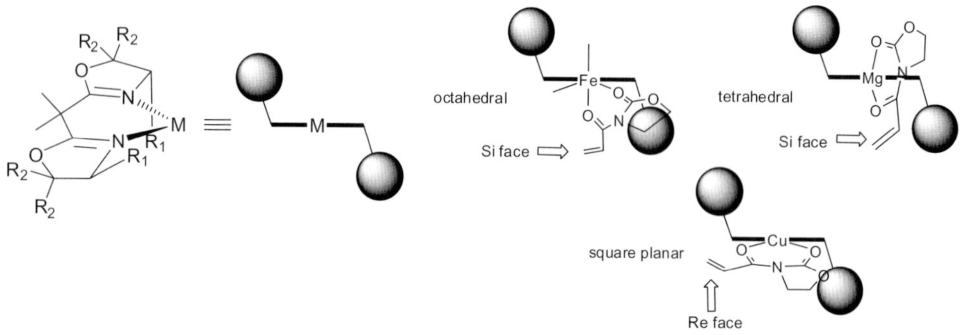

Figure 6.15 Proposed variations of face discrimination with different Box–metal complexes.

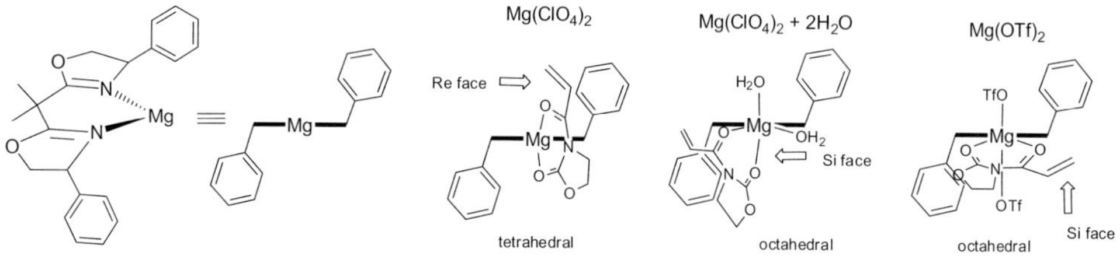

Figure 6.16 Proposed effect of anion and coordinating water molecules on enantioselectivity of Box–Mg complexes.

Table 6.4 Variation of enantioselectivity in Diels–Alder reactions with the counterion and/or additives in Box–Mg complexes.

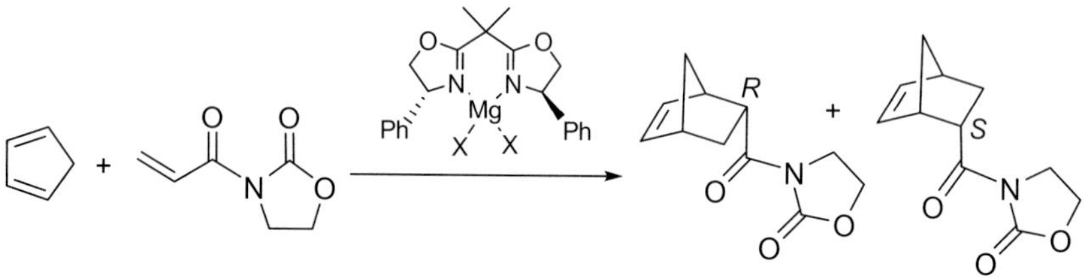

Counterion	Additive	% ee (major isomer)
ClO_4^-	–	73 (*S*)
	2 H_2O	73 (*R*)
	2 MeOH	42 (*R*)
	$HOCH_2CH_2OH$	58 (*R*)
	2 TMU	51 (*R*)
TfO^-	–	88 (*R*)
	2 H_2O	86 (*R*)
	2 TMU	88 (*R*)

the dienophile, in a similar way to that observed in the case of the square-planar Cu complex (Figure 6.15). The geometry of this intermediate does not change by the addition of external ligands, as shown by the same results in the presence of water or TMU.

As can be seen, these results have been explained according to proposed models, in many cases without additional experimental or theoretical evidences apart from the sense of the chiral induction.

6.4.3
The Poorly Understood Effect of Surface

Amorphous silica has been used as a support for complexes bearing triflate anions due to the capacity to form hydrogen bonds between the surface silanol groups and the fluorine atoms of triflate. This type of immobilization has been shown to affect the enantioselectivity in the case of complexes using a Box with phenyl groups [27] (Figure 6.17).

With the three metals tested (Cu, Zn, and Mg), a reversal in enantioselectivity was observed in the immobilized catalysts with respect to the homogeneous ones, for example, from 60% ee (*S*) in solution to 30% ee (*R*) with the heterogeneous catalyst in the case of the Mg catalyst. This reversal was ascribed to a change in the coordinating ability of the anion, as described above for $Mg(ClO_4)_2$ and $Mg(OTf)_2$ complexes in solution, but in this case the reduction in the coordinating ability of the anion must be produced by the hydrogen bonds between silanols and triflates.

However, this explanation cannot be applied to the copper catalyst, as the reversal is not observed in solution. In this case, new interactions of unknown origin must affect the geometry of the complex and the corresponding TS, modifying the relative energies.

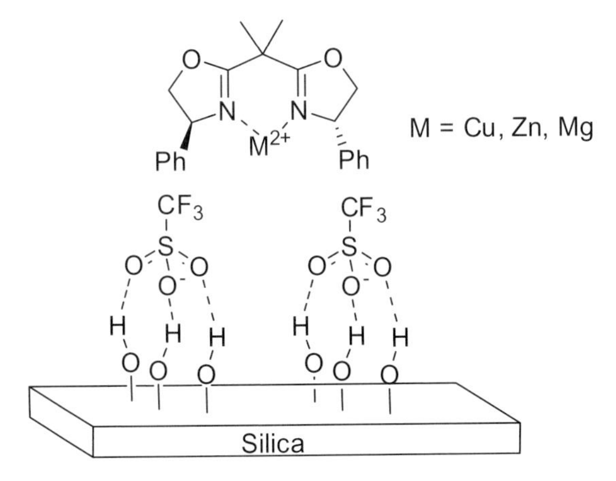

Figure 6.17 PhBox–metal complexes immobilized on silica by hydrogen bond interactions.

6.4.4
Similar But Not the Same: Control of Induction Sense with Different Lanthanides

We have seen how the change from a square-planar geometry to a tetrahedral or octahedral disposition may change completely the sense of the asymmetric induction. This situation is even more complicated when metals with higher coordination numbers are involved. Pyridinebis(oxazoline) complexes with lanthanide metals are also chiral Lewis acids able to catalyze Diels–Alder reactions. The sense of the enantioselection does not depend only on the metal, but also ligands of the same absolute configuration with different bulky substituents lead to opposite enantiomers with the same lanthanide metal (Table 6.5) [28].

From those results it is clear that both ligands, with the same absolute configuration and the same metal, lead to major products of opposite configuration. Furthermore, the induction sense changes from scandium to the rest of lanthanides. Based on X-ray structures, a model was proposed for the ligand with R = Ph (Figure 6.18). Scandium, with a coordination number of 7, is complexed to pybox ligand in equatorial, whereas dienophile docks with the exocyclic carbonyl group in the apical position, in such a way that phenyl group efficiently shields the *Si* face. However, lanthanum, with a coordination number of 9, keeps two triflates in apical positions and the dienophile coordinates in the equatorial plane, living *Si* face more accessible to the attack of diene. However, the explanation for the reversal in enantioselectivity when using the ligand with isopropyl groups requires the coordination of dienophile in a completely different orientation,

Table 6.5 Variation of enantioselectivity in Diels–Alder reactions with the shape of the chiral pocket in pyridinebis(oxazoline)–metal complexes.

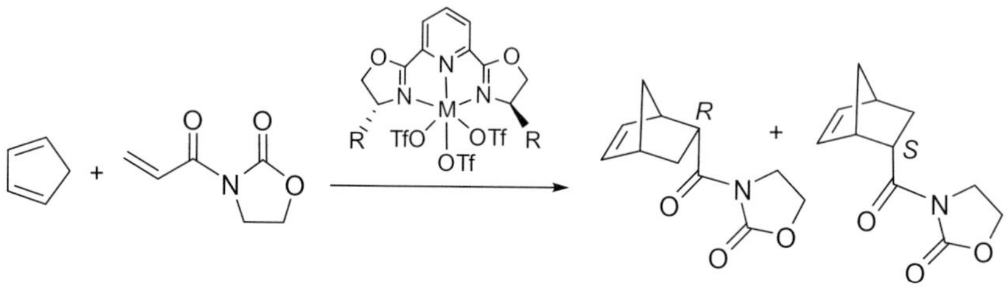

		% ee (major isomer)	
Metal	Ionic radius (Å)	R = ^{i}Pr[a]	R = Ph[b]
Sc	0.870	84 (*R*)	20 (*S*)
Yb	0.985	0	66 (*R*)
Eu	1.068	58 (*S*)	38 (*R*)
La	1.160	17 (*S*)	78 (*R*)

a) With 4 Å MS.
b) Without 4 Å MS.

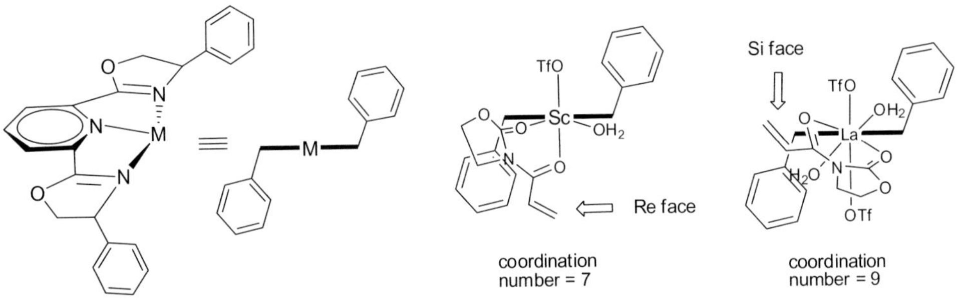

Figure 6.18 Proposed models to explain the change of the sense of chiral induction with different lanthanide metals.

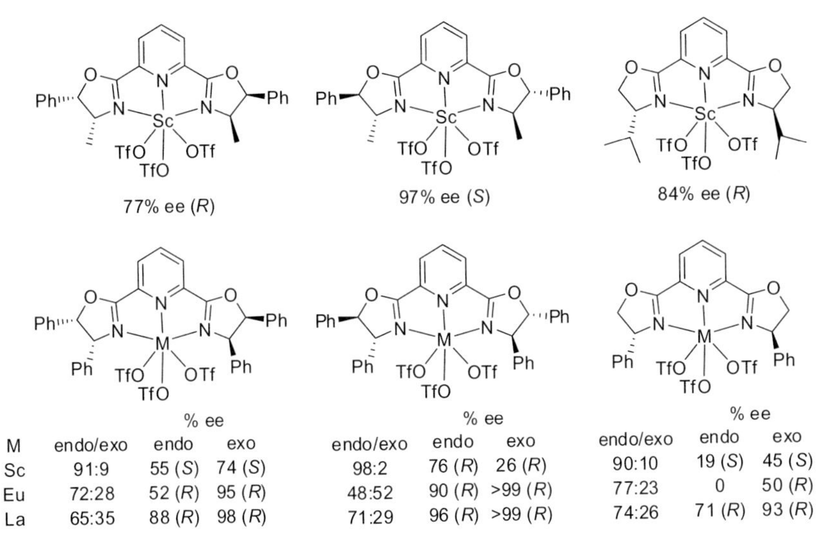

M	endo/exo	% ee endo	% ee exo	endo/exo	% ee endo	% ee exo	endo/exo	% ee endo	% ee exo
Sc	91:9	55 (*S*)	74 (*S*)	98:2	76 (*R*)	26 (*R*)	90:10	19 (*S*)	45 (*S*)
Eu	72:28	52 (*R*)	95 (*R*)	48:52	90 (*R*)	>99 (*R*)	77:23	0	50 (*R*)
La	65:35	88 (*R*)	98 (*R*)	71:29	96 (*R*)	>99 (*R*)	74:26	71 (*R*)	93 (*R*)

Figure 6.19 Variations in enantioselectivity and sense of chiral induction with different lanthanide metals and pybox ligands substituted in position 5.

probably due to the role of water, that is, it is excluded in that case by the use of 4 Å MS.

In this kind of system, distal electronic and steric effects have been demonstrated. In the case of electronic effects, substitution in position 4 of the pyridine ring with electron-donating groups is detrimental, whereas substitution with electron-withdrawing groups is positive, probably due to a more electrophilic character of Sc(III) that would bind more tightly the dienophile, improving the discriminating capacity of the ligand [29]. The other distal effect is that of substitution in position 5 of the oxazoline ring (Figure 6.19). In principle, this position is quite far from the reaction center, but it has a decisive influence in the stereochemical course of the reaction [30]. The presence of a phenyl group in that position is able to efficiently shield the attack to the Sc-coordinated dienophile, in conjunction with a small methyl group in position 4. In fact, the phenyl group in *cis* of the methyl group leads to almost the same enantioselectivity as the isopropyl group, and in *trans* reverts completely the induction sense up to almost complete enantioselection. The case with a phenyl group in position 4 is even more complex, as it depends on the metal and it also provokes important variations in the *endo/exo* selectivity. Again the presence of the second *cis*-phenyl in position 5 improves the results obtained with one single phenyl, either in one induction sense (Sc) or in the opposite (Eu and La). The presence of the *trans*-phenyl controls the induction sense, which is thus independent of the metal, with very high values for La and Eu. The explanation for this complicated pattern of results is not straightforward.

In the case of lanthanides, the presence of water or coordinating anions, acting as ligands, may dramatically change the stereochemical course of the reaction. The addition of other nonchiral ligands can be used to get both enantiomers using the same chiral ligand, in this case BINOL [31]. When 3-acetyl-1,3-oxazolidin-2-one is added, the 2*S* *endo* cycloadduct is preferentially obtained (Table 6.6), whereas the 2*R* *endo* cycloadduct is the major one when 3-phenylacetylacetone is used as additional ligand.

The existence of two binding sites due to the coordination number of Yb has been proposed as the origin of this change in enantioselection. 3-Acetyl-1,3-oxazolidin-2-one would compete with the dienophile for the site A, favoring the formation of 2*S* enantiomer, whereas 3-phenylacetylacetone would block this site, imposing the coordination of dienophile to site B, leading to the 2*R* enantiomer.

Table 6.6 Variation of enantioselectivity in Diels–Alder reactions with additional achiral ligands in BINOL–Yb complexes.

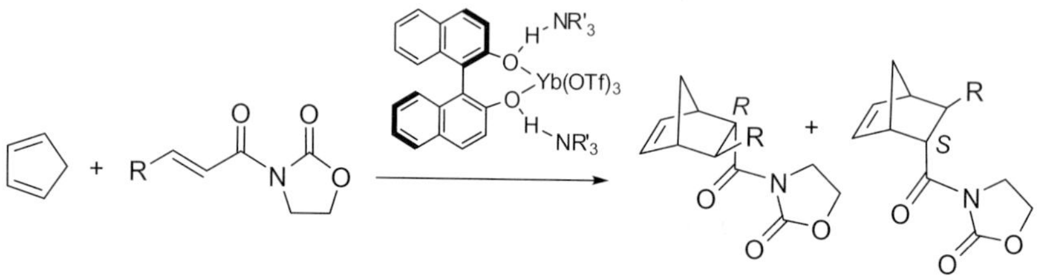

Added ligand	NR_3'	R	% ee
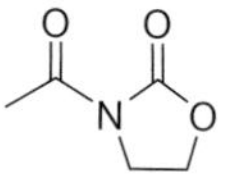	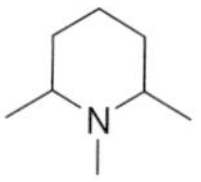	Me	93 (2*S*)
		Ph	83 (2*S*)
		nPr	86 (2*S*)
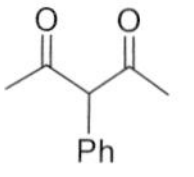 Ph	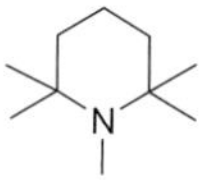	Me	81 (2*R*)
		Ph	83 (2*R*)
		nPr	80 (2*R*)

However, this effect is less simple than exposed here, as the amine plays an important role, probably due to the transmission of the axial chirality of BINOL to the amines, which efficiently shield one or another face of the dienophile as an effective part of the chiral pocket. In addition to this effect, the presence of 3-phenylacetylacetone introduces a nonlinear effect on enantioselectivity, indicating the possible role of aggregates.

6.4.5 Chiral Relay Effects

The concept of "chiral relay" was introduced by Davies in chiral auxiliary-controlled reactions [32]. This strategy is based on the use of conformationally flexible protecting groups that are inserted between the stereogenic center and the prochiral reactive center. Due to steric interactions with the stereogenic center, the conformationally flexible group adopts a defined conformation that efficiently shields one face of the reactive center. By this process, the chiral information is relayed and even amplified, thus enabling an efficient control of the diastereoselectivity.

Later on, several authors have used this concept in reactions catalyzed by chiral Lewis acids, in which the chiral information comes from a chiral catalyst and not from a chiral auxiliary directly bonded to one of the reagents [33]. In this strategy, the chiral Lewis acid would convert an achiral template into a chiral auxiliary, so that in most cases both the chiral catalyst and the template will influence the stereochemical course of the reaction. If the template is "structured" in such a way that it matches the chiral catalyst, the result will be the amplification of the enantioselectivity. However, in a mismatched scenario, the selectivity will be reversed in comparison to that obtained by the use of the chiral Lewis acid alone.

Two methodologies can be followed to transfer the chirality to the template. In the first one, a conformationally flexible template is used, and complexation with the chiral catalyst locks the template into a chiral conformation. In the second one, complexation of an enantiotopic group generates a new stereogenic unit.

Following the first approach, pyrazolidinones (Table 6.7) have been used to substitute the commonly employed oxazolidinones [34]. The tetrahedral N(1) atom of the template inverts rapidly, but in the presence of the chiral catalyst, it preferentially exists in one of the forms, acting as a new stereogenic center. As it is close to the reactive center, it strongly influences the stereoselectivity of the reaction. The results show a correlation between the enantioselection and the size of the relay group (Table 6.7). The same effect is also shown by other Lewis acids able to adopt square-planar geometry [35], which points to a model with the Box occupying two coordination sites and with bidentate coordination of the dienophile in s-*cis* conformation.

The proposed model (Figure 6.20) places the relay group shielding one face of the crotonate, reinforcing the role of one isopropyl group of the Box. The authors have not a clear explanation of the preferred positions of the fluxional substituents and they propose that when the substituent is placed on the opposite side of the dienophile, both sides are efficiently shielded and the conformation is not reactive.

Table 6.7 Chiral relay effect in enantioselective in Diels–Alder reactions with pyrazolidinone derivatives.

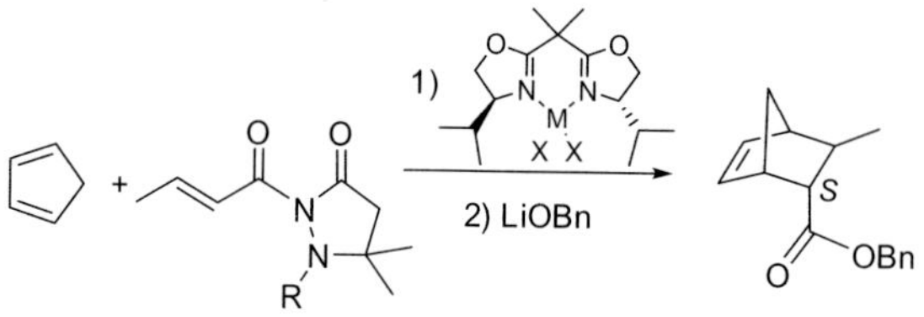

MX_2	R	% ee
$Cu(OTf)_2$	H	8
	Et	56
	Bn	71
	CH_2-1-Napht	92
$Cu(ClO_4)_2$	Bn	86
$Pd(ClO_4)_2$	Bn	96
$Mg(ClO_4)_2$	Bn	23
$Zn(OTf)_2$	Bn	45
$Yb(OTf)_3$	Bn	3

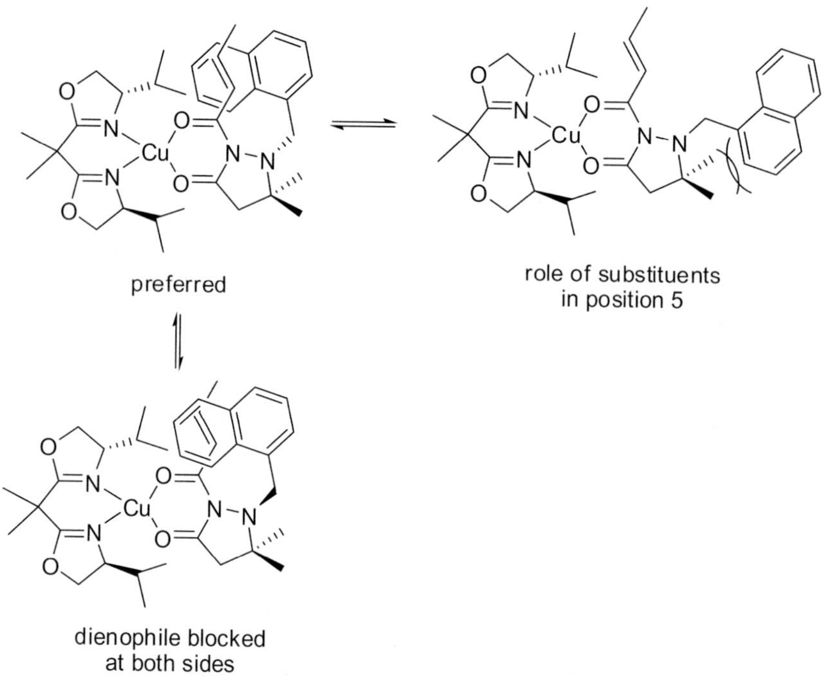

Figure 6.20 Proposed model for chiral relay in Diels–Alder reaction of pyrazolidinone derivatives catalyzed by Box–Cu complexes.

Thus, the conformation with isopropyl and fluxional group at the same side of dienophile is more reactive and, in Curtin–Hammett conditions, directs the reaction. Moreover, the substituents in position 5 of pyrazolidinone force the conformational equilibrium of the fluxional substituent.

Another example uses 4-substituted 1,3-benzoxazol-2(3*H*)-ones as templates [36]. Under chelate control with a chiral Lewis acid, the acryloyl group cannot be coplanar with the aromatic ring but strongly twisted, generating two possible diastereomeric conformers that differ in the absolute configuration of the chirality axis in the template (Figure 6.21). The bulkiness of R modifies the twisting angle, which seems to be optimal around 45°. As observed in other cases, the anion and the hydration degree of the Mg salt affect the results by changing the coordination from tetrahedral to octahedral, obtaining even a reversal in the enantioselectivity from 70% ee (*R*) for R=H to 88% ee (*S*) for R=Bn, making even more difficult the rationalization of the results.

In the second strategy, a new stereogenic center is generated by complexation of an enantiotopic group. A first example is the Diels–Alder reaction of *ortho*-substituted *N*-arylmaleimides (Figure 6.22) [37]. Both carbonyl groups of

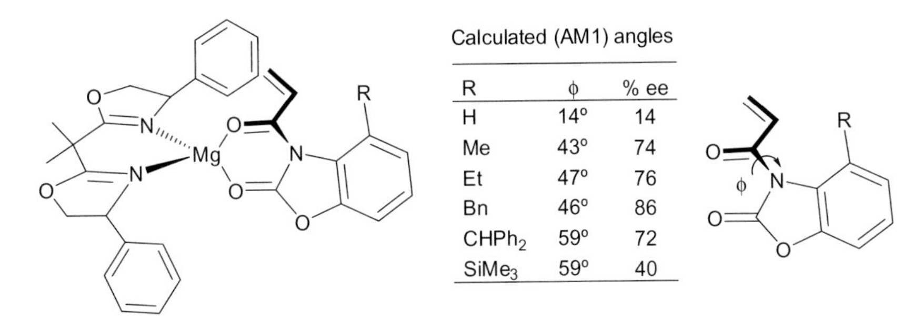

R	ϕ	% ee
H	14°	14
Me	43°	74
Et	47°	76
Bn	46°	86
$CHPh_2$	59°	72
$SiMe_3$	59°	40

Figure 6.21 Proposed model for chiral relay effect with 4-substituted 1,3-benzoxazol-2(3*H*)-ones.

R	% ee
H	62
Me	90
^{t}Bu	95

Figure 6.22 Generation of a new stereogenic center by complexation of an enantiotopic group.

maleimide are enantiotopic, and complexation with a chiral Lewis acid produces two diastereomeric complexes. Increasing the size in the *ortho* position has a very positive effect on enantioselectivity that can be envisaged as a cumulative effect of ligand control and chiral relay.

6.4.6
Subtle Changes in TADDOLate Geometry: Substitution and Immobilization

TADDOLs (α,α,α′,α′-tetraaryl-1,3-dioxolane-4,5-dimethanols) and their derivatives constitute one of the most successful families of chiral ligands that, coordinated to a large variety of metals, have produced excellent results in many different enantioselective reactions in which they have been used as chiral catalysts [38]. In particular, Ti-TADDOLates have acted as efficient chiral Lewis acids in Diels–Alder reactions. As in many of the other examples, 3-enoyl-1,3-oxazolidin-2-ones were identified as suitable dienophiles, able to form intermediate chelate complexes.

The comparison of several TADDOLs in the Diels–Alder reaction between cyclopentadiene and (*E*)-3-butenoyl-1,3-oxazolidin-2-one (Table 6.8, entries 1–6) showed that enantioselectivity depends not only on the nature of the aromatic substituents, whose influence was expected due to the role in the construction of the chiral pocket, but also on the nature of the distal substituents R_1 and R_2 [39]. The effect is significant in case of comparing phenyl and 2-naphthyl aromatic groups, but it is really dramatic in the case of 1-naphthyl group, given that 2*R endo* cycloadduct is obtained as major product in contrast with 2*S* obtained with the other ligands.

An even deeper influence of the dioxolane substitution was found in studies devoted to the immobilization of TADDOLs by covalent bonding to polymers [40]. TADDOLs bearing 3,5-dimethylphenyl aromatic groups (Table 6.8, entries 7–9, 11–14) lead 2*R endo* cycloadduct as the major enantiomer when only one of the substituents of the dioxolane ring is an aromatic group, including the case of TADDOL grafted to a Merrifield resin (P1 support). The reaction is not enantioselective when both substituents in the dioxolane ring are aromatic and the presence of two methyl groups produces a reversal in the enantioselection, leading to 2*S endo* cycloadduct. This influence is particular for 3,5-dimethylphenyl groups and it has not been detected for other aromatic groups. The existence of some kind of interaction between 3,5-dimethylphenyl groups and the aromatic substituent of the dioxolane ring was made evident by the comparison between the effect of a flexible (P1) and a rigid monolithic (P2) polymeric support. Other flexible TADDOL-containing polymers do not show any similar reversal effect on the enantioselection. In those cases, the polymerization position plays a crucial role on enantioselectivity, and very low values were obtained when the polymer was linked to two aromatic groups of the α,α′-positions [41].

The explanation for these results is not straightforward. From X-ray structures of several TADDOL ligands, the existence of an intramolecular H-bond in the most stable conformers has been considered as a good model for the Ti–chelate

Table 6.8 Effect of TADDOL substituents on the enantiomeric ratio of Diels–Alder reaction.

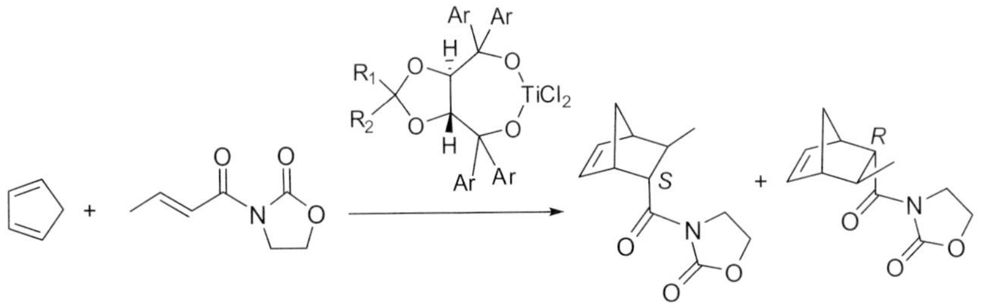

Entry	R_1	R_2	Ar	2S/2R
1	Me	Me	Ph	72:28
2	Me	Me	2-Napht	94:6
3	Ph	Me	Ph	94:6
4	Ph	Me	2-Napht	71:29
5	Ph	H	Ph	69:31
6	Ph	Ph	Ph	90:10
7	3-(P1)O-C_6H_4–[a]	H	3,5-diMePh	38:62
8	3-(P2)O-C_6H_4–[b]	H	3,5-diMePh	59:41
9	3-BnO-C_6H_4–	H	3,5-diMePh	31:69
10	3-BnO-C_6H_4–	H	Ph	67:33
11	Ph	H	3,5-diMePh	31:69
12	Ph	Me	3,5-diMePh	38:62
13	Ph	Ph	3,5-diMePh	50:50
14	Me	Me	3,5-diMePh	91:9
15	Ph	Ph	Ph	90:10
16	Me	Me	Ph	72:28

a) P1 = Merrifield resin (1% cross-linking).
b) P2 = Monolithic polymer obtained by copolymerization of TADDOL monomer (R_1 = 3-(4-vinylbenzyloxy)phenyl) and divinylbenzene (monomer/DVB ratio = 40:60).

complexes as the H-bond falls nearly along the C_2 axis, in the same position occupied by Ti in the complex (Figure 6.23). The substitution at the ketal carbon must influence the conformation around the C-aryl bond, modifying in a subtle way the chiral pocket around the metal position. The dramatic change in enantioselectivity from phenyl or 2-naphthyl aromatic groups to 1-naphthyl groups can be also explained by the difference in disposition of the fused aromatic ring following the same model (Figure 6.23). Whereas in 2-naphthyl the second fused aromatic ring is placed far from the catalytic center, in 1-naphthyl the fused ring extends forward in the quasiequatorial position but back in the quasiaxial position, producing a strong difference in shielding properties of both aromatic groups.

The dramatic variations in enantioselectivity, produced by slight changes in the ligand structure, show that the mechanism determining the stereochemical

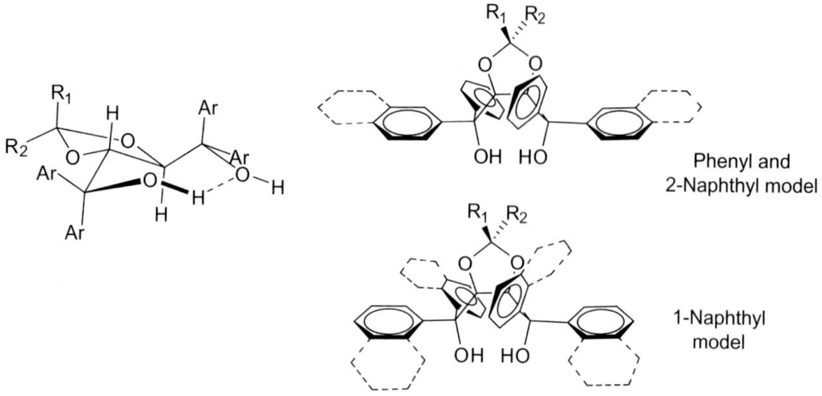

Figure 6.23 Models based on TADDOL conformations to explain the effect of α-aryl substituents.

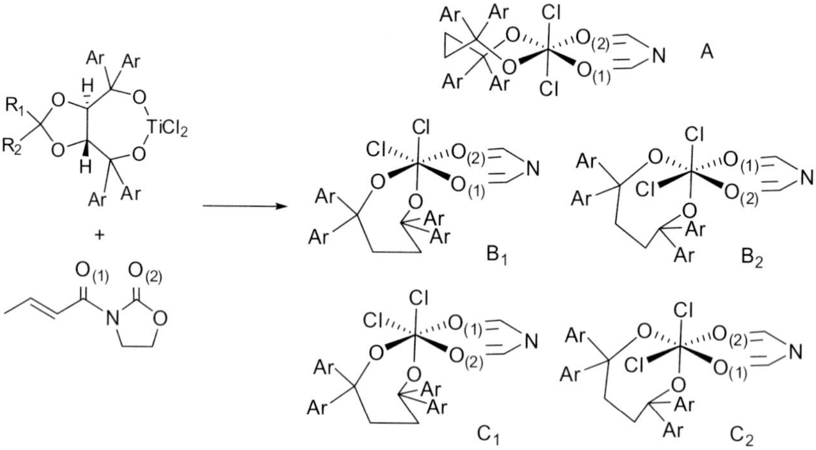

Figure 6.24 Possible TADDOL–Ti–dienophile intermediate complexes.

outcome of the reaction is not simple. In fact, under Curtin–Hammett conditions, selectivity depends on the energy differences between the TS leading to the different products. Better explanations should be obtained by considering the catalyst–dienophile intermediate complexes and several studies have been devoted to this point, with some controversial degree.

The coordination of the commonly used dienophiles can lead to five diastereomeric complexes, and X-ray and NMR studies of different TADDOLate–$TiCl_2$–dienophile complexes [42, 43] have shown that the complex bearing the two chlorine atoms in relative *trans* position (species A in Figure 6.24) is the most abundant, and hence the most stable. The main controversy comes from the relative reactivity of these intermediates. In fact, in a Curtin–Hammett scenario, it is the most reactive intermediate and not the most stable that determines the stereochemical result of the reaction. Whereas some authors proposed also a higher reactivity of species A, theoretical calculations seemed to indicate a higher degree of Lewis acid activation in the case of intermediates B [44].

With regard to enantioselectivity, the hypothesis favoring intermediate A was not able to explain the experimental results, both the high enantioselectivity obtained in some cases and the variations in the sense of asymmetric induction with TADDOLs bearing 3,5-dimethylphenyl substituents. Molecular mechanics and molecular dynamic calculations, using the MM2 force field [42b], showed that relative energies of B_1 and B_2 intermediates (Figure 6.24) are determined by the substitution pattern of dioxolane ring for those TADDOLs bearing 3,5-dimethylphenyl substituents. As both intermediates show shielding of a different face of the C=C double bond, they lead to different cycloadducts, and the energy changes may be the origin of the enantioselectivity changes experimentally observed. Differences in the substitution of the dioxolane ring provoke energy differences in good qualitative agreement with the experimental results (Table 6.8), and the existence of π-stacking interaction between one of the 3,5-dimethylphenyl groups and a phenyl group in the dioxolane ring has been proposed as responsible for the energy approach between B_1 and B_2 given that this interaction would be present in both intermediates (Figure 6.25).

It is clear that the exact mechanism of this reaction is still controversial and more complicated than expected. It is possible that the relative reactivity of the different intermediates changes from one to another TADDOL ligand. In fact, the subtle conformational changes and the existence of interactions between the different groups may modify the relative energy of the different diastereomeric TS.

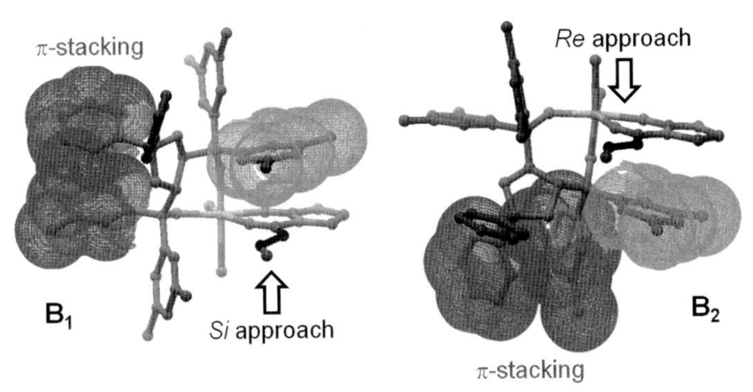

Figure 6.25 Effect of a possible π-stacking in the TADDOL–Ti catalyzed Diels–Alder.

6.5
Case of Study (3): Salen-Based Catalysts

6.5.1
The Structural Variations of Salen Ligands and Complexes

Salen stands for bis(salicylidene)ethylenediamine, whose chiral derivatives have been used as ligands for a large variety of metal complexes able to catalyze different enantioselective reactions. These ligands and their complexes present an ample array of possible structural variations (Figure 6.26). Salicylidene moiety can be substituted in different positions, although the most usually are C3 and C5 ones (R_3 and R_5, respectively, in Figure 6.26). In most cases, those substituents are bulky ones, and even R_3 may contain a stereogenic element, either a carbon atom or an axis. The main (and mostly the only) chirality source of the ligand is the diamine moiety that generally presents a C_2-symmetry axis (R = R′), and this symmetry is extended to the whole chiral ligand when $R_3 = R_3'$ and $R_5 = R_5'$. Once the complex is formed, other elements such as the anion and its coordinating ability, or the presence of additional chiral or achiral ligands (L in Figure 6.26) are factors that may influence significantly the diastereo- and enantioselectivity obtained in the catalytic reaction. Salen complexes of Mn, Ru, V, Ti, Al, Co, Cr, Cu, Zn, or Zr have been described as catalysts for reactions such as epoxidation, aziridination, epoxide ring opening, cyclopropanation, sulfoxidation, hetero-Diels–Alder, sulfimidation, conjugate addition, Baeyer–Villiger, etc. [45]. In this section we will analyze only two significant cases: epoxidation and cyclopropanation.

6.5.2
Effects of the Structural Variations in Epoxidation Reactions Catalyzed by Salen–Mn Complexes

Chiral salen–Mn complexes were described as catalysts for enantioselective epoxidation reactions in the early 1990s [46], but the mechanism for this reaction is still under debate and the contribution of each structural parameter is not fully understood.

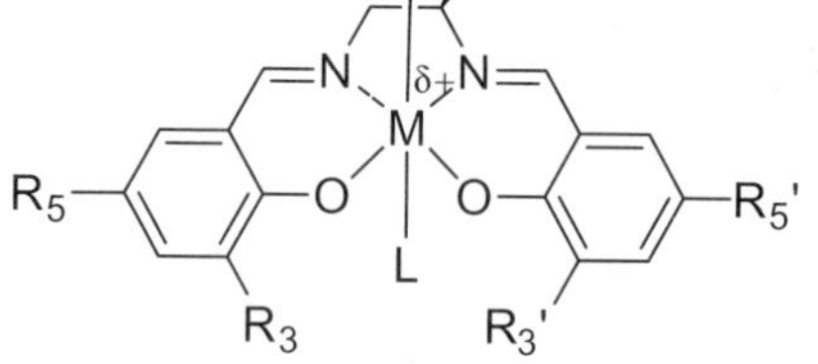

Figure 6.26 General structure of salen–metal complexes.

Which seems clear is that the intermediate is an $Mn^{V}=O$ species, and alkene must approach to the oxo group. The main role of the R_3 and R_5 groups is to hinder the approach of the alkene through a number of possible trajectories, allowing only the approach by the zone under the influence of the stereogenic centers of the chiral diamine (Figure 6.27). The approach of the alkene parallel to the main plain of the complex also explains the strong preference of this catalytic system for *cis*-alkenes, able to place both substituents far away from the complex, minimizing in this way the steric interaction. Moreover, the best results are obtained in the epoxidation of alkenes conjugated with aryl groups. This fact has been explained by a possible π–π repulsion of the approaching alkene and the aromatic rings of the salicylidene moieties. In fact, this repulsion can be increased by extending the aromatic system with naphthyl groups (Figure 6.27), which led to better results [47].

However, this scheme showed to be too simplistic, as some features of the reactions remained unexplained. In fact, the complex can adopt different conformations depending on that of the ethylenediamine-metal five-member chelate, either half-chair or envelope, leading to the so-called stepped and umbrella conformations (Figure 6.28) [48]. The two-stepped conformations become diastereomeric by the presence of stereogenic centers in the ethylenediamine moiety, and several experimental results seem to indicate the preference for the stepped conformation that places the two substituents in pseudoequatorial positions.

It is also remarkable that the reversal of enantioselectivity obtained when the R substituent of the diamine moiety is a carboxylate group, able to coordinate to Mn, forcing in this way the stepped conformation with the R groups in the pseudoaxial position (Figure 6.28). Moreover, the preference for one of the two-stepped conformations may not be due to salen substituents. The use of a chiral axial ligand would produce the same type of diastereodifferentiation, with a preference for one of the two conformers. Such effect was confirmed by the moderate enantioselectivity (73% ee) obtained in epoxidation using achiral salen ligand with bulky R_3 and R_5 substituents (*tert*-butyl) and enantiopure sparteine as axial ligand. In fact, the donor character of the axial ligands and/or anion seems to be responsible for the distortion degree in the preferred conformations, as shown by the

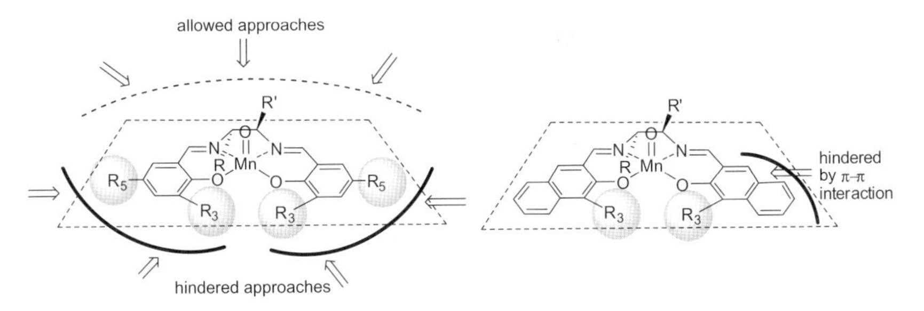

Figure 6.27 Proposed role of the different substituents in the approach restrictions to salen–Mn catalysts.

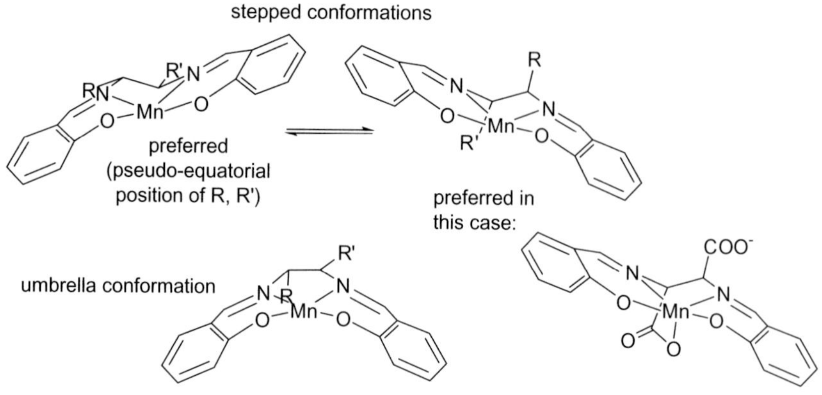

Figure 6.28 Stepped and umbrella conformations of salen–Mn complexes.

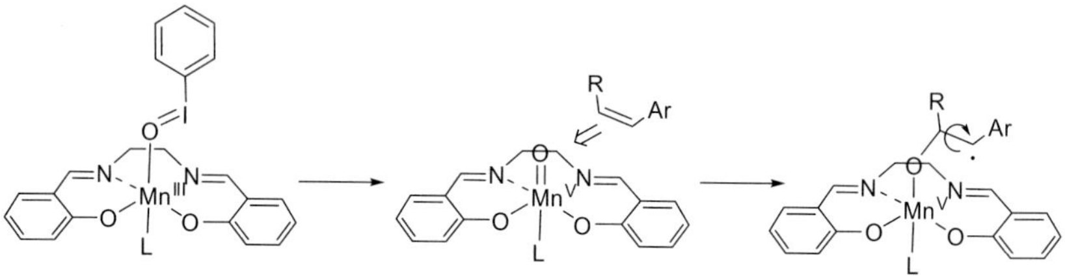

Figure 6.29 Possible intermediates in the epoxidation reaction catalyzed by salen–Mn complexes.

recent studies in model compounds [49], and hence for the observed differences in enantioselectivity.

The general picture of the selectivity control is even more complicated than exposed until now due to several additional factors. In some cases, a dependence of enantioselectivity on the nature of the hypervalent iodine oxidants (such as PhIO, C_6F_5IO, and MesIO) has been observed [50]. The only possible explanation for this fact is the participation of a new oxidizing species based on coordination of the oxidant to salen-Mn, acting in this case as a Lewis acid without oxygen transfer (Figure 6.29).

The existence of this new oxidation pathway is highly dependent on the nature of the donor ligand (L) and the oxidant. Moreover, the classical reaction pathway involves the formation of a radical upon addition of Mn=O to the alkene (Figure 6.29), with free rotation around the single C–C bond. This explains the observed yield of *trans*-epoxide from *cis*-alkene, but the variable amounts of *trans*-epoxide with metal (Cr > Mn), anion, oxidant, temperature, and donor ligand are more difficult to explain. In fact, the proper choice of all those parameters allowed obtaining high yields and enantioselectivities of *trans*-epoxides from *cis*-alkenes with salen–Cr complexes and triphenylphosphine oxide as donor ligand [51].

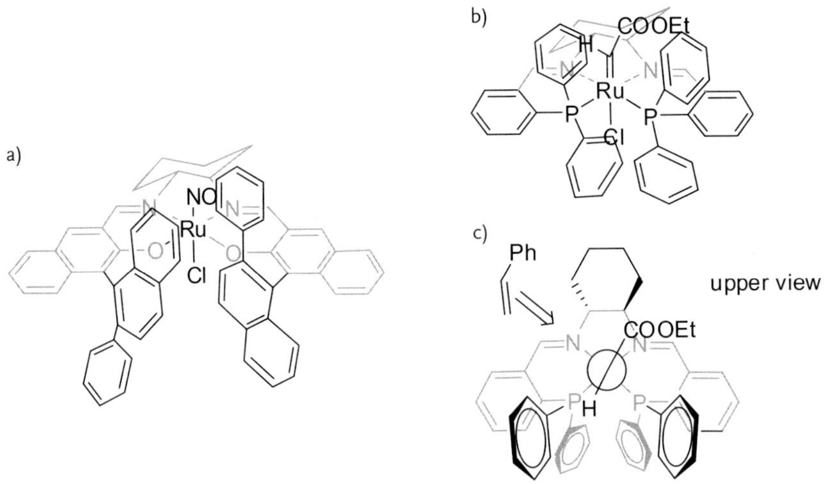

a)

b)

c)

upper view

Figure 6.30 Proposed models for *cis* selectivity in salen–Ru catalyzed cyclopropanation.

6.5.3
Control of the Sense of Asymmetric Induction in Salen–Ru Complexes

Probably, the most dramatic effect on enantioselectivity observed with salen–metal complexes is the reversal of induction sense in the case of Ru-catalyzed cyclopropanation. Salen–Co(III) complexes had shown high efficiency in enantioselective cyclopropanation when R_3 substituents were absent. However, the use of Ru instead of Co allowed the presence of bulky R_3 substituents with dramatic influence in the final results. A salen ligand derived from binaphthyl units (Figure 6.30) led to the best results of enantioselectivity, with an important match–mismatch effect in the two sources of chirality, the cyclohexanediamine and the binaphthyl units. The use of (*R*)-binaphthyl and (*S*)-diamine led to low *trans*-preference in the cyclopropanation of styrene with *tert*-butyl diazoacetate and 51% ee in the *trans* isomers. On the contrary, the combination of (*R*)-binaphthyl and (*R*)-diamine led to a high *cis*-preference, not usual with most of homogeneous catalysts, and up to 89% ee in the *cis* isomers [52]. These results in the absence of solvent were even improved by the use of THF, with 96:4 *cis*/*trans* selectivity and 99% ee in the *cis* isomers. Regarding the mechanism for the high *cis* preference and the high enantioselectivity, a mechanism has been proposed using an analogous PNNP ligand (Figure 6.30b) [53]. The key point for the high *cis* preference is the conformation of the Ru–carbene intermediate that places the H in the hindered zone of the chiral ligand, and styrene approaches with its phenyl group also far from that zone (Figure 6.30c). The presence of the cyclohexanediamine moiety is the main responsible factor for the energy difference between the two *cis* TS (>7 kcal/mol), in agreement with the very high enantioselectivity.

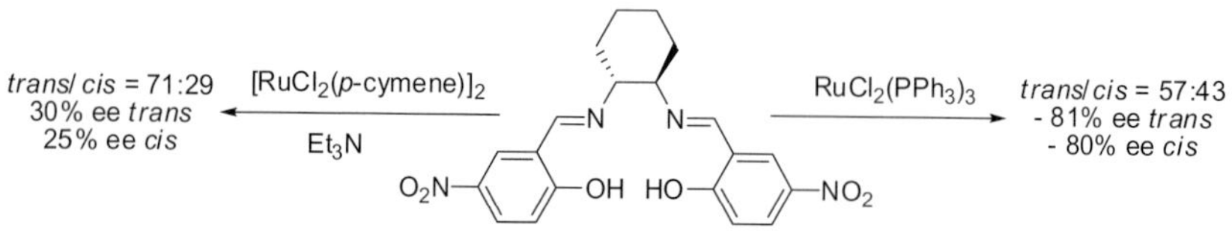

trans/*cis* = 71:29
30% ee *trans*
25% ee *cis*
$[RuCl_2(p\text{-cymene})]_2$
Et_3N
$RuCl_2(PPh_3)_3$
trans/*cis* = 57:43
- 81% ee *trans*
- 80% ee *cis*
O_2N OH HO NO_2

Figure 6.31 Reversal of enantioselectivity with the Ru precursor.

With the salen–Ru complex, a complete reversal of enantioselectivity was observed when solvents such as THF or ethyl acetate were changed by diisopropyl ether or hexane, obtaining up to 83% ee of the *cis* isomer with opposite absolute configuration. The proposed explanation for this dramatic solvent effect was the poor solubility of the complex in the latter type of solvents. The true catalyst in such case aggregates, whereas in THF, the complete solution of the complex led to monomeric species. A similar effect was observed in the case of using less-hindered salen and some donor ligands (Figure 6.31). The use of triphenylphosphine as donor ligand leads, in the case of a salen ligand with withdrawing nitro groups, to a reversal in the enantioselectivity. The proposed mechanism is the formation of the Ru–carbene intermediate by the breakage of an N–Ru bond, instead of the usual Ru–L bond proposed for the rest of the salen and donor ligands.

6.6
Case of Study (4): Multifunctional Catalysis

6.6.1
Cooperative Effects

Some enantioselective catalytic reactions require the simultaneous activation of the two partners involved, generally speaking a nucleophile and an electrophile, in a phenomenon known as cooperative effect. The activation of only one reagent is not enough to produce the reaction in high yield, and usually enantioselectivity is also very low. Enzymes play the same role with the presence of several catalytic centers (Brønsted and Lewis acids and bases) that simultaneously participate in the catalytic process. Apart from catalysts using some organic functionality, either acid or base, as one of the cooperative centers, in the case of processes requiring two metal centers, four types of catalytic systems can be considered (Figure 6.32): homo- or heterobimetallic catalysis in an inter- or intramolecular way. A chiral environment around both metals seems to be crucial to obtain high enantioselectivities, probably due to the need of a strict control on the geometry of the TS, in theory better controlled in case of intramolecular systems by the link between both metals or complexes. Homobimetallic systems present the limitation of the same metal having to activate both reagents, although this limitation is less important in case of intramolecular systems, as the electronic and steric

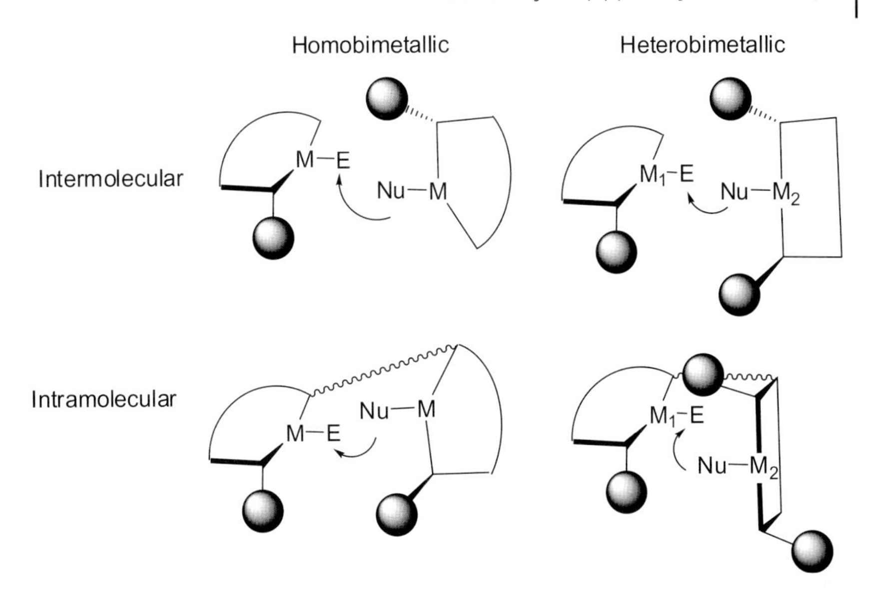

Figure 6.32 Types of catalytic systems with cooperative effects.

environment of both metals can be different enough to selectively activate one or another reagent.

6.6.2 Intermolecular Homobimetallic Catalysis

Different salen–metal complexes have shown cooperative effects in several enantioselective reactions. One example is salen–Cr as catalyst for the enantioselective opening (desymmetrization) of *meso*-epoxides with azide (Figure 6.33) [54]. The same complex is able to coordinate both azide and epoxide, but kinetic evidences point to an intermolecular transfer of azide to coordinated epoxide (Figure 6.33).

A similar case has been described for the conjugate addition of cyanide to α,β-unsaturated imides catalyzed by salen–Al complexes [55]. In this case, spectroscopic evidences show the formation of two different species in solution, probably salen–Al–CN and salen–Al-imidate (Figure 6.34). The second-order dependence of reaction rate on catalyst concentration indicates the existence of a cooperative homobimetallic intermolecular mechanism, although this double coordination requires a large amount of catalyst.

6.6.3 Intermolecular Heterobimetallic Catalysis

The same metal may not be the best to activate both nucleophile and electrophile. This is the case of the above-mentioned conjugate addition of cyanide to

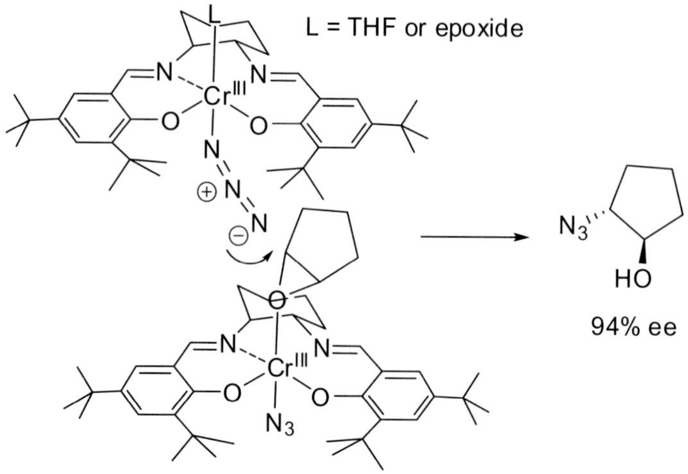

L = THF or epoxide

94% ee

Figure 6.33 Proposed mechanism of intermolecular desymmetrization of meso-epoxides.

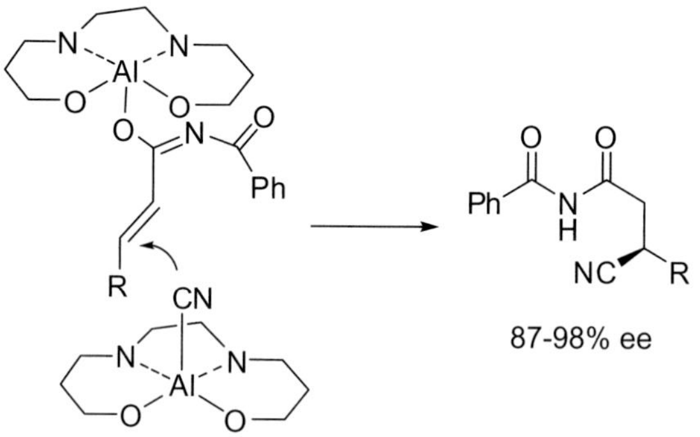

87-98% ee

Figure 6.34 Homobimetallic intermolecular mechanism of conjugate addition of cyanide.

α,β-unsaturated imides, given that aluminum catalysts are not the optimum for cyanide activation. The same authors demonstrated that the combination of salen–Al complex with a pybox–$ErCl_3$ complex was able to efficiently catalyze the same reaction (Figure 6.35) with much lower amount of catalysts (2% salen–Al and 3% pybox–Er instead of 10% or even 15% salen–Al) [56], due to the specialization of each type of complex, in this case the ability of lanthanide complexes to activate cyanide. Activation of imide was proven to be also necessary by the almost no conversion obtained with pybox–Er alone. Both catalysts are involved in the TS, and match in the chirality of both catalysts is necessary in order to obtain high enantioselectivity. In fact, with both (*S*) catalysts, 96% ee was obtained (R = nPr), whereas enantioselectivity was reduced if (*R*)-pybox (72% ee) or a nonchiral analogous (84% ee) were used. The same effect was observed if a

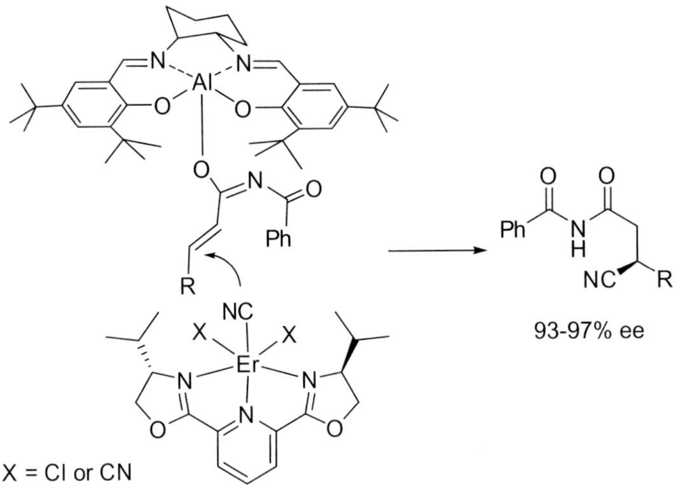

X = Cl or CN

93-97% ee

Figure 6.35 Heterobimetallic intermolecular mechanism of conjugate addition of cyanide.

nonchiral salen ligand was used in combination with the enantiopure pybox (78% ee).

6.6.4
Intramolecular Homobimetallic Catalysis

Shibasaki and coworkers are probably the authors that have most widely explored the intramolecular bimetallic catalytic systems [57]. In the case of homobimetallic systems, the same metal has to play two different roles in the activation of both reaction partners. Several examples of this type of systems have been described with linked-BINOL ligands (Figure 6.36). In this way, the ligand has up to five coordination centers able to form bimetallic complexes. When 1 mol of this ligand reacted with 2 mol of Et_2Zn, an oligomeric species was formed, in which one phenol group of each ligand remained free and the oxygen of the linker played an important role in the coordination of Zn (Figure 6.36) [58].

However, when this system was used as catalyst in the aldol reaction between an aldehyde and the α-hydroxyketone shown in Figure 6.36, the true intermediate of the reaction was more complicated, as shown by cold spray ionization-mass spectrometry, formed by Zn, linked-BINOL, and hydroxyketone in 7 : 3 : 4 ratio. The key parameters seem to be the control of the relative position of the Zn atoms to carry out the cooperative effect, demonstrated by the poor results obtained with BINOL, the participation of the linker oxygen in the coordination of Zn, as the linker without heteroatom also gave poor results, the hydroxyl in α-position of the substrate that stabilizes the enolate, and finally, the presence of the *ortho*-methoxy group that helps to fix the relative position of the enolate to the intermediate complex (Figure 6.37).

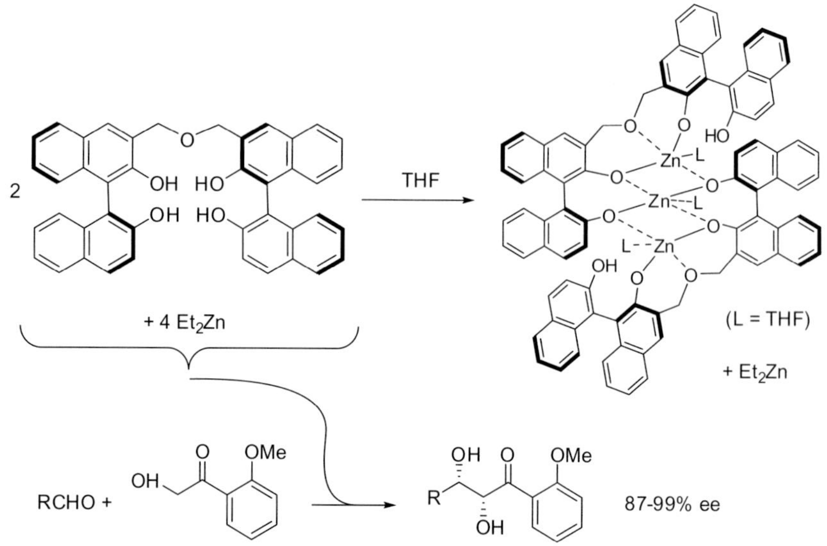

Figure 6.36 Structure of the major (linked-BINOL)$_2$Zn$_3$(THF)$_3$ species in solution and catalytic enantioselective aldol reaction.

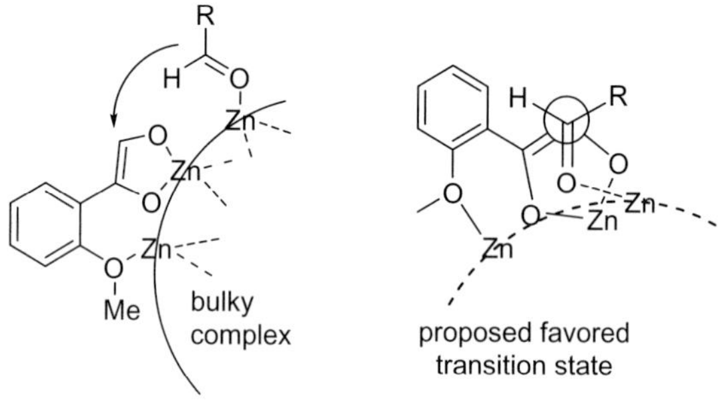

Figure 6.37 Proposed role of the multicenter Zn homometallic catalyst.

Homomultimetallic complexes of Zn, Y, In, and La have been used in reactions with other nucleophiles, such as α,β-unsaturated carbonyl compounds (Michael reactions), *N*-diphenylphosphinoyl-imines or *N*-tosyl-imines (Mannich reactions), and other electrophiles such as *N*-(2-hydroxyacetyl)pyrrole or dialkyl malonate.

One general conclusion from these results was the possibility of getting good results with simpler modified BINOL ligands without C_2-symmetry. In fact, the systematic study of several modified BINOLs showed that the only requirements were the presence of a second aromatic groups linked to BINOL by a linker with a heteroatom, the functionalization of this second aromatic group with an additional phenol in position 2′, and the presence of a substituent in position 3′

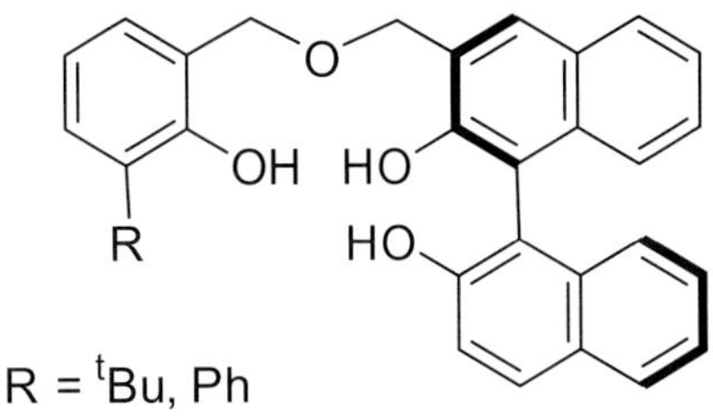

R = ^{t}Bu, Ph

Figure 6.38 Minimum structure required to obtain high enantioselectivity with linked-BINOL ligands.

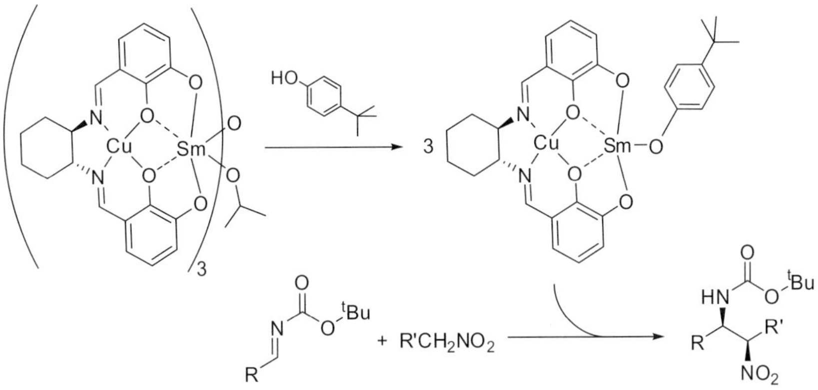

Figure 6.39 Trimeric structure of Cu–Sm bimetallic system, formation of the monomeric catalytic species and enantioselective nitro-Mannich reaction.

(Figure 6.38). The axial chirality in this second group and the presence of a fourth phenol are not necessary.

6.6.5
Intramolecular Heterobimetallic Catalysis

In the case of heterobimetallic systems, the relative position of both metals has to be controlled by the correct design of the chiral ligand with the suitable coordination centers. As an example, nitro-Mannich reactions have been catalyzed by Cu–Sm bimetallic complexes (Figure 6.39) [57]. The control of the relative positions of both metals was carried out with a hexadentate ligand, able to form trimeric complexes in the presence of $Cu(OAc)_2$ and $Sm(O^{i}Pr)_3$, using one μ-isopropoxy and one μ-oxo ligands (Figure 6.39). However, the active species is a monomeric complex formed by reaction with 4-*tert*-butylphenol (Figure 6.39). Formation of a Sm-nitronate by deprotonation of the nitro compound and coordination of Boc group to Cu bring both partners together in a suitable position to react with 83–98% ee depending on R′. The same ligand is able to form bimetallic

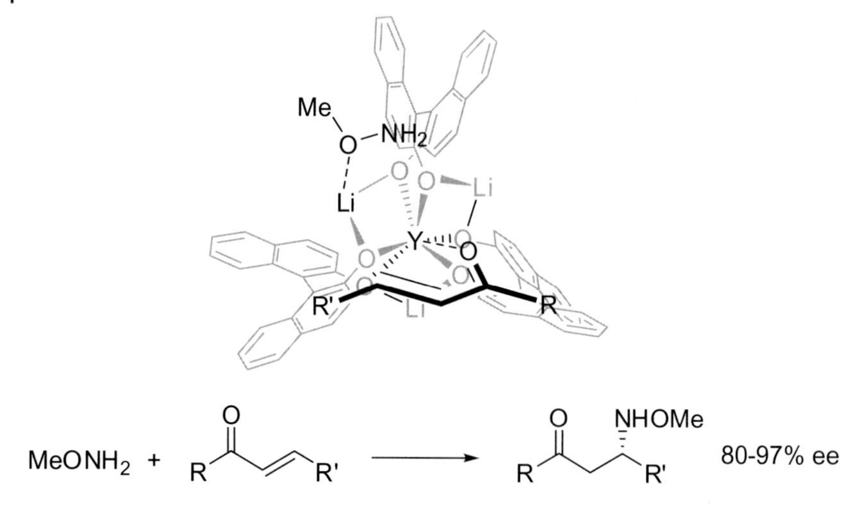

Figure 6.40 Y-Li-BINOL system and enantioselective *aza*-Michael addition.

complexes with other metals, expanding in this way the applicability of this kind of system. As an example, recently, the Pd–La complex was described to be optimum to catalyze nitro-aldol reactions between aldehydes and nitro compounds [59].

In the previous examples, one of the roles played by the catalysts is that of the base, able to deprotonate one partner, whereas one of the metals efficiently coordinates the formed enolate and the other one acts as a Lewis acid. However, in other cases, both metals can play the role of Lewis acid, and in such case, the choice of metals is really important. An example is the *aza*-Michael addition of methoxyamine to enones (Figure 6.40). A rare-earth metal, Y in this case, is used to coordinate the enone, whereas methoxyamine is coordinated through an alkaline metal, Li in this case, with BINOL as a chiral ligand. Complexes include Y, Li, and ligand in a 1 : 3 : 3 ratio (Figure 6.40), the cooperative effect is demonstrated by the lack of efficiency of Y–BINOL and Li–BINOL systems, and the poor performance of the Y–K–BINOL system shows the specificity of the Y–Li pair for this reaction. The simultaneous coordination of methoxyamine to Li and enone to Y brings both reagents together in a suitable fashion to react with high efficiency and enantioselectivity.

6.7 Conclusions

Chiral catalysis is the tool selected by the nature to transfer chirality and the process is as complex as nature itself. Given that only small differences in energy are involved in enantioselectivity, a couple kcal mol^{-1} often makes all the differences between success and failure, subtle factors have a decisive influence. In the nanoenvironment of a chiral catalyst, there are many elements that can influence decisively the stereochemical course of a reaction. Some of them such as close substituents, counterions, auxiliary ligands, or solvent molecules can be in the immediate neighborhood of the catalytic center (usually a metal), and its influence is easier to ascertain, at least in principle. Other elements such as remote substituents, support, or even other cocatalytic species are usually located farther apart from the catalytic center, but they can still exert a strong influence by inducing conformational changes and, in general, by determining preferential reaction channels. In that regard, asymmetric catalysis should be considered a nanometric phenomenon in which all the actors taking part in the reaction are important.

Deep knowledge on the mechanism of the stereodifferentiation in a catalytic process is rather unusual, but in those cases where a good knowledge of the steric requirements of the chiral pocket is achieved, either through mechanistic studies or by trial-error experiments, this knowledge can be effectively used to design new ligands taking advantage of the different nanoenvironmental factors, to improve the catalytic results.

References

1 (a) Ojima, I. (ed.) (2000) *Catalytic Asymmetric Synthesis*, John Wiley & Sons, Inc., New York.
(b) Noyori, R. (1994) *Asymmetric Catalysis in Organic Synthesis*, John Wiley & Sons, Inc., New York.
(c) Jacobsen, E.N., Pfaltz, A., and Yamamoto, H. (eds). (1999) *Comprehensive Asymmetric Catalysis*, Springer, Berlin.

2 (a) Salomon, R.G. and Kochi, J.K. (1973) *J. Am. Chem. Soc.*, **95**, 3300.
(b) Díaz-Requejo, M., Belderrain, T.R., Nicasio, M.C., Prieto, F., and Pérez, P.J. (1999) *Organometallics*, **18**, 2601.

3 Dai, X. and Warren, T.H. (2004) *J. Am. Chem. Soc.*, **126**, 10085. Only one Cu–carbene complex has been isolated and characterized by X-ray diffraction analysis.

4 Fraile, J.M., García, J.I., Martínez-Merino, V., Mayoral, J.A., and Salvatella, L. (2001) *J. Am. Chem. Soc.*, **123**, 7616.

5 Rasmussen, T., Jensen, J.F., Èstergaard, N., Tanner, D., Ziegler, T., and Norrby, P.-O. (2002) *Chem. Eur. J.*, **8**, 177.

6 García, J.I., Jiménez-Osés, G., Martínez-Merino, V., Mayoral, J.A., Pires, E., and Villalba, I. (2007) *Chem. Eur. J.*, **13**, 4064.

7 Aratani, T. (1985) *Pure Appl. Chem.*, **57**, 1839. Aratani's catalysts, industrially used in the production of chrisantemic acid derivatives.

8 Reiser, O. (1993) *Angew. Chem. Int. Ed.*, **32**, 547.

9 Cornejo, A., Fraile, J.M., García, J.I., Gil, M.J., Herrerías, C.I., Legarreta, G., Martínez-Merino, V., and Mayoral, J.A. (2003) *J. Mol. Catal. A*, **196**, 101.

10 Lyle, M.P.A., Draper, N.D., and Wilson, P.D. (2006) *Org. Biol. Chem.*, **4**, 877.

11 Nishiyama, H., Soeda, N., Naito, T., and Motoyama, Y. (1998) *Tetrahedron Asymmetry*, **9**, 2865.

12 Cornejo, A., Fraile, J.M., García, J.I., Gil, M.J., Martínez-Merino, V., and Mayoral, J.A. (2005) *Angew. Chem. Int. Ed.*, **44**, 458.

13 (a) García, J.I., Mayoral, J.A., Pires, E., and Villalba, I. (2006) *Tetrahedron Asymmetry*, **17**, 2270.
(b) Werner, H., Vicha, R., Gissibl, A., and Reiser, O. (2003) *J. Org. Chem.*, **68**, 10166.

14 Fraile, J.M., García, J.I., Gissbl, A., Mayoral, J.A., Pires, E., Reiser, O., Roldán, M., and Villalba, I. (2007) *Chem. Eur. J.*, **13**, 8830.

15 Burguete, M.I., Fraile, J.M., Garcia, J.I., Garcia-Verdugo, E., Herrerias, C.I., Luis, S.V., and Mayoral, J.A. (2001) *J. Org. Chem.*, **66**, 8893.

16 (a) Evans, D.A., Woerpel, K.A., Hinman, M.M., and Faul, M.M. (1991) *J. Am. Chem. Soc.*, **113**, 726.
(b) Fraile, J.M., García, J.I., Mayoral, J.A., and Tarnai, T. (1999) *J. Mol. Catal. A*, **144**, 85.

17 (a) Fraile, J.M., García, J.I., Mayoral, J.A., Tarnai, T., and Harmer, M.A. (1999) *J. Catal.*, **186**, 214.
(b) Fernández, M.J., Fraile, J.M., García, J.I., Mayoral, J.A., Burguete, M.I., García-Verdugo, E., Luis, S.V., and Harmer, M.A. (2000) *Top. Catal.*, **13**, 303.
(c) Fraile, J.M., García, J.I., Herrerías, C.I., Mayoral, J.A., and Harmer, M. (2004) *J. Catal.*, **221**, 532.

18 Fraile, J.M., García, J.I., Gil, M.J., Martínez-Merino, V., Mayoral, J.A., and Salvatella, L. (2004) *Chem. Eur. J.*, **10**, 758.

19 Fernández, A.I., Fraile, J.M., García, J.I., Herrerías, C.I., Mayoral, J.A., and Salvatella, L. (2001) *Catal. Commun.*, **2**, 165.

20 García, J.I., López-Sánchez, B., Mayoral, J.A., Pires, E., and Villalba, I. (2008) *J. Catal.*, **258**, 378.

21 Fraile, J.M., García, J.I., Jiménez-Osés, G., Mayoral, J.A., and Roldán, M. (2008) *Organometallics*, **27**, 2246.

22 (a) Sibi, M.P. and Liu, M. (2001) *Curr. Org. Chem.*, **5**, 719–755.
(b) Desimoni, G., Faita, G., and Jørgensen, K.A. (2006) *Chem. Rev.*, **106**, 3561.

23 (a) Corey, E.J., Imai, N., and Zhang, H.Y. (1991) *J. Am. Chem. Soc.*, **113**, 728.
(b) Corey, E.J. and Isihara, K. (1992) *Tetrahedron Lett.*, **33**, 6807.

24 (a) Evans, D.A., Murry, J.A., von Matt, P., Norcross, R.D., and Miller, S.J. (1995) *Angew. Chem. Int. Ed. Engl.*, **34**, 798.
(b) Evans, D.A., Miller, S.J., Lectka, T., and von Matt, P. (1999) *J. Am. Chem. Soc.*, **121**, 7559. (c) Evans, D.A., Kozlowski, M.C., and Fedrow, J.S. (1996) *Tetrahedron Lett.*, **37**, 7481.

25 Ghosh, A.K., Mathivanan, P., and Cappiello, J. (1996) *Tetrahedron Lett.*, **37**, 3815.

26 (a) Desimoni, G., Faita, G., and Righetti, P.P. (1996) *Tetrahedron Lett.*, **37**, 3027.
(b) Desimoni, G., Faita, G., Invernizzi, A.G., and Righetti, P. (1997) *Tetrahedron*, **53**, 7671.
(c) Carbone, P., Desimoni, G., Faita, G., Filippone, S., and Righetti, P. (1998) *Tetrahedron*, **54**, 6099.

27 Wang, H., Liu, X., Xia, H., Liu, P., Gao, J., Ying, P., Xiao, J., and Li, C. (2006) *Tetrahedron*, **62**, 1025.

28 Desimoni, G., Faita, G., Guala, M., and Pratelli, C. (2003) *J. Org. Chem.*, **68**, 7882.

29 Wang, H., Wang, H., Liu, P., Yang, H., Xiao, J., and Li, C. (2008) *J. Mol. Catal. A*, **285**, 128.

30 (a) Desimoni, G., Faita, G., Guala, M., and Laurenti, A. (2004) *Eur. J. Org. Chem.*, 3057.
(b) Desimoni, G., Faita, G., Guala, M., Laurenti, A., and Mella, M. (2005) *Chem. Eur. J.*, **11**, 3816.

31 (a) Kobayashi, S. and Ishitani, H. (1994) *J. Am. Chem. Soc.*, **16**, 4083. (b) Kobayashi, S., Ishitani, H., Hachiya, I., and Araki, M. (1994) *Tetrahedron*, **50**, 11623.

32 (a) Bull, S.D., Davies, S.G., Epstein, S.W., Leech, M.A., and Ouzman, J.V.A. (1998) *J. Chem. Soc. Perkin Trans. 1*, 2321.
(b) Bull, S.D., Davies, S.G., Fox, D.J., and Sellers, T.G.R. (1998) *Tetrahedron Asymmetry*, **9**, 1483.
(c) Bull, S.D., Davies, S.G., and Ouzman, J.V.A. (1998) *Chem. Commun.*, 659.

33 Corminboeuf, O., Quaranta, L., Renaud, P., Liu, M., Jasperse, C.P., and Sibi, M.P. (2003) *Chem. Eur. J.*, **9**, 29.

34 Sibi, M.P., Venkatraman, L., Liu, M., and Jasperse, C.P. (2001) *J. Am. Chem. Soc.*, **123**, 8444.

35 Sibi, M.P., Stanley, L.M., Nie, X., Venkatraman, L., Liu, M., and Jasperse, C.P. (2007) *J. Am. Chem. Soc.*, **129**, 395.

36 Quaranta, L., Corminboeuf, O., and Renaud, P. (2002) *Org. Lett.*, **4**, 39.

37 (a) Corey, E.J. and Lee, D.-H. (1994) *J. Am. Chem. Soc.*, **116**, 12089.
(b) Corey, E.J. and Letavic, M.A. (1995) *J. Am. Chem. Soc.*, **117**, 9626.

38 Seebach, D., Beck, A.K., and Heckel, A. (2001) *Angew. Chem. Int. Ed.*, **40**, 92.

39 Seebach, D., Dahinden, R., Marti, R.E., Beck, A.K., Plattner, D.A., and Kühnle, F.N.M. (1995) *J. Org. Chem.*, **60**, 1788.

40 (a) Altava, B., Burguete, M.I., Escuder, B., Luis, S.V., Salvador, R.V., Fraile, J.M., Mayoral, J.A., and Royo, A.J. (1997) *J. Org. Chem.*, **62**, 3126.
(b) Altava, B., Burguete, M.I., Fraile, J.M.,

García, J.I., Luis, S.V., Mayoral, J.A., Royo, A.J., and Vicent, M.J. (1997) *Tetrahedron Asymmetry*, **8**, 2561.
(c) Altava, B., Burguete, M.I., Fraile, J.M., García, J.I., Luis, S.V., Mayoral, J.A., and Vicent, M.J. (2000) *Angew. Chem. Int. Ed.*, **39**, 1503.

41 Seebach, D., Marti, R.E., and Hintermann, T. (1996) *Helv. Chim. Acta*, **79**, 1710.

42 (a) Gothelf, K.V., Hazell, R.G., and Jørgensen, K.A. (1995) *J. Am. Chem. Soc.*, **117**, 4435.
(b) Gothelf, K.V. and Jørgensen, K.A. (1995) *J. Org. Chem.*, **60**, 6847.

43 Haase, C., Sarko, C.R., and DiMare, M. (1995) *J. Org. Chem.*, **60**, 1777.

44 García, J.I., Martínez-Merino, V., and Mayoral, J.A. (1998) *J. Org. Chem.*, **63**, 2321.

45 Cozzi, P.G. (2004) *Chem. Soc. Rev.*, **33**, 410.

46 (a) Jacobsen, E.N., Zhang, W., Muci, A.R., Ecker, J.R., and Deng, L. (1991) *J. Am. Chem. Soc.*, **113**, 7063.
(b) Irie, R., Noda, K., Ito, Y., Matsumoto, N., and Katsuki, T. (1991) *Tetrahedron Asymmetry*, **32**, 1055.

47 Sasaki, H., Irie, R., Hamada, T., Suzuki, K., and Katsuki, T. (1994) *Tetrahedron*, **50**, 11827.

48 Katsuki, T. (2004) *Chem. Soc. Rev.*, **33**, 437.

49 Kurahashi, T. and Fujii, H. (2008) *Inorg. Chem.*, **47**, 7556.

50 Collman, J.P., Zeng, L., and Brauman, J.I. (2004) *Inorg. Chem.*, **43**, 2672.

51 Daly, A.M., Renehan, M.F., and Gilheany, D.G. (2001) *Org. Lett.*, **3**, 663.

52 (a) Uchida, T., Irie, R., and Katsuki, T. (1999) *Synlett*, 1163.
(b) Uchida, T., Irie, R., and Katsuki, T. (1999) *Synlett*, 1793.

53 Bachmann, S., Furler, M., and Mezzetti, A. (2001) *Organometallics*, **20**, 2102.

54 Hansen, K.B., Leighton, J., and Jacobsen, E.N. (1996) *J. Am. Chem. Soc.*, **118**, 10924.

55 Sammis, G.M. and Jacobsen, E.N. (2003) *J. Am. Chem. Soc.*, **125**, 4442.

56 Sammis, G.M., Danjo, H., and Jacobsen, E.N. (2004) *J. Am. Chem. Soc.*, **126**, 9928.

57 Matsunaga, S. and Shibasaki, M. (2008) *Bull. Chem. Soc. Jpn.*, **81**, 60.

58 Kumagai, N., Matsunaga, S., Kinoshita, T., Harada, S., Okada, S., Sakamoto, S., Yamaguchi, K., and Shibasaki, M. (2003) *J. Am. Chem. Soc.*, **125**, 2169.

59 Handa, S., Nagawa, K., Sohtome, Y., Matsunaga, S., and Shibasaki, M. (2008) *Angew. Chem. Int. Ed.*, **47**, 3230.

7
Selective Catalysts for Petrochemical Industry
Shape Selectivity in Microporous Materials

Stian Svelle and Morten Bjørgen

7.1
Overview of Petrochemical Industry and Refinery Processes

The objective of this brief introductory section is to give a very short and simple overview of the refinery and petrochemical industries with respect to raw materials, intermediates, and products. Our goal is merely to allow the reader to recognize the context of the reactions relying on catalyst shape selectivity discussed in later sections.

7.1.1
Primary Raw Materials for the Petrochemical Industry

The world depends heavily on fossil hydrocarbon resources for energy and as raw material for petrochemical products such as polymers, nitrogen fertilizers, and fine chemicals. By far, the major part of these resources is utilized as transportation fuels and for power generation. More than 80% of the worldwide energy consumption is derived from coal, oil, and natural gas [1, 2]. Of these three, oil is the largest contributor. As a consequence of the high level of integration between energy production and petrochemistry, oil and natural gas are also crucial for the production of a large variety of petrochemicals, and it is therefore of interest to consider the future perspective with respect to these raw materials. Most estimates suggest that the overall reserves of fossil fuels will last from 200 to 300 years [2]. However, the increasing energy demand, and thus also the petrochemical industry, can only be based so heavily on liquid petroleum for a few more decades (proven reserves of oil will last 40 years at present rate of consumption). Natural gas is more abundant, and the reserves will outlast those of oil (70 years at current consumption) [2]. The reserves of coal are even larger, representing about 170 years of supply at the current rate of consumption [2]. In addition, there are large reserves of unconventional oil sources such as tar sands and oil shale [2]. Also, biomass, which in the broadest sense encompasses all biological material that can be used as fuel or for chemicals otherwise produced from petroleum sources, is a renewable energy source with increasing potential in the energy

Selective Nanocatalysts and Nanoscience, First Edition. Edited by Adriano Zecchina, Silvia Bordiga, Elena Groppo.

market worldwide. Its main advantage is that biomass can be converted to liquid, solid, and gaseous fuels (besides heat and power) [3]. This has led to the biorefinery concept, which is analogous to today's petroleum refineries that produce multiple fuels and products from petroleum. In summary, the outline of the petrochemical industry and the catalytic reactions involved and presented in this chapter will most likely not be dramatically altered over the next decades, despite the increased focus on nonhydrocarbon energy sources in our society. However, a shift toward more environmentally friendly hydrocarbon-based processes will inevitably continue to occur, and this will be inherently linked to the development of improved catalysts based on fundamental chemical insight.

7.1.2
Processing of Petroleum and Natural Gas

The petrochemical industry is a huge field that encompasses vast amounts of chemicals, which are vital for our present day standard of living. Among the petrochemical products and derivatives thereof, we find plastics, fertilizers, paints, insecticides, pharmaceuticals, and so on. Most of these diverse products are derived from three types of intermediates (numbers in parentheses are European production in 2009 in kilotons) [4]:

- Lower alkenes: ethene (20 000), propene, (15 000), and butadiene (2000).
- Aromatics (BTX): benzene (5100), toluene (1600), and xylenes (2600).
- Synthesis gas: a mix of CO and H_2 of varying composition; syngas.

Currently, liquid petroleum and natural gas constitute the dominant raw materials of these three groups of intermediates for the petrochemical industry.

The petrochemical plant in many cases is highly integrated with the refinery, and to distinguish between the two is not always useful from the chemist's perspective. Moreover, many of the processes occurring in the refinery, for instance, catalytic cracking and hydrocracking (see Section 7.4.2), are highly relevant to those seeking insight into shape-selective catalysis by microporous materials. Briefly, crude petroleum is processed predominantly into fuels in the refinery, but also various hydrocarbon intermediates (Figure 7.1). Key petrochemical intermediates such as alkenes and aromatics are produced in cracking and catalytic reforming processes, respectively. Of course, several other processes are also employed for the production of these intermediates such as thermal steam cracking of LPG or naphtha for the production of light alkenes.

Utilization of natural gas (mainly composed of methane) for petrochemical production usually proceeds by conversion into syngas, most often in the catalytic steam-reforming reaction. Syngas may also be prepared from coal, which is the historical source, or from biomass as a potential future raw material (see Figure 7.2). Syngas is then converted mainly into ammonia and methanol. Also, the conversion of syngas into diesel oil in the Fischer–Tropsch process is commercially feasible and expected to be of increasing importance. Ammonia is used for nitrogen fertilizers such as urea and ammonium salts. Other applications are nitric acid, amines, nitriles, and other nitrogen-containing compounds for

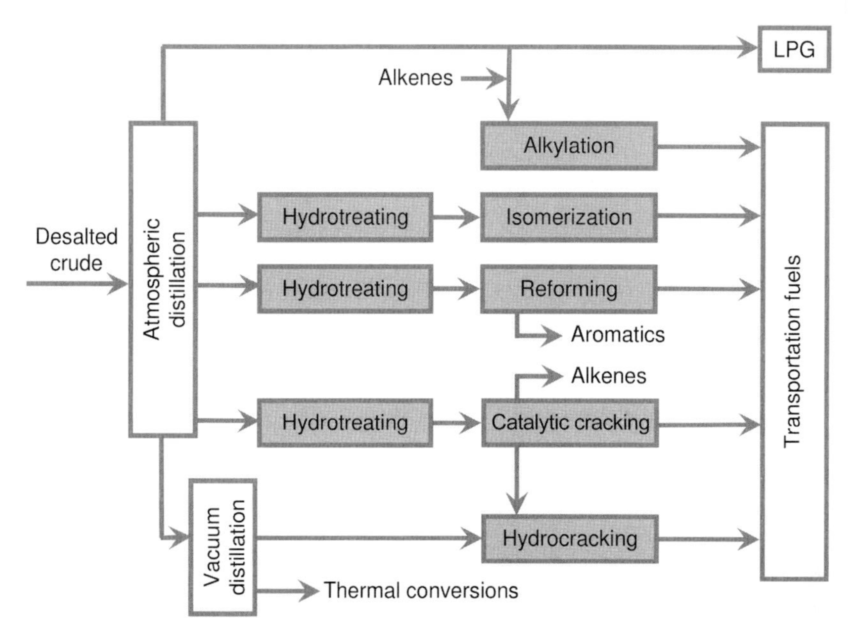

Figure 7.1 Simplified diagram of catalytic processes (gray boxes) in an oil refinery. Thermal/separation processes and straight run production are not included [5,6].

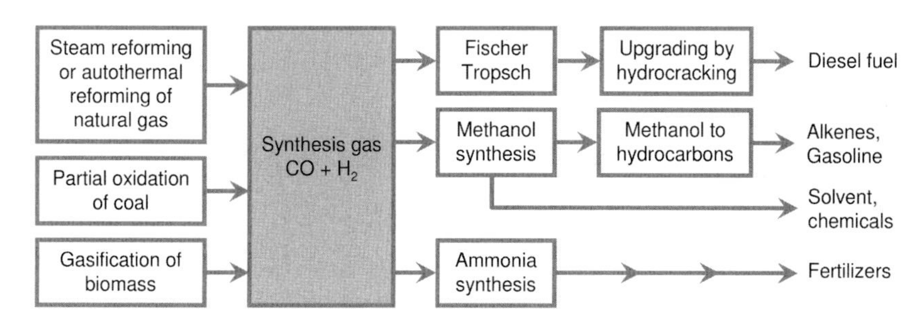

Figure 7.2 Box diagram for the production of some petrochemical products from synthesis gas. Presently, natural gas is the main source of synthesis gas, but coal and biomass are likely to become important raw materials in the future. Such a shift in raw material will, however, mostly affect the processes represented by the boxes to the left in the diagram, whereas existing technology may be employed with moderate modification for the further conversion of the synthesis gas.

fine chemicals and pharmaceuticals. Methanol is a convenient solvent, but it is also extensively used in making other chemicals [5]: It is the feedstock for petrochemical processes such as the manufacture of formaldehyde, acetic acid, higher alcohols, ethers, and so on. An application of increasing commercial interest, due to the current high price of oil relative to gas (and coal), is the conversion of methanol into ethene and propene or gasoline range hydrocarbons (see Section 7.4.1).

7.2 Catalysis in the Petrochemical Industry

In this section, we will emphasize the indispensability of industrial catalysis, which cannot be overestimated. For example, support of the present world population would be impossible without synthetic fertilizers produced in catalytic reactions, and the correlation between ammonia production and world population is clear. Moreover, we will attempt to point out the challenges and importance of further catalyst improvements, in particular with respect to the selectivity of the catalytic reactions.

7.2.1 The Importance of Catalysis

Both living organisms and most industrial chemical processes rely heavily on catalysis and could not have reached current levels of sophistication without its implementation, either as enzymes in living organisms or as homogeneous or, in particular, heterogeneous catalysts found in industry. A catalyst may be defined as *a substance that increases the rate at which a chemical system approaches equilibrium without being stoichiometrically consumed in the process*. Thus, the presence of a catalyst allows a reaction to proceed more efficiently or under milder conditions than would otherwise be possible. It is often (and correctly) emphasized that a catalyst will affect only the rate at which chemical equilibrium is approached and not the overall thermodynamics and the equilibrium concentrations. The catalyst will therefore not allow the formation of thermodynamically unfavorable products, and the maximum yields will be thermodynamically limited. However, in many cases, in particular within heterogeneous catalysis, the presence of the catalyst allows the formation of products that would otherwise not be formed altogether. Hence, comparison of the two situations, that is reaction with or without catalyst, is quite often of little interest.

It can be estimated that 85–90% of the products in the overall chemical industry are manufactured in catalytic processes [6]. Of the top 10 organic chemicals manufactured in Europe (by mass), all are produced in catalytic, petrochemical processes, and shape selectivity is relevant for several of these [4]. In addition to their use in the chemical industry and fuel production in refineries, catalysts are essential for flue gas cleaning. A prominent example is the catalytic reduction of nitrogen oxides in automotive exhaust and industrial off-gas. Two other aspects pertaining to the importance of catalysis become evident if we consider the energy consumption and waste by-production accompanying large-scale industrial processes. As mentioned above, the catalyst may allow the reaction to occur at mild conditions (lower pressures and temperatures), and clearly the more active a catalyst, the milder the reaction conditions. Thus, the energy consumption may be drastically reduced by replacing a stoichiometric process with a catalytic process and further gains may be made by additional catalyst improvement. In the ammonia synthesis, the early coal-based catalytic Haber–Bosch process was about a factor of five less energy demanding than the noncatalytic electric discharge process, whereas modern, steam-reforming-based ammonia synthesis is twice as efficient as Haber–Bosch process [5, 6]. Bearing in mind that the ammonia synthesis accounts for about 1% of the total global energy consumption [6], the economic and environmental impacts of such process improvements become evident. Within the field of green chemistry, the aim is to design processes that reduce by-products, thereby eliminating costly end-of-the-pipe treatments, to avoid using hazardous reactants and to reduce the use of energy and other resources. Clearly, catalytic processes will meet these targets more readily than stoichiometric reactions. Indeed, in 2010, more than 40% of the scientific publications containing the concept "green chemistry" also contained the concept "catalysis"[1].

7.2.2
Catalyst Selectivity

We are now at a position where we can fully appreciate the importance of catalyst selectivity. If there are several possible products of a catalytic reaction, the product selectivity for each product may be superficially defined as the fraction (or percentage) of that product with respect to the total amount of product. Hence, a selective catalyst or selective catalytic process refers to the situation where the desired product is produced in a satisfactory excess with respect to the relevant by-products. It should also be noted that different catalysts may display selectivity toward different products for a given reactant. A classical example is seen for ethanol, which is selectively dehydrogenated into acetaldehyde and hydrogen over Cu metal, dehydrated into ethylene and water over alumina, whereas zeolite H-ZSM-5 displays a high selectivity to a gasoline mixture and water. We note that in the final example, selectivity is considered toward a group of compounds rather than a single species. Another prominent example is the conversion of syngas (Figure 7.3): Over Ni

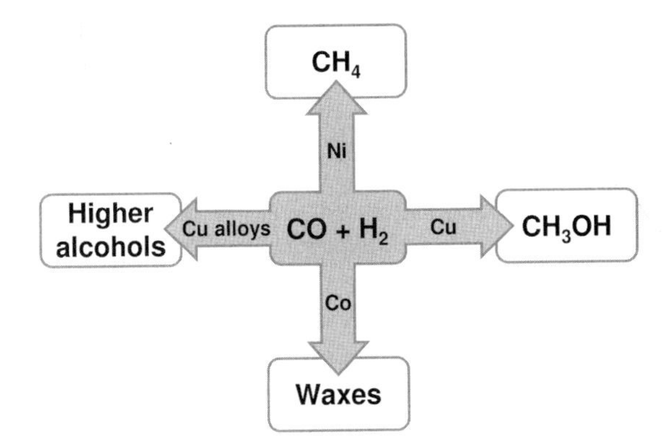

Figure 7.3 Depending on the catalyst, synthesis gas can be selectively converted into a wide variety of products.

1) Search carried out using SciFinder, American Chemical Society, 2010.

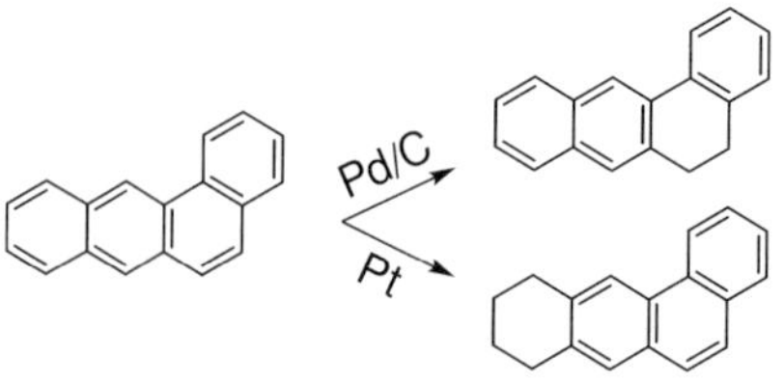

Figure 7.4 Regioselecitivity in the hydrogenation of fused aromatic rings. Depending on the catalyst, either a terminal or an internal ring is selectively hydrogenated [7].

catalysts, methane is formed selectively; over Cu-based materials, methanol is the main product; over Co-based catalysts, heavy waxes are formed; whereas over alloyed Cu catalysts, the selectivity toward higher alcohols is enhanced. Clearly, catalyst selectivity is a high priority issue for green catalysis, and it is a major challenge to improve the selectivity of existing catalysts or to develop new catalysts with superior selectivity characteristics compared to the existing catalysts.

The concept of selectivity can be further subdivided into chemo-, regio-, and stereoselectivity. The two systems discussed above (Figure 7.3) are examples of chemoselectivity: A reactant may give chemically different products depending on the catalyst employed. A reaction is called regioselective if it takes place preferentially at only one of two or more possible reaction sites (functional groups), and a prominent example is found for the hydrogenation of fused aromatic rings, such as benz[a]anthracene (Figure 7.4) [7]: Over Pd/C catalysts, the internal ring is hydrogenated with 97% selectivity, whereas over Pt catalysts, the terminal ring is hydrogenated with a selectivity of 95%. Stereoselectivity is the preferential production of one stereoisomer over another, leading to an enantiomeric excess. Examples from homogeneous organometallic catalysis are abundant, and one example is stereoselective polymerization leading to tactic or syndiotactic polymer chains, due to the possibility of creating chiral active sites by the use of chiral ligands coordinated to the metal. Examples from traditional heterogeneous catalysis are less common, due to the difficulties of controlled synthesis of chiral solids. However, recent progress in the technique of anchoring or grafting organometallic complexes to solid supports, thereby creating a solid catalyst, has led to the development of several stereoselective heterogeneous catalysts [8, 9]. Also, the rapidly expanding class of metal organic framework compounds (MOFs), which are synthesized at comparatively mild conditions, has resulted in examples of heterogeneous stereoselective catalysis [10–14].

Thus far, we have been concerned with the classification of the various types of selectivity, that is, chemo-, regio-, and stereoselectivity and not considered the catalyst properties that give rise to selectivity. Clearly, this is both a complex and essential issue for understanding and developing improved catalysts. The local, atomic scale environment around the active site, in particular electronically and geometrically, is decisive to obtain any type of selectivity [15], although it is usually not easy to completely disassociate electronic and geometric effects (see Figure 7.5) [15–17]. It is now straightforward to realize that surface site heteroge-

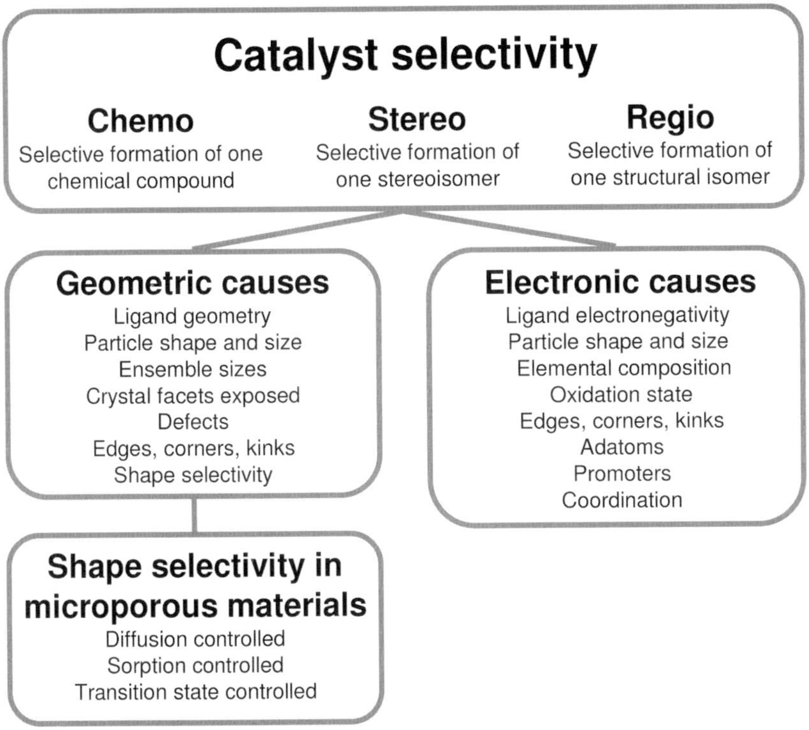

Catalyst selectivity

Chemo
Selective formation of one chemical compound

Stereo
Selective formation of one stereoisomer

Regio
Selective formation of one structural isomer

Geometric causes
Ligand geometry
Particle shape and size
Ensemble sizes
Crystal facets exposed
Defects
Edges, corners, kinks
Shape selectivity

Electronic causes
Ligand electronegativity
Particle shape and size
Elemental composition
Oxidation state
Edges, corners, kinks
Adatoms
Promoters
Coordination

Shape selectivity in microporous materials
Diffusion controlled
Sorption controlled
Transition state controlled

Figure 7.5 The three types of product selectivity effected by a catalyst (top box) may as a general simplification be attributed to either geometric or electronic causes (which sometimes cannot be meaningfully distinguished). Shape selectivity is a prominent geometric effect found for microporous catalysts that may give rise to product selectivity. Shape selectivity may be associated with diffusion, adsorption properties, or steric constraints imposed on transition states or key reaction intermediates.

neity should be avoided. Within the field of heterogeneous catalysis, this idea has led to the concept of single-site heterogeneous catalysts [18–21], whereas in the field of homogeneous catalysis, this is more readily achieved due to the unambiguous nature of the catalytic organometallic molecular complex. The electronic environment for heterogeneous catalysts is affected by factors such as the nature of the metal site, for example, element, oxidation state (or degree of oxidation/reduction), exposed crystal facets, crystal irregularities (leading to edges, kinks, and corners), alloying, the presence of electronic promoters, and so on. Geometric effects are evident in homogeneous and enzymatic catalysis, where the exclusion and stabilization of the species involved in reaction on the active sites are manipulated by the choice of ligands and the folding of the macromolecular structure. The most obvious geometric cause of selectivity for heterogeneous catalysts is *shape selectivity*, which is affected by the presence of regular pores, channels, and cavities of molecular dimensions that encapsulate the active sites. This sterically restricted environment surrounding the active sites will determine which species are allowed to partake and be formed in a reaction. The term shape selectivity is most often reserved for crystalline, so-called molecular sieve materials that possess microporosity (pore dimensions up to 2 nm), and we shall explore these materials and their selective properties in Section 7.3.

7.3
Microporous Materials and Shape Selectivity

Porosity, which can be defined as the ratio of the total pore volume to the apparent volume of the material particles, will lead to increases in the exposed and accessible surface area of a material, by orders of magnitude. As heterogeneous catalysis is intrinsically a surface phenomenon, porosity is a key property of almost all catalytic materials. An alternative way to create a large surface area is to synthesize materials composed of nanoparticles, which will lead to substantial external surfaces. However, such surfaces do not in general give rise to shape selectivity, which, as we shall see, is facilitated by the internal surface in porous materials. It is customary to classify porous materials according to pore width: Materials with pores with widths not exceeding 2 nm are called microporous; those with widths exceeding 50 nm are called macroporous, whereas materials with pores of intermediate size (2–50 nm) are called mesoporous. Among the microporous materials, we find zeolites and zeotypes, activated carbons, various amorphous oxides (such as silica and titania), modified clays, MOFs, and so on. Illustrations of two non-zeolite microporous materials are given in Figure 7.6. Microporous materials find widespread application as adsorbents, in ion-exchange, for separation pur-

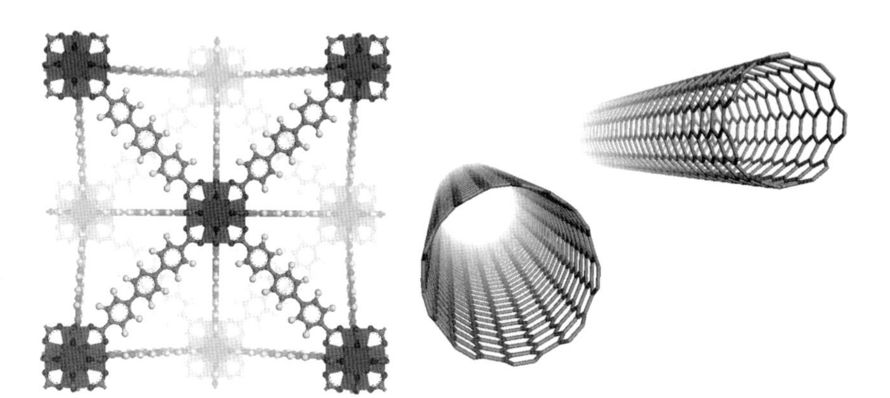

Figure 7.6 Examples of non-zeolite microporous materials. Left panel: A metal organic framework compound (UiO-68 composed of Zr_6-based cornerstones and with terphenyl dicarboxylate as linker, creating a porous framework with accessible pore diameters ~1 nm and a theoretical BET surface area of 4170 m^2/g). Right panel: Microporous carbon nanotubes with accessible inner diameter of between 0.6 and 0.87 nm.

poses, and in catalysis. Among the microporous solids, zeolites (and zeotypes) have a pivotal role in the catalytic processing of petrochemicals and refining of oil [22, 23].

7.3.1 Zeolites and Zeotypes

The history of zeolites began in 1756 when the Swedish mineralogist Cronstedt discovered a suite of well-formed crystals in northern Sweden [24]. Nearly 200 years after this initial discovery, it was noted that dehydrated zeolite crystals would adsorb small organic molecules and reject larger ones [25]. This phenomenon was described as "molecular sieving" [26], and this concept is inherently linked to shape-selective catalysis.

In the following section, we will distinguish between zeolites, defined as aluminosilicates, and zeotypes, which are materials of similar construction (in many cases identical) that contain other elements in tetrahedral framework positions. Thus, zeolites are a class of materials built up from aluminum, silicon, and oxygen. They also contain charge-balancing cations and adsorbed water. Structurally, zeolites may be viewed as crystalline materials based on a three-dimensional network of TO_4 tetrahedra, where T is Si or Al, connected by sharing oxygen atoms at each tetrahedral corner. As illustrated in Figure 7.7, these primary structural units form larger secondary building units that are combined to form three-dimensional framework structures. Zeolites are traditionally classified as 8-, 10-, or 12-ring structures, based on the number of tetrahedral atoms (T-atoms) comprising the circumference of the pores. Figure 7.8 displays the framework and channel structure of zeolites beta (a), mordenite (b), ZSM-5 (c), and ZSM-11 (d), which will be considered in the following sections. Moreover, there is a substantial effort directed toward the synthesis of novel, extra-large pore zeolites, that is, materials with pores larger than 12-rings [27], such as ITQ-33, which comprises 18-ring pores [28–30]. Typical pore dimensions are 0.3–0.45 nm for small-pore zeolites, 0.45–0.60 nm for medium-pore zeolites, and up to 0.8 nm for large-pore zeolites [31].

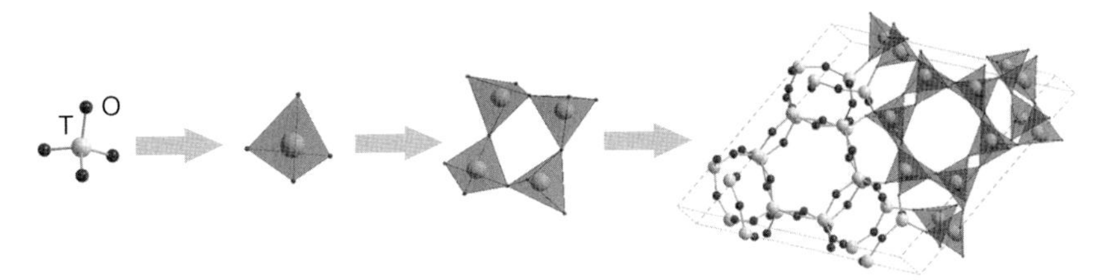

Figure 7.7 Zeolites are built up from corner sharing TO_4 tetrahedra, where T is Si or Al, thus forming three-dimensional lattices with internal porosity. For zeotypes, T may be a wide variety of elements. The example shown is the chabazite structure, which is isostructural with the SAPO-34 zeotype catalyst employed in the conversion of methanol to alkenes.

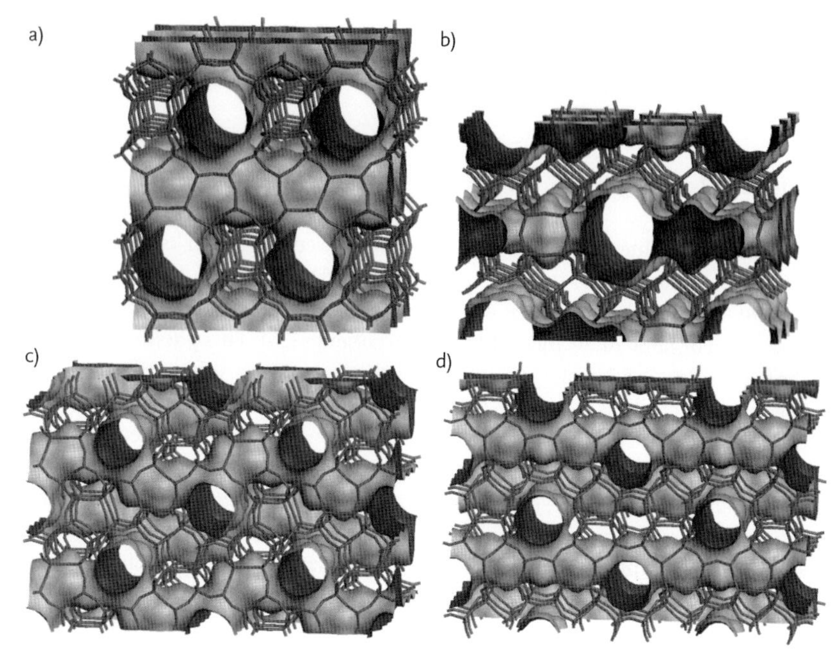

Figure 7.8 Zeolite pore structures. (a) Zeolite beta has straight and zig-zag 12-ring channels. (b) Mordenite has straight 12-ring channels intersected by highly tortuous 8-ring channels leading to the formation of side pockets. (c) Zeolite ZSM-5 has straight and zig-zag 10-ring channels with identical cavities at each intersection. (d) Zeolite ZSM-11 has straight 10-ring channels with two distinct types of cavities at the intersections; one is larger than that in ZSM-5, the other is comparable to the ZSM-5 intersections. Beta, ZSM-5, and ZSM-11 have intersecting channels leading to a material in which diffusivity may occur in all three directions, whereas mordenite is often considered to have a one-dimensional pore system.

7.3.2
Catalytic Sites in Zeolites and Zeotypes

A zeolite built up exclusively from tetravalent Si and O would constitute an overall electrically neutral lattice with chemical formula SiO_2. Each substitution of Si(IV) with trivalent Al(III) will therefore result in one negative electric charge in the framework, and these negative charges must be balanced by cations located within the pores of the zeolite in order to maintain overall electric neutrality. The charge-compensating cations may in principle be any cation, and if they are protons, the zeolite becomes a solid Brønsted acid, giving rise to the use of zeolites as acidic catalysts (see Figure 7.9). The acidic strength of the sites can be varied considerably and this may result in substantially different catalytic properties: Other trivalent cations from elements such as Ga, B, and Fe may be incorporated as T-atoms and this typically gives rise to lower acidic strength. In addition, AlPO materials, built up from strictly alternating P(V) and Al(III) T-atoms, will

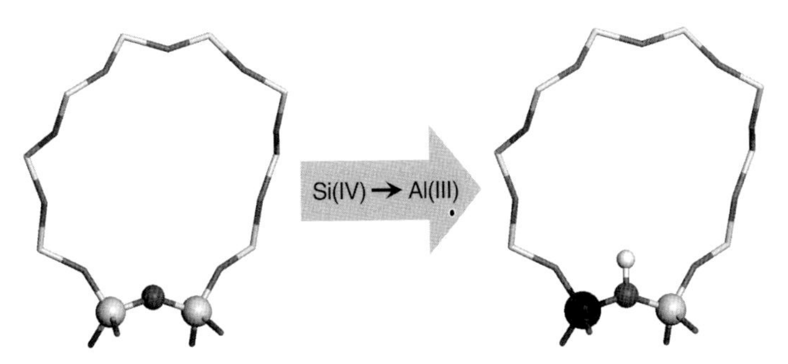

Figure 7.9 A Brønsted acidic site in a zeolite is formed when one Si^{IV} (light gray) is replaced by an Al^{III} (black) and the resulting charge deficit in the zeolite framework is counterbalanced by a proton to ensure overall electrical neutrality.

upon substitution of P(V) by Si(IV) give an acidic SAPO zeotype with a slightly lower acidic strength than the isostructural zeolite, as has been demonstrated for the isostructural SAPO-34 and H-SSZ-13 materials [32–34]. Finally, the presence of nonframework species, such as positively charged Al–O moieties of less-defined nature, in the micropores has been associated with sites of Lewis acidic character [35].

Zeolites with basic active sites are obtained by employing alkaline or earth alkaline cations as charge-compensating species [31]. Impregnation methods have also led to stronger basic sites caused by the formation of alkaline- or alkaline oxide clusters in the micropores.

In addition to acidic and basic catalysis, transition metal-containing zeolites display a rich redox chemistry of industrial significance. The perhaps most prominent example is the titanosilicate TS-1, which is used as catalyst in epoxidation reactions [36, 37]. This possibility of isomorphous substitution of T-atoms in both zeolites and zeotypes opens up for a vast number of microporous catalysts with potential redox catalytic application. In addition to having metals in framework positions, zeolites and zeotypes may be loaded with transition metals by ion-exchange or impregnation. In the latter case, the function of the microporous material may be described more as a traditional (active) support. Prominent examples are Cu-loaded ZSM-5 and in particular beta, which are promising systems for de-NO_x applications [38]. Ga- and Mo-containing zeolites display activity in the aromatization of alkanes (Cyclar process) and are also widely investigated in the aromatization of methane mixed with higher alkenes and aromatics [39–41]. Finally, Pt-containing mordenite is currently employed in commercial hydroconversion applications [42]. It should be emphasized, however, that in many cases, the distinction between zeolite and zeotype catalysts with metals in framework position and as extra-framework species is gradual, as elements originally present as labile T-atoms after synthesis tend to migrate from these positions and become extra-framework material upon pretreatment, reaction, or regeneration.

7.3.3
Zeolites in Petrochemistry and Refining

Zeolites and zeotypes may be fine-tuned to possess several unique properties that constitute the basis for their widespread applications:

- a wealth of crystalline structures, which give rise to a variety of pore systems of molecular dimensions with sharply defined shape and size;
- large adsorption capacities and surface areas, typically several hundred meter square per gram BET areas;
- Brønsted acidic properties;
- basic properties;
- variable framework elemental composition by isomorphous substitution, giving rise to redox activity;
- mobile, exchangeable cations;
- tunable electrostatic properties, by adjusting framework elemental composition and cation content;
- substantial thermal stability, normally above 600 °C;
- metals deposited within the voids, giving rise to bifunctionality; and
- shape-selective properties (see Section 7.3.4).

It is these properties that form the basis for the use of zeolites as catalysts in petrochemistry and refining. It is also noteworthy, but beyond the scope of this chapter, that zeolites are employed as adsorbents and in separation processes. The sharply defined size of the pore apertures makes zeolites particularly suited for applications such as the separation of normal- and isoparaffins [42, 43]. The use of zeolite A as a desiccant, or selective water remover, is familiar to most chemists.

The first use of zeolites as catalysts occurred in 1959 when zeolite Y was used as an isomerization catalyst by Union Carbide [42]. In 1962, Mobil Oil introduced the use of synthetic zeolite X as a cracking catalyst [44]. Presently, the major employment of zeolites within catalytic chemistry is as acid-cracking catalysts. It can be estimated that more than 30% of the world's gasoline production from crude oil involves the use of zeolite catalysts [45], and Faujasite-type zeolites (X and Y) account for at least 95% of the zeolite catalysis market, by volume [44]. An extensive list of commercialized zeolite-facilitated processes based on shape-selective catalysis has been compiled by Degnan [46].

7.3.4
Zeolites as Shape-Selective Catalysts

Previously, we have qualitatively discussed selectivity and shape selectivity as an underlying cause of selectivity in microporous catalysts. Let us now consider selectivity in narrow pores in a more quantitative manner. For pores of near molecular dimensions, it has been emphasized that even very minor differences in molecular size and shape may have dramatic impact on the diffusivity. Diffu-

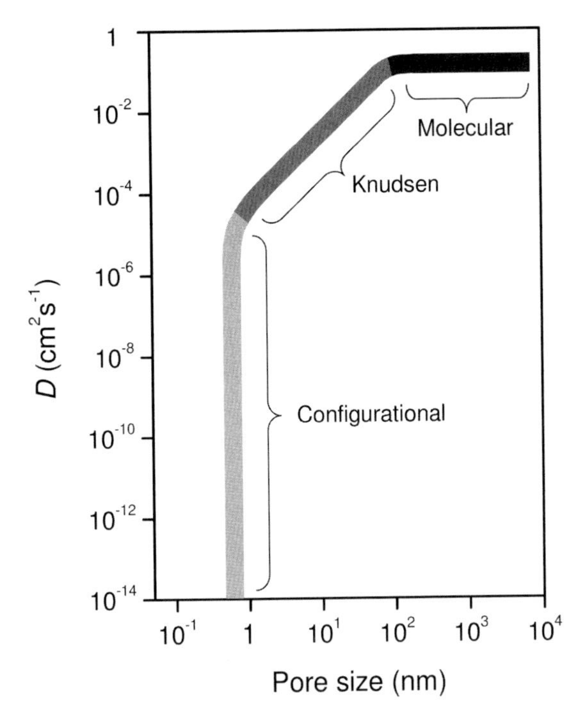

Figure 7.10 The dependency of the diffusivity on the size of the pore in which diffusion takes place. When the pores become very narrow, as is the case in microporous zeolites, the diffusivity will be extremely sensitive to the pore diameter, or vice versa, the kinetic diameter of the diffusing molecule.

sion in such pores, which is even more restricted than Knudsen diffusion (when collisions with the walls are more frequent than intermolecular collisions), proceeds via so-called configurational diffusion and may be associated with substantial energy barriers. This is illustrated in the often encountered and very informative Figure 7.10. According to Wheeler [47], it is convenient to distinguish between three kinetically different types of selectivity, as these are affected differently by the pore structure.

Type I selectivity pertains to the situation where we wish to consider the ability of the catalyst to selectively convert one compound in a mixture of compounds, as defined by Eq. (7.1a) and (b) below:

$$R_1 \xrightarrow{k_1} P_1 \tag{7.1a}$$

$$R_2 \xrightarrow{k_2} P_2 \tag{7.1b}$$

For two first-order reactions, the selectivity of reaction (7.1a), which we may consider to be the desired reaction, is defined as the ratio between the rate constants; $S = k_1/k_2$. When the reactions take place in narrow pores, it is important to realize that the fastest reaction (7.1a) will be slowed down by mass transport limitations more severely than the slow reaction (7.1b). This is because R_1 is converted at the highest rate and will be consumed before reaching the innermost internal surface, meaning that a smaller fraction of the internal surface (and active sites) will be available to the faster reaction. Thus, narrow pores will reduce Type I selectivity if R_1 and R_2 are affected similarly by the pore structure, that is, if they have the same kinetic diameter. If R_1 and R_2 have different kinetic diameters, the effects of Type I selectivity will be even more severe, as we will see below.

Type II selectivity involves the discrimination of parallel reactions in which a single reactant is converted into two products via separate paths (Eq. (7.2a) and (b))

$$R \xrightarrow{k_1} P_1 \tag{7.2a}$$

$$R \xrightarrow{k_2} P_2 \tag{7.2b}$$

A selective catalyst will favor one reaction over the other, and the selectivity is defined as $S = k_1/(k_1 + k_2)$ for first-order reactions. Given that the reactions are of the same order, the selectivity will be unaffected when we impose diffusion limitations on the reactant in narrow pores. If the reactions are of different order, the selectivity will depend on pore size and diffusion limitations, as the decrease in partial pressure of R will affect the higher order reaction more severely.

Type III selectivity is found with consecutive reactions, where the desired product is an intermediate, which may undergo further reactions into undesired products (Eq. (7.3)).

$$R \xrightarrow{k_1} P_1 \xrightarrow{k_2} P_2 \tag{7.3}$$

High selectivity toward P_1 is achieved when k_1 is much larger than k_2. Type III selectivity is easily distinguished from Type II selectivity since in the latter case, the selectivity is independent of conversion, whereas the selectivity will decline with increasing conversion for Type III reactions. It can be rigorously shown that diffusion limitations imposed by narrow pores will lead to a reduced selectivity. This is also the intuitive result, as diffusion limitations force P_1 to linger for a longer time within the microporous crystal, thereby increasing probability of conversion into P_2.

Shape selectivity may be defined as the preferred production of one compound (or a group of compounds) due to spatial influence imposed by the catalyst on the species involved in the reaction. Shape selectivity may be classified into several different categories. Three categories of shape selectivity were identified relatively early and have become well established [48, 49], whereas several other much more material-specific concepts regarding shape selectivity have originated more recently and are less established. Haag distinguishes between three underlying causes leading to shape selectivity: diffusion controlled, sorption controlled, and transition state controlled [50] and we will adopt this in the following text.

Reactant shape selectivity occurs when the pore openings of a zeolite are such that certain reactants are excluded, whereas others are allowed to diffuse into the catalyst crystal. A classical example is the dehydration of *iso*-butanol and *n*-butanol.

Over Ca-exchanged zeolite A (5A), which has pore openings of approximately 0.49 nm [32], only the unbranched alcohol is dehydrated, as the branched *iso*-butanol is too large to enter the pore system. In contrast, over Ca-exchanged zeolite X (10X), the substantially larger pores allow rapid dehydration of both butanol isomers [51, 52]. Reactant shape selectivity is a case of Type I selectivity, as discussed above, and the underlying cause of the effect can unequivocally be ascribed to differences in diffusivity.

Product shape selectivity refers to the situation where several products are formed inside fairly large voids or cages of the microporous catalyst, but only those smaller than a certain size are permitted to leave the catalyst through small-pore apertures. The larger product molecules are trapped inside the zeolite cages unless further reactions occur, resulting in the formation of smaller molecules. A prominent example is the conversion of methanol to hydrocarbons (MTH) over acidic zeolites/zeotypes: Over the acidic H-ZSM-5 catalyst (0.51×0.55 nm straight pores intersected by 0.53×0.56 nm sinusoidal pores, see Figure 7.8), molecules as large as 1,2,4,5-tetramethylbenzene are detected in the gas phase. The acidic H-SAPO-34 zeotype (with cages (about 0.7×1.0 nm) connected by small windows (0.38×0.38 nm)) displays even more pronounced product shape selectivity; ethene and propene are the main products and among the butenes and pentenes, only linear isomers are allowed to escape the porous structure, even though much larger molecules, such as methyl benzenes, reside within the cages (see Figure 7.11). Product shape selectivity is a case of Type II selectivity, and, as was the case for reactant shape selectivity, the selectivity is caused by varying diffusivities.

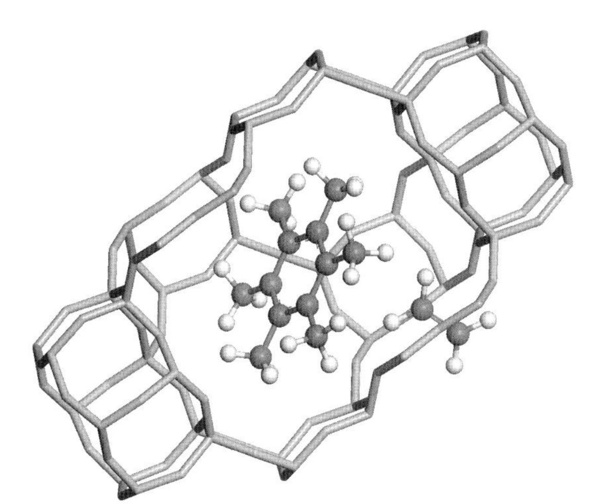

Figure 7.11 Hexamethylbenzene, which is an intermediate in the hydrocarbon pool mechanism, is locked inside the SAPO-34 cages during the conversion of methanol to hydrocarbons. Only smaller molecules, such as ethene (shown) and propene, may diffuse through the narrow 8-ring windows and enter the gas phase. Some framework atoms have been removed for clarity.

Transition state shape selectivity takes place when the zeolite topology is such that there are spatial (steric) restrictions prohibiting the progress of reaction via a bulky transition state, whereas other reactions, proceeding via geometrically smaller transition states, are allowed. Transition state shape selectivity applies not only to the actual transition state but also to key reaction intermediates. The concept was coined by Csicsery and observed in the transalkylation of alkylbenzenes over H-mordenite; the rate of formation of 1,3,5-substituted trialkylbenzenes was found to be surprisingly low, and it was concluded that the reaction was inhibited because there was not enough space to form the large 1,1-diphenylalkane-type intermediate [53]. Transition state shape selectivity is conceptually more complex than reactant and product shape selectivity and may be harder to assess than the two other types, but it is helpful to realize that, in contrast to product shape selectivity, transition state shape selectivity will not depend on crystal size, meaning that the effect is independent of diffusion pathway length.

Molecular traffic control is believed to occur when the microporous material has two different, but interconnected pore systems of different geometry, such as pore diameter, tortuosity, or pore shape (the presence of side pockets, etc.). Reactants may reach the active sites via one channel type, and the products leave via the other. Molecular traffic control was originally proposed by Derouane and Gabelica [54] to account for the lack of rate-limiting counterdiffusion effects in the MTH reaction over H-ZSM-5. They proposed that the hydrocarbon products diffuse out through the straight channels and that the reactants enter through the sinusoidal channels. Methanol would then be transformed into products in the relatively spacious channel intersections. Clark *et al.* [55] employed molecular dynamics to investigate the phenomenon and formulated four criteria that must be met for the effect to occur: (i) In the narrow pore, the small molecule has a higher diffusivity than the large molecule, (ii) in the wide pore, the large molecule has a higher diffusivity than the small molecule, (iii) the large molecule has a higher diffusivity in the wide pore than in the narrow pore, and (iv) the small molecule has a higher diffusivity in the narrow pore than in the wide pore. Criterion (ii) seems counterintuitive, but the large molecule may in fact diffuse faster than the small molecule in the wide pores when the fit with the pore is optimal and the molecule becomes supermobile or “floating” [56]. Again, differences in diffusivities are the underlying cause of the phenomenon.

Inverse shape selectivity was originally described by Santilli *et al.* [57]. They observed high yields of mono- and dibranched C_6 isomers in hydrocracking of *n*-C_{16} over the isostructural H-SSZ-24 and SAPO-5 (same framework type, unidirectional circular pores with diameter 0.73 nm) catalysts. This was illuminated using computational techniques and adsorption measurements at low pressures, and it was concluded that adsorption and thus also the formation of branched C_6 isomers was particularly favorable due to snug interaction with the pore walls for materials with channels in the range 0.65–0.74 nm. Clearly, this is related to the idea of “floating” molecules as proposed by Derouane in his explanation of molecular traffic control [56]. Baron and coworkers [58] employed tracer chromatography for adsorption of mixtures of linear and branched C_4–C_8 alkanes on SAPO-5, and it was found that at low loading, specific branched molecules are adsorbed preferentially over the corresponding linear isomers. In agreement with Santilli *et al.*, Baron and coworkers found that at low loading the interaction with the branched alkanes with the pore wall results in a higher adsorption enthalpy compared to the linear chains, which outweighs the negative contribution of the adsorption entropy to the adsorption-free energy. Smit and Maesen and coworkers have simulated the adsorption of branched and linear C_6 alkanes as a function of pore diameter at high pressure, which is more similar to the experimental conditions of the hydrocracking reaction [59, 60]. A better stacking efficiency was found for the branched isomers relative to linear isomers in pores of size 0.70–0.75 nm. This entropic effect occurs only at high loadings. These results have challenged the early ideas behind the concept of inverse shape selectivity and led to a more refined understanding of shape selectivity caused mainly by sorption phenomena [61].

The external surface of microporous catalysts, or sites located in close proximity of the external surface, may also affect the overall selectivity of a given reaction. This has given rise to three related categories of shape selectivity. These are pore mouth shape selectivity, key lock shape selectivity, and the "nest" effect. Pore mouth shape selectivity has been suggested to account for the preferential isomerization of *n*-alkanes into 2-methyl-branched alkanes over Pt/H-ZSM-22 (0.46×0.57 nm straight channels) [62, 63]. This is believed to proceed by adsorption of only parts of the alkane and that the isomerization reaction occurs in immediate proximity of the crystallite surface. Key lock shape selectivity addresses the same reaction over the same catalyst, but is used to explain the tendency toward cracking or chain isomerization in the central part of the long-chain alkanes [64, 65]. The two ends of an alkane molecule may partly penetrate into two different pores, or one end along the terminating crystal surface, thus favoring reactivity around the center of the molecule. The "nest" effect is derived mostly from investigations of the alkylation of benzene with ethene or propene over H-MCM-22 (cages with pockets (about 0.71×1.8 nm) connected by windows (0.41×0.51 nm and 0.45×0.55 nm). Termination of the H-MCM-22 crystals leads to exposure of pockets or "nests" (half a cage) that cover the surface of the crystals and do not lead into the interior of the crystal [66]. The reaction is then believed to occur in these pockets (nests) without diffusion limitations, but under steric control, leading to suppressed alkene oligomerization.

In addition, there are several more controversial effects related to the diffusivity and adsorption of relatively long-chain alkanes that have bearing on our understanding of shape selectivity. The most prominent example is the so-called window-effect, which was proposed by Chen *et al.* [67]. A gap (window) in the C_5–C_8 range in the product distribution in the hydrocracking over erionite zeolites was found. This has later been proposed to be caused by unfavorable diffusivity of molecules curled up in cages without penetrating out of the windows connecting these cages. The effect has been predicted to appear in microporous structures where larger cages are connected by small-pore openings (~0.4 nm) [68, 69]. This and other effects such as single-file diffusion, incommensurate diffusion, molecular path control, and levitation effects have recently been reviewed by Dubbeldam and Snurr [70].

7.4 Selected Examples of Shape-Selective Catalysis by Zeolites/Zeotypes

This section will review three recent studies that have led to substantial new insight into the role of shape selectivity in catalytic reactions relevant to the petrochemical industry. For researchers in the field of catalysis, quantum chemistry and molecular mechanics continue to become more powerful and available even to nonexperts. It is therefore our intention to emphasize the potential of interplay between experiment and rapidly evolving computational methods.

7.4.1 Industrial Relevance of the Conversion of Methanol to Hydrocarbons

Methanol can be produced via syngas (see Figure 7.2), and may subsequently be converted into alkenes or gasoline over acidic zeolites/zeotypes. Such processes are generically known as the MTH reaction. The MTH reaction was discovered by Chang and coworkers for the H-ZSM-5 catalyst, which selectively yields an aromatics rich hydrocarbon mixture terminated at C_{10} when operated at long contact times [71, 72]. The methanol-to-gasoline reaction was commercialized in New Zealand, where gasoline production (600 000 ton/year) was started in 1986, but was later shutdown due to drops in the price of oil relative to that of methanol [73]. The Topsøe integrated gasoline synthesis process, also based on H-ZSM-5, was demonstrated on pilot scale in the mid-1980s [74]. This process was, however, never further developed due to the situation in the global energy market. Later on, more attention was paid to the methanol-to-olefins reaction for the production of polymer grade ethene and propene from methanol. The SAPO-34-based UOP/Norsk Hydro technology was proven in a demo plant in the 1990s [75]. Lurgi developed a process aiming at converting methanol into propene over H-ZSM-5 and the process was demonstrated in a joint Lurgi–Statoil demo plant. Currently, intense commercialization efforts for MTH technologies are ongoing [76].

7.4.2 Shape Selectivity in the Conversion of Methanol to Hydrocarbons

Many aspects of shape selectivity are encountered in the MTH reaction. As mentioned above, product shape selectivity is prominent, especially for the H-SAPO-34 catalyst. Also, molecular traffic control was first described for the MTH reaction [54]. However, this section covers more recent insights regarding transition state shape selectivity for the formation of light alkenes over H-ZSM-5 and H-beta (three-dimensional intersecting channels of diameter 0.76×0.64 nm, see Figure 7.8). An overall, conceptual representation of the generally accepted reaction mechanism for the conversion of methanol to light alkenes, the hydrocarbon pool mechanism, is given in Figure 7.12. Note that alkenes are not formed by direct coupling of methanol molecules [77–79]. The hydrocarbon pool mechanism was introduced by Dahl and Kolboe in the early 1990s [80–83], and stipulates that methanol is added to aromatic reaction centers, which are adsorbed or even

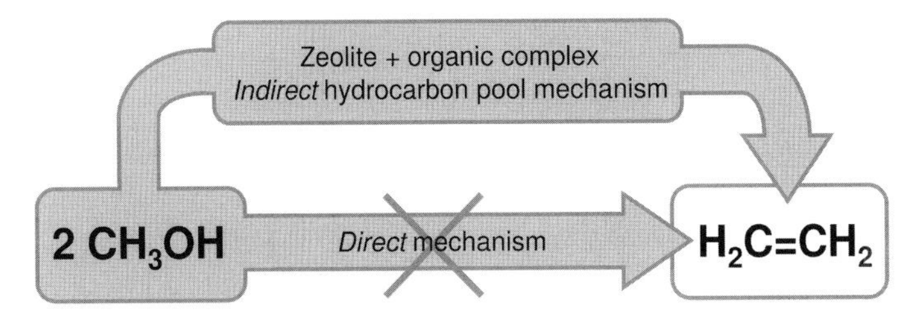

Figure 7.12 Light alkenes (exemplified by ethene) are not formed by *direct* coupling of methanol or dimethylether molecules (bottom pathway) in the methanol to hydrocarbons reaction. Rather, alkenes are created from species comprising an organic complex (methylbenzenes) in interaction with an acidic zeolite proton confined within the micropores (top pathway). This *indirect* pathway is referred to as the *hydrocarbon pool* mechanism.

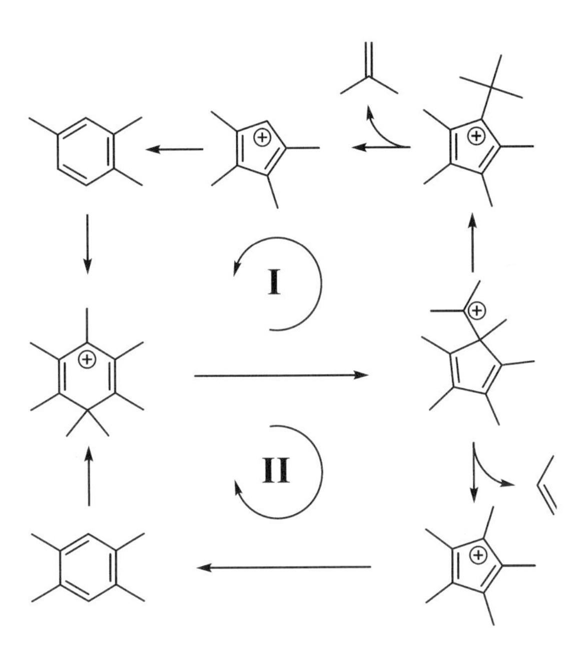

Figure 7.13 Isobutene (cycle I) and propene (cycle II) are believed to be formed from methylbenzenes and their cationic counterparts via ring contractions, methyl shift (for isobutene), dealkylations, and ring expansion, according to the so-called paring reaction.

trapped in the microporous catalysts, in methylation steps and that light alkenes are split off in subsequent rearrangement steps. The experimental evidence in favor of the hydrocarbon pool mechanism is substantial, and summarizing publications exist [84, 85]. A rather extensive series of studies have identified multiply methylated benzenes, such as hexamethylbenzene (hexaMB), or their protonated counterparts as the hydrocarbon pool species responsible for alkene formation, as illustrated in Figure 7.13 [86–90].

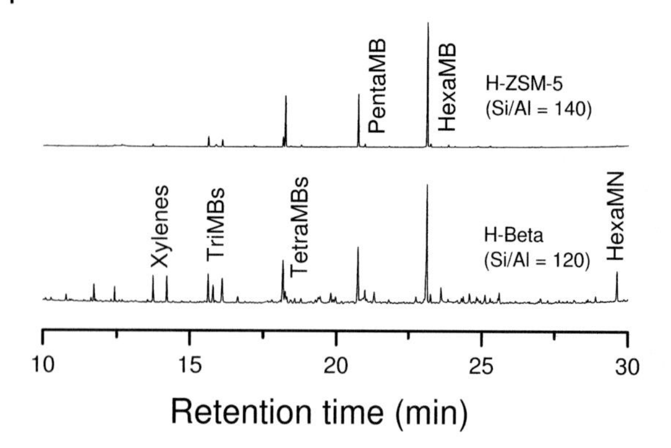

Figure 7.14 GC-MS analyses (total ion chromatograms) of the hydrocarbons retained within the zeolite voids after 20 min of methanol conversion at 350 °C over H-ZSM-5 (top) and H-beta (bottom). Methylbenzenes, which are intermediates for alkene formation in the hydrocarbon pool mechanism, are dominant in both H-ZSM-5 and H-beta, but larger compounds are formed and retained within the 12-ring pores in H-beta.

In a comparative investigation of large-pore zeolite H-beta and medium-pore H-ZSM-5, which are topologically described above, in the MTH reaction, the propene to ethene ratio was 21 for H-beta and merely 3 for H-ZSM-5 [91]. Thus, H-beta is seven times more selective toward propene relative to ethene when compared to H-ZSM-5. A mechanistic understanding of this substantial difference in selectivity is of importance for potential methanol-into-propene applications, and this difference may be attributed to shape selectivity with respect to the reaction intermediates (transition state shape selectivity). The chemical identity of the hydrocarbons located within the zeolite crystals, which are possible candidates as intermediates, was determined by rapidly quenching the reaction, dissolving the catalyst samples in hydrofluoric acid, extraction, and analysis by GC-MS. The results for the two catalysts are shown in Figure 7.14. Clearly, the composition of the retained hydrocarbons is similar for both topologies and methyl-substituted benzenes are dominant. HexaMB is most abundant for both catalysts, followed by pentamethylbenzene/tetramethylbenzenes, trimethylbenzenes, and *p*/*m*-xylene. Hydrocarbons larger than hexaMB were not found to be present in noticeable amounts in the H-ZSM-5 samples, whereas the large-pore beta zeolite allows the formation of larger molecules. Methylbenzenes up to 1,2,4,5-tetramethylbenzene is typically observed in the effluent from H-ZSM-5 samples, so only penta- and hexaMB were considered to be trapped inside H-ZSM-5. For H-beta, however, also penta- and hexaMB were observed among the gas phase products, and this was attributed to the less-restricted space in beta compared to H-ZSM-5. It may be also noted, as a digression, that the substantial amount of hexamethylnaphtalene formed in large-pore H-beta is in contrast to the clean cutoff in molecular size seen at hexaMB for H-ZSM-5. This reflects the limit to available space within the channels of the two materials, and is related to the outstanding coking resistance of H-ZSM-5, which may be caused by transition state shape selectivity [90, 92].

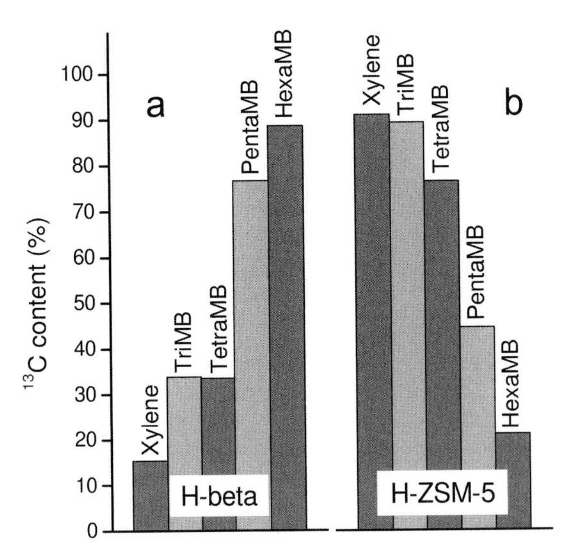

Figure 7.15 Total ^{13}C contents in confined methylbenzene reaction intermediates after 18 min of ^{12}C-methanol reaction followed by 2 min of ^{13}C-methanol reaction at 350 °C over H-beta (a) and H-ZSM-5 (b). The organics were made available for GC-MS analysis by dissolving the zeolite in HF. For H-beta, the higher methyl benzenes, i.e., penta- and hexaMB, display the fastest incorporation of ^{13}C from the methanol feed and are thus the most important intermediates in the hydrocarbon pool mechanism, whereas the opposite trend with respect to the number of methyl groups is seen for the H-ZSM-5 catalyst [90]. Reproduced with permission.

The role of the identified methylbenzenes as reaction intermediates was investigated by transient isotope analysis. Normal ^{12}C methanol was reacted for a predetermined time over both catalysts to build up a ^{12}C hydrocarbon pool. In order to follow the reaction pathway of methanol via these intermediates, a switch to ^{13}C methanol feed was performed. The degree of incorporation of labeled carbons into the methylbenzenes located in the catalyst pores after 2 min of ^{13}C methanol reaction is shown in Figure 7.15. Interestingly, in H-beta, a clear trend is observed: The highest rate of carbon incorporation from methanol, and thus the highest reactivity, is seen for the aromatics with the most methyl substituents. For H-ZSM-5, however, the opposite trend is observed as the xylenes, tri- and tetramethylbenzene now have the highest rate of ^{13}C incorporation. This shows that different compounds are predominant hydrocarbon pool species in H-beta and H-ZSM-5. The low reactivity of the higher methylbenzenes in H-ZSM-5 was corroborated by Lesthaeghe *et al.* using computational methods [93, 94]. It was shown that the formation of penta- and hexaMB in H-ZSM-5 is strongly influenced by steric limitations, implying a lack of sufficient space to undergo the necessary reaction steps for alkene formation and thus a minor role as reaction intermediate in this catalyst. It was therefore concluded that the *higher* methylbenzenes are intermediates in alkene formation in H-beta, whereas the *lower* methylbenzenes play the analogous role in H-ZSM-5. Haw and coworkers [95] have studied the alkene selectivities in the methanol-to-olefin reaction over H-SAPO-34, and it was found that the ethene selectivity is related to the number of methyl groups on the benzene rings trapped in the cages of the catalyst. Based on the correlation of ^{13}C NMR spectra of the hydrocarbons confined inside the H-SAPO-34 cages to the effluent composition, it was specifically concluded that propene is favored by methylbenzenes with four to six methyl groups, whereas ethene is predominantly formed from methylbenzenes with two or three methyl groups. NMR spectra and the ethene and propene selectivity as a function of the average number of methyl groups per ring are shown in Figure 7.16. Armed with this insight from H-SAPO-34, it could be stated that the hydrocarbon pool mechanism operating in H-ZSM-5, which is based on the lower methylbenzenes, yields predominantly ethene, whereas for H-beta, the higher homologues are active and propene is favored.

The observation that different species function as reaction intermediates in the hydrocarbon pool mechanism due to differences in catalyst topology, which in turn leads to different alkene selectivities, explains the large difference in ethene to propene ratio observed for the two catalysts. This example of transition state shape selectivity accounts for the very low ethene yield over H-beta compared to H-ZSM-5, and is an example of how fundamental studies might have implications for industrial applications.

7.4.3
Industrial Relevance of Hydroconversion Reactions

Many important catalytic reactions for the processing of hydrocarbon-based feedstocks with hydrogen take place in the refinery. Among these we find various hydrotreating processes for the removal of heteroatoms (such as sulfur, oxygen, nitrogen, and metals) and hydroconversion reactions. Hydroconversion reactions comprise hydrocracking, where the purpose is to convert heavy fractions such as vacuum gas oils into lighter products, mostly transportation fuels, and hydroisomerization, where linear pentanes and hexanes are isomerized into branched products of similar molecular weight with increased octane number. Hydrocrack-

Figure 7.16 Top panel: ^{13}C CP/MAS NMR spectra showing the loss of methyl groups as a function of time from methylbenzenes trapped in the H-SAPO-34 nanocages at 400 °C. For each spectrum, methanol was reacted for a fixed time at 400 °C, followed by an abrupt cutoff in the methanol feed and flushing with He at the reaction temperature for 0–60 min before recording a spectrum of the material retained inside the pores. Based on the relative intensities of peaks corresponding to methyl carbons (20 and 25 ppm) and aromatic carbons (130 and 134 ppm), the average number of methyl groups per aromatic ring (M_{ave}) may be calculated. Bottom panel: Ethene and propene selectivity as a function of the average number of methyl groups per aromatic ring located in the pores during flushing. Clearly, ethene is favored by methylbenzenes with two or three methyl groups, while propene is favored with four or more methyl groups [94]. Reproduced with permission.

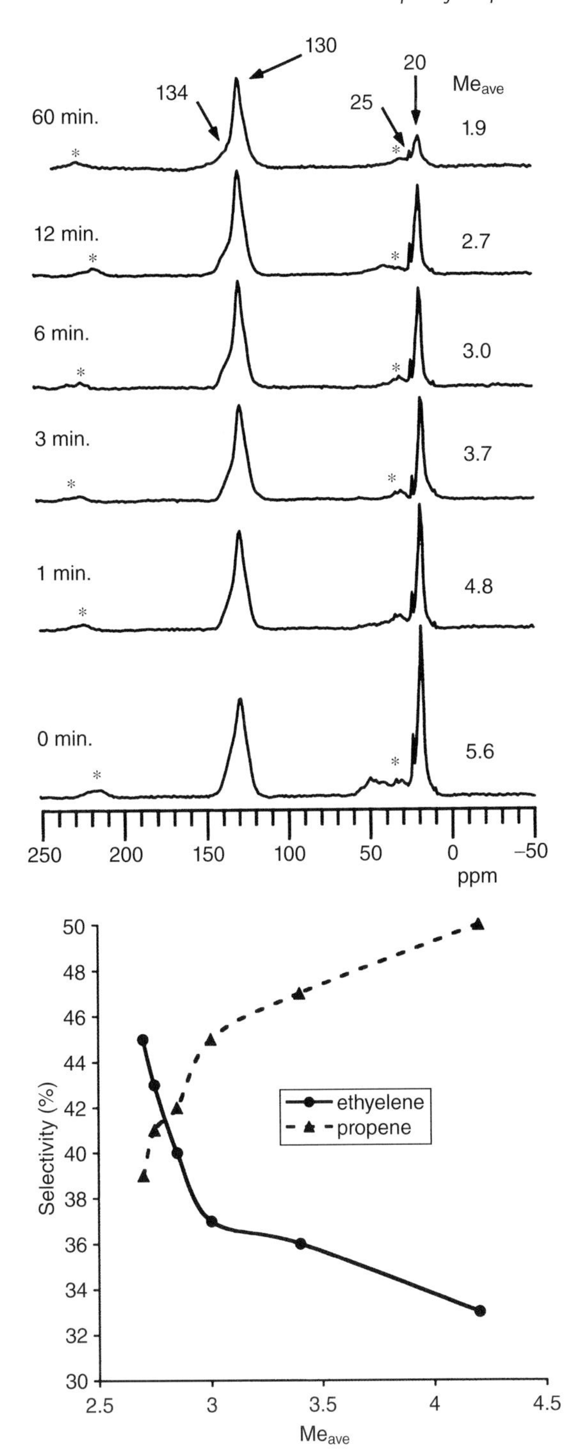
130
20
134
Meave
60 min.
25
1.9
12 min.
2.7
6 min.
3.0
3 min.
3.7
1 min.
4.8
0 min.
5.6
250
200
150
100
50
0
–50
ppm
Selectivity (%)
ethyelene
propene
Meave

ing is of increasing relevance due to increasing market demands for transportation fuels, more stringent environmental restrictions (sulfur and other pollutants are removed during hydrocracking), and the fact that there is a growing need for processing increasingly heavier crude oils [5]. Similarly, hydroisomerization is a process that may contribute to increased octane numbers and reduced sulfur/aromatics contents in gasoline. For both hydrocracking and hydroisomerization, high surface area bifunctional catalysts with a hydrogenation/dehydrogenation function and an acidic function are needed. Many metals in combination with different oxides are in commercial use and have been studied in the laboratory. The hydrogenation/dehydrogenation activity may be provided by, for example, Co, Mo, W, Ni, V, Pd, or rare earth elements, whereas the acidic function can be afforded by materials such as amorphous silica-alumina, γ-alumina, or zeolites [42, 96]. Among the zeolite-based systems, zeolite Y prevails, although other topologies have been investigated, for example, L, mordenite, omega, X, beta, ZSM-22, and ZSM-5 [42, 63, 64].

7.4.4
Shape Selectivity in Hydrocracking

In order to discuss the shape selectivity phenomena in the hydrocracking reaction, we must take a closer look at the reaction mechanism. As mentioned, hydrocracking and hydroisomerization catalysts are bifunctional with a metal function and an acidic function. In hydrocracking, a metal hydrogenation function is required to remove heteroatoms and to form fully hydrogenated products. However, the same catalyst component also facilitates dehydrogenation to yield the alkene intermediates required in both hydrocracking and hydroisomerization. These alkenes subsequently interact with acidic sites to form carbenium ion-like reaction intermediates, which undergo cracking and isomerization steps. After return of a proton to regenerate the acidic site, the cracked or isomerized species are hydrogenated by the metal function in order to yield alkane products. Shape selectivity may occur in the hydrocracking reaction when the acidic function is provided by a zeolite and the cracking and isomerization reactions take place within the micropores. When using an industrial feedstock, consisting of paraffinic and polyaromatic compounds, it is straightforward to realize that reactant shape selectivity based solely on size exclusion may occur. However, most scientific studies employ model feedstocks, such as decane (C_{10}) and hexadecane (C_{16}), implying that reactant shape selectivity based on strict exclusion by size is less relevant. Also, as mentioned above, Martens and coworkers [63, 65] have studied hydrocracking over ZSM-22-based catalysts and emphasized the importance of reactions occurring in the very outermost parts of the zeolite channels, that is, pore mouth and key lock shape selectivity. In this section, however, we focus on a theoretical approach toward understanding shape selectivity, wherein the stability (Gibbs free energy) of adsorbed reactants, intermediates, and products is considered. Clearly, knowledge of the relative Gibbs free energy of every species involved in a catalytic chemical reaction will allow a detailed understanding of the chemistry, and these quantities are most often not readily determined experimentally, except for the overall reaction.

Smit and Maesen have employed so-called configurational bias Monte Carlo (CBMC) simulations to calculate the free energies of adsorption of alkanes in zeolites [60, 97]. Briefly, in the CBMC scheme, molecules are grown atom by atom in such a way that the empty channels inside the zeolites are found. This approach is 1–2 orders of magnitude more efficient than traditional Monte Carlo or docking procedures. When using such traditional approaches, the molecules loaded into the zeolite cell have a large possibility of ending up in positions where prohibitive overlap between the molecule and framework occurs. This inevitably leads to the rejection of the majority of the simulation steps. Further, the zeolite framework is kept rigid and built up exclusively from Si and O, meaning that acidic sites are not taken into account. The hydrocarbons are modeled by the so-called united atoms approach, where each CH_x unit in the hydrocarbon is considered to be a single interaction center. Alkanes are considered to model the chemistry well. This approach allows the computation of adsorption isotherms and the Gibbs free energy of the adsorbed molecules. The salient feature in this context is that CBMC allows simulation of larger molecules at higher pressures than otherwise feasible.

Smit and Maesen have investigated shape selectivity during hydrocracking over ZSM-11- and ZSM-5-based catalysts, for which detailed product distributions and interesting differences have been reported [98]. The ZSM-11 zeolite shown in Figure 7.8 is structurally similar to ZSM-5. It also consists of two sets of intersecting 10-ring channels, thus leading to a three-dimensional pore structure. In contrast to ZSM-5, ZSM-11 has two identical *straight* channels (0.53×0.54 nm) leading to two different types of intersections, one of which is larger than the single intersection type found in ZSM-5 (Figure 7.8). Despite structural similarities, the two catalysts display markedly different product selectivities in the hydrocracking of a decane model feed: ZSM-11 has a higher selectivity (a factor 2) toward isobutane relative to linear butanes than ZSM-5 [98]. Based on traditional experimental techniques, it is not straightforward to assess these differences in selectivity properties between ZSM-5 and ZSM-11. Smit and Maesen employed the CBMC approach outlined above to investigate this peculiar selectivity difference. The Gibbs free energy of all intermediates leading from a linear C_{10} feed via monobranched to potential dibranched cracking precursors was calculated for both structures (see Figure 7.17b), thus charting the free energy landscape. The free energy of a specific species relative to that of C_{10} is taken as a measure of the probability of involvement in reactions leading to cracking products. It is important to realize that fast cracking requires a specific arrangement of the methyl groups on the C_8 backbone in the dibranched species. In a formal representation, particularly unstable carbenium ions are thus avoided on C–C scission. Moreover, diffusion simulations indicate that the diffusion rates of the dibranched species are vanishingly small, meaning that they must undergo either cracking reactions or isomerization reactions leading to fewer branches [59, 99]. Figure 7.18 shows the free energy of several representative reaction intermediates relative to the free

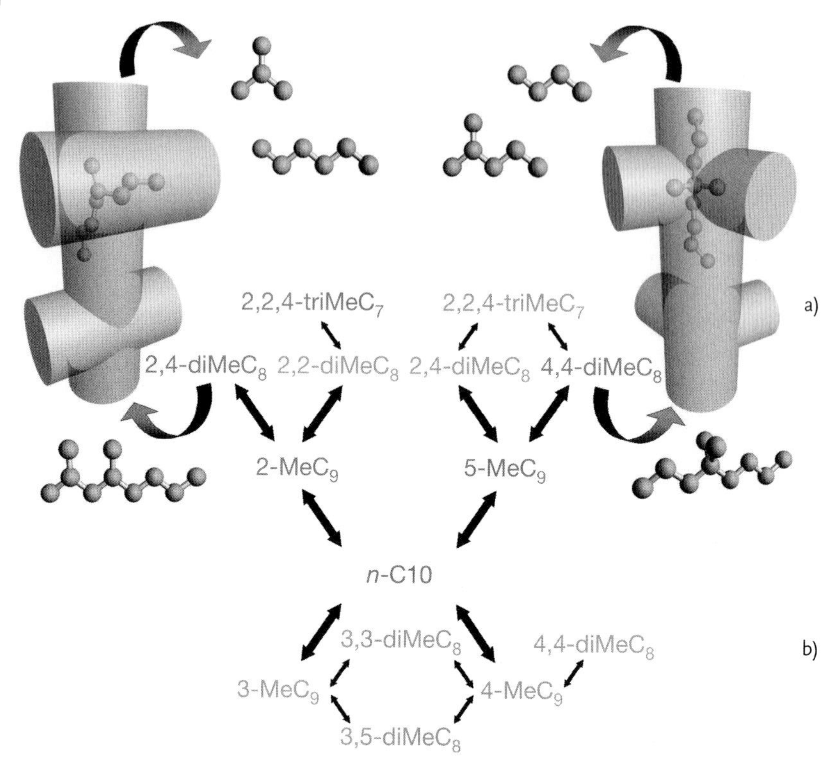

Figure 7.17 Reaction scheme for the hydrocracking of n-C_{10} in ZSM-11 based (left) and ZSM-5 based (right) catalysts. In ZSM-11, 2,4-diMe-C_8 constitutes a perfect match with the pore structure and is the prominent cracking precursor, resulting in a preferred formation of isobutane, whereas in ZSM-5, 4,4-diMe-C_8 is a better geometrical match and n-butane is the preferred C_4 product [59]. Reproduced with permission.

energy of *n*-C_{10}. Clearly, monobranched isomers are readily formed in both catalysts without any significant energetic penalty (or gain). For the dibranched cracking precursors however, the situation is quite different. For ZSM-11, the formation of 2,4-dimethyl-C_8 is significantly favored relative to the other representative cracking precursor by as much as 16 kJ mol^{-1}. On the other hand, for ZSM-5, the situation is reversed: 4,4-dimethyl-C_8 is favored by 8 kJ mol^{-1}. The reason for this behavior is seen in Figure 7.17. In ZSM-5, the geminal arrangement of the two methyl groups on the carbon chain in 4,4-dimethyl-C_8 perfectly matches the geometry of an intersection. For ZSM-11, on the other hand, 2,4-dimethyl-C_8 provides the perfect geometrical match. Thus, for an optimum stability and therefore importance as a reaction intermediate, the shape of the molecule should be *commensurate* with the zeolite topology. The pivot in the argument presented by Smit and Maesen is the intuitive fact that 2,4-dimethyl-C_8 (favored in ZSM-11) will give isobutane and *n*-hexane on cracking, whereas cracking of 4,4-dimethyl-C_8 (favored in ZSM-5) will give *n*-butane and 2-methylpentane. Thus, the experimentally

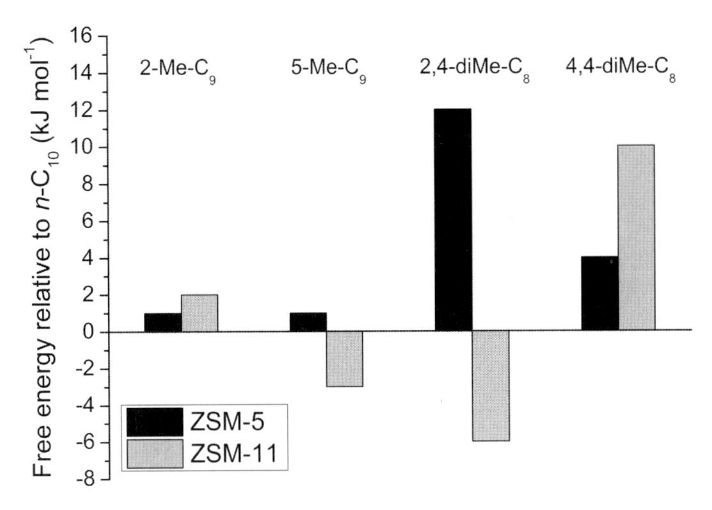

Figure 7.18 The free-energy differences between the reaction intermediates and the n-C_{10} reactant in ZSM-11 and ZSM-5. The two topologies energetically favor different cracking precursors, and thus different cracking products [59].

observed product selectivities may be rationalized by the detailed insight about the free energy landscape in a very gratifying manner.

This case is a prominent example of the quantitative and detailed nature of data, in this instance, thermochemical data, that can only be accessed with theoretical methods. The aspects of shape selectivity in the hydrocracking reaction discussed above are clearly related to the adsorption of the various species involved in the pathway. In terms of classification according to the types of shape selectivity defined above, the insight provided by the simulations of Smit and Maesen is obviously related to both product- and transition state shape selectivity. Smit and Maesen prefer the term reaction *intermediate shape selectivity* [60].

7.4.5
Industrial Relevance of Carbonylation Reactions

Carbonylation reactions offer the possibility to form C–C bonds by adding CO to various organic substrates such as alcohols or olefins. The products formed from carbonylation reactions are typically organic carbonyls (e.g., aldehydes and carboxylic acids). Rh-, Pt-, Ni-, Co-, Pd-, and Ru-based organometallic complexes are well known to catalyze carbonylation reactions, often with rather high selectivities [100]. The Rh-based Monsanto process has been the leading technology for industrial production of acetic acid from methanol since the 1960s, but mainly due to a significantly higher plant throughput, the recently deployed Ir-based BP Cativa process is attractive [101]. Another alternative for production of acetic acid from methanol is the Acetica technology that employs a resin-supported Rh-based catalyst rather than homogeneous metal organic catalysts. However, an alternative process for methanol carbonylation without the need for any costly noble metals is naturally highly desired. Thus, there is a clear incentive to develop alternative, preferably heterogeneously catalyzed carbonylation processes. In the following section, we describe some shape selectivity aspects of carbonylation of dimethyl ether (DME) to methyl acetate. Methyl acetate is a precursor to acetic acid that, in turn, is central for the production of for example, pharmaceuticals, plastics, and textiles. DME may be formed from methanol by a simple dehydration step carried out over an acidic heterogeneous catalyst. Today, carbonylation of methanol accounts for the major part of acetic acid production and about 1 million tons of acetic acid were produced in Europe in 2009 [4].

7.4.6
Shape Selectivity in Carbonylation

Cheung *et al.* [102] reported the highly selective low-temperature carbonylation of DME into methyl acetate over Brønsted acid zeolites. This was a significant breakthrough, as the catalysts were halide free and without any noble metal functionality (see above). The zeolite catalysts investigated were H-mordenite, H-ferrierite, H-ZSM-5, H-USY, and H-beta. Remarkably, H-mordenite was found to be order(s) of magnitude more active than any of the other materials. The carbonylation reaction was found to be zero order with respect to DME and first order with respect to CO (see Figure 7.19a and b). This implies that the active sites were saturated with DME-derived intermediates (most likely surface-bound methoxy groups, as determined by stoichiometric titration of Brønsted sites by DME), and that the kinetically relevant step of this reaction involves a reaction between CO, either from the gas phase or in an adsorbed state. However, the remarkable activity differences among the various topologies were not accounted for at the time, and further investigations of the nature of the CO-bindings sites were announced [102]. Subsequently, a wide range of H-mordenite samples of various Al content and

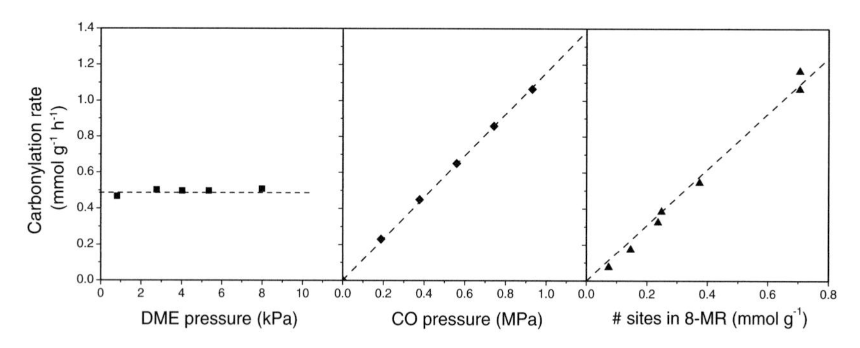

Figure 7.19 Left panel: DME carbonylation rate as a function of DME partial pressure. Middle panel: DME carbonylation rate as a function of CO partial pressure. Right panel: DME carbonylation rate measured as a function of the density of Brønsted sites located exclusively in the 8-ring channels of H-mordenite catalysts [101, 104].

degrees of Na exchange were investigated as DME carbonylation catalysts [103]. Again, a puzzling observation was made, as it was seen that H-mordenite samples with similar densities of acid sites catalyzed the methyl acetate formation at varying rates. Further, the rate of carbonylation was found to increase more than linearly with the density of acid sites, suggesting that the distribution of Al among framework positions could play a role [103]. As shown in Figure 7.8b, the mordenite structure comprises 12-ring channels with intersecting 8-ring channels leading to the formation of so-called side pockets. Bhan *et al.* [104] employed FTIR spectroscopy and elegantly demonstrated that the signal representing the ν(OH) modes of the Brønsted sites in H-mordenite is composed of two contributions. Upon successively increasing the coverage of *n*-hexane, only the high-frequency component of the Brønsted band was eroded, whereas the low-frequency component was left intact. These two components are ascribed to Brønsted sites in the 12-ring channels (high frequency) and the 8-ring channels (low frequency). The *n*-hexane adsorption experiment thus shows that there is a substantial difference in accessibility among the two groups of Brønsted sites. Armed with this insight, and the possibility of distinguishing between the two groups of acid sites, it was shown that the carbonylation rates increased linearly with the number of Brønsted sites in the eight-ring channels (Figure 7.19c). It appeared that the surface methoxy groups confined within the eight-ring channels displayed the greatest reactivity toward CO, because of specific stabilization of the ionic, acetyl-like transition state involved in the rate-determining step, that is, the insertion of CO into the methoxy group [104]. To further probe the particular stabilizing environment in the H-mordenite side pockets, H/D isotope exchange between CD_4 and the Si(OH)Al Brønsted sites was employed to probe the chemical and steric properties of the Brønsted sites in H-mordenite [105]. For this reaction, which has substantially smaller steric requirements and a less charged transition state, no such preference for reaction in the H-mordenite side pockets was found. Indeed, among H-mordenite, H-ferrierite, and H-ZSM-5, a very minor fivefold difference in H/D exchange rate constants was found, in stark contrast with the 100-fold rate difference among these catalysts in the carbonylation reaction [105].

The work of Iglesia and coworkers described in the preceding paragraph on the zeolite-catalyzed carbonylation of DME is an example of transition state selectivity influencing a reaction occurring in a zeolite containing Brønsted sites encapsulated within subnanometer channels. The work extends the scope of shape selectivity concepts beyond those reflecting size exclusion and preferential adsorption [104, 105].

7.5 Summary and Outlook

In this chapter, the immense importance of hydrocarbon-based petrochemical production, today and in the coming decades, is outlined. Overall, the petrochemical and refining industries will most likely to a considerable extent maintain its current configuration. However, increased environmental considerations and the desire of an improved and more efficient utilization of increasingly heavier and contaminated feed stocks will lead to an intense focus on improved processes, in particular with respect to product selectivity. This must be achieved by improved engineering solutions and *catalyst design*. As we have seen, catalysis is deeply involved in the chemical industry, and the properties of the catalyst are decisive for most petrochemical processes. One prominent class of catalysts in the petrochemical industry is the microporous zeolites. It is the well-defined porosity of these catalysts that is determined by the crystal structure that provides the foundation for the successful and widespread use of zeolite-based catalysts. The presence of regular molecular sized channels allows the control of reactants, intermediates/transition states, and products by selective discrimination on the nanometer length scale. It is the interplay between species involved in the reactions and the surrounding zeolite framework that determines the reaction selectivity. Such unique and intricate events are commonly referred to as shape selectivity. We have reviewed three case studies, two relying mostly on experimental approaches and one mostly on theoretical methods. These studies emphasize the importance of molecular level understanding of shape selectivity in order to understand and perhaps even predict the actual outcome of a zeolite-catalyzed reaction. In all three cases, it is the very specific geometric nature of the interaction between the zeolite nanostructure and a wide variety of potential reaction intermediates that dictates the reaction pathway and thus the reaction selectivity.

One way to pursue in order to meet the increasingly more stringent standards for the petrochemical processes outlined above is to attain an even deeper fundamental description of the phenomena giving rise to selectivity. For zeolite-based processes, these phenomena are various forms of shape selectivity. In two of the presented case studies, we have seen how sophisticated experiments may offer knowledge relevant for industrial applications. In the second case, we have seen how a theoretical approach can provide similar information. However, it is the adamant opinion of the authors that if these two approaches were applied to a single problem, the resulting intellectual outcome would by far surpass the sum of the individual contributions. Thus, by integrating several different approaches it might be possible to meet future challenges and demands for petrochemistry. Two equally important aspects have not been covered in this chapter. Materials characterization and synthesis of novel and improved materials should be united with reaction studies. Materials synthesis has traditionally been somewhat of an art, but a significant progress toward fundamental insights also here has become possible by the advent of high throughput methods and more powerful techniques applied *in situ* during synthesis.

References

1 Smoot, L.D. (1998) Progress in energy and combustion. *Science*, **24**, 409–501.

2 Olah, G.A., Goeppert, A., and Prakash, G.K.S. (2006) *Beyond Oil and Gas: The*

Methanol Economy, Wiley-VCH Verlag GmbH, Weinheim.
3 Nilsen, M.H., Antonakou, E., Bouzga, A., Lappas, A., Mathisen, K., and Stöcker, M. (2007) *Microporous Mesoporous Mater.*, **105**, 189–203.
4 (2010) Facts & figures of the chemical industry. *Chem. Eng. News*, **88**, (**27**), 54–62.
5 Moulijn, J.A., Makkee, M., and van Diepen, A. (2001) *Chemical Process Technology*, Wiley-VCH Verlag GmbH, Weinheim.
6 Chorkendorff, I. and Niemantsverdriet, J.W. (2003) *Concepts of Modern Catalysis and Kinetics*, Wiley-VCH Verlag GmbH, Weinheim.
7 Fu, P.P., Lee, H.M., and Harvey, R.G. (1980) *J. Org. Chem.*, **45**, 2797–2803.
8 Gertosio, V., Santini, C.C., Taoufik, M., Bayard, F., Basset, J.M., Buendia, J., and Vivat, M. (2001) *J. Catal.*, **199**, 1–8.
9 Ryndin, Y.A., Santini, C.C., Prat, D., and Basset, J.M. (2000) *J. Catal.*, **190**, 364–373.
10 Lin, W. (2005) *J. Solid State Chem.*, **178**, 2486–2490.
11 Wu, C.D., Hu, A., Zhang, L., and Lin, W. (2005) *J. Am. Chem. Soc.*, **127**, 8940–8941.
12 Mantion, A., Massüger, L., Rabu, P., Palivan, C., McCusker, L.B., and Taubert, A. (2008) *J. Am. Chem. Soc.*, **130**, 2517–2526.
13 Qiu, S. and Zhu, G. (2009) *Coord. Chem. Rev.*, **253**, 2891–2911.
14 Corma, A., Garcia, H., and Xamena, F.X.L. (2010) *Chem. Rev.*, **110**, 4606–4655.
15 Raimondi, F., Scherer, G.G., Kötz, R., and Wokaun, A. (2005) *Angew. Chem. Int. Ed.*, **44**, 2190–2209.
16 Bond, G.C. (1987) *Heterogeneous Catalysis: Principles and Applications*, 2nd edn. Oxford University Press, Oxford.
17 Uzio, D. and Berhault, G. (2010) *Catal. Rev. Sci. Eng.*, **52**, 106–131.
18 Thomas, J.M. (2008) *J. Chem. Phys.*, **128**, 182502–182519.
19 Thomas, J.M. and Raja, R. (2010) *Top. Catal.*, **53**, 848–858.
20 Thomas, J.M., Raja, R., Gai, P.L., Gronbeck, H., and Hernandez-Garrido, J.C. (2010) *ChemCatChem*, **2**, 402–406.
21 Thomas, J.M., Raja, R., and Lewis, D.W. (2005) *Angew. Chem. Int. Ed.*, **44**, 6456–6482.
22 Thomas, J.M. and Thomas, W.J. (1997) *The Principles and Practice of Heterogeneous Catalysis*, Wiley-VCH, Weinheim.
23 Vermeiren, W. and Gilson, J.-P. (2009) *Top. Catal.*, **52**, 1131–1161.
24 Mumpton, F.A. (1978) *Natural Zeolites. Occurrence, Properties, Use*, Pergamon, Oxford.
25 Weigel, O. and Steinhoff, E. (1925) *Z. Kristallogr.*, **61**, 125–154.
26 McBain, J.W. (1932) *The Sorption of Gases and Vapors by Solids*, Rutledge & Sons, London.
27 Jiang, J.X., Yu, J.H., and Corma, A. (2010) *Angew. Chem. Int. Ed.*, **49**, 3120–3145.
28 Corma, A., Diaz-Cabanas, M.J., Jorda, J.L., Martinez, C., and Moliner, M. (2006) *Nature*, **443**, 842–845.
29 Lobo, R.F. (2006) *Nature*, **443**, 757–758.
30 Moliner, M., Diaz-Cabanas, M.J., Fornes, V., Martinez, C., and Corma, A. (2008) *J. Catal.*, **254**, 101–109.
31 Dyer, A. (1988) *An Introduction to Zeolite Molecular Sieves*, John Wiley & Sons, Inc., New York.
32 Gale, J.D., Shah, R., Payne, M.C., Stich, I., and Terakura, K. (1999) *Catal. Today*, **50**, 525–532.
33 Bordiga, S., Regli, L., Cocina, D., Lamberti, C., Bjørgen, M., and Lillerud, K.P. (2005) *J. Phys. Chem. B*, **109**, 2779–2784.
34 Bordiga, S., Regli, L., Lamberti, C., Zecchina, A., Bjørgen, M., and Lillerud, K.P. (2005) *J. Phys. Chem. B*, **109**, 7724–7732.
35 Li, S., Zheng, A., Sul, Y., Zhang, H., Chen, L., Yang, J., Ye, C., and Deng, F. (2007) *J. Am. Chem. Soc.*, **129**, 11161–11171.
36 Fan, W., Duan, R.G., Yokoi, T., Wu, P., Kubota, Y., and Tatsumi, T. (2008) *J. Am. Chem. Soc.*, **130**, 10150–10164.
37 Gleeson, D., Sankar, G., Catlow, C., Richard, A., Thomas, J.M., Spano, G., Bordiga, S., Zecchina, A., and Lamberti,

C. (2000) *Phys. Chem. Chem. Phys.*, **2**, 4812–4817.
38 Smeets, P.J., Sels, B.F., van Teeffelen, R.M., Leeman, H., Hensen, E.J.M., and Schoonheydt, R.A. (2008) *J. Catal.*, **256**, 183–191.
39 Cejka, J. and van Bekkum, H. (2005) *Zeolites and Ordered Mesoporous Materials: Progress and Prospects*, Elsevier, The Netherlands.
40 Sarioglan, A., Savasci, O.T., Erdem-Senatalar, A., Tuel, A., Sapaly, G., and Taarit, Y.B. (2007) *J. Catal.*, **246**, 35–39.
41 Ma, D., Shu, Y., Zhang, C., Zhang, W., Han, X., Xu, Y., and Bao, X. (2001) *J. Mol. Catal. A*, **168**, 139–146.
42 Guisnet, M. and Wilson, J.P. (2002) *Zeolites for Cleaner Technologies, Catal. Sci. Ser. 3*, Imperial College Press, London.
43 Dyer, A. and Amin, S. (2001) *Microporous Mesoporous Mater.*, **46**, 163–176.
44 van Bekkum, H., Flanigen, E.M., Jacobs, P.A., and Jansen, J.C. (2001) *Introduction to Zeolite Science and Practice*, Elsevier Science, The Netherlands.
45 Corma, A. (1995) *Chem. Rev.*, **95**, 559–614.
46 Degnan, T.F. (2003) *J. Catal.*, **216**, 32–46.
47 Wheeler, A. (1951) *Adv. Catal.*, **3**, 250–328.
48 Song, C., Garces, J.M., and Sugi, Y. (eds) (2000) *Shape Selective Catalysis. Chemicals Synthesis and Hydrocarbon Processing, ACS Symp. Ser. 738*, American Chemical Society, Washington, DC.
49 Chen, N.Y., Degnan, T.F., and Smith, C.M. (1994) *Molecular Transport and Reaction in Zeolites*, VCH Publishers Inc., New York, NY.
50 Haag, W.O. (1994) *Stud. Surf. Sci. Catal.*, **84**, 1375–1394.
51 Weisz, P.B. and Frilette, V.J. (1960) *J. Phys. Chem.*, **64**, 382–383.
52 Chen, N.Y., Garwood, W.E., and Dwyer, F.G. (1989) *Shape Selective Catalysis in Industrial Applications*, Marcel Dekker, New York, Basel.
53 Csicsery, S.M. (1984) *Zeolites*, **4**, 202–213.
54 Derouane, E.G. and Gabelica, Z. (1980) *J. Catal.*, **65**, 486–489.
55 Clark, L.A., Ye, G.T., and Snurr, R.Q. (2000) *Phys. Rev. Lett.*, **84**, 2893–2896.
56 Derouane, E.G., Andre, J.M., and Lucas, A.A. (1988) *J. Catal.*, **110**, 58–73.
57 Santilli, D.S., Harris, T.V., and Zones, S.I. (1993) *Microporous Mater.*, **1**, 329–341.
58 Denayer, J.F.M., Ocakoglu, A.R., Martens, J.A., and Baron, G.V. (2004) *J. Catal.*, **226**, 240–244.
59 Smit, B. and Maesen, T.L.M. (2008) *Nature*, **451**, 671–678.
60 Smit, B. and Maesen, T.L.M. (2008) *Chem. Rev.*, **108**, 4125–4184.
61 Schenk, M., Calero, S., Maesen, T.L.M., van Benthem, L.L., Verbeek, M.G., and Smit, B. (2002) *Angew. Chem. Int. Ed.*, **41**, 2500–2502.
62 Parton, R., Uytterhoeven, L., Martens, J.A., Jacobs, P.A., and Froment, G.F. (1991) *Appl. Catal.*, **76**, 131–142.
63 Souverijns, W., Martens, J.A., Froment, G.F., and Jacobs, P.A. (1988) *J. Catal.*, **174**, 177–184.
64 Martens, J.A., Souverijns, W., Verrelst, W., Parton, R., Froment, G.F., and Jacobs, P.A. (1995) *Angew. Chem. Int. Ed.*, **34**, 2528–2530.
65 Claude, M.C. and Martens, J.A. (2000) *J. Catal.*, **190**, 39–48.
66 Lawton, S.L., Leonowicz, M.E., Partridge, R.D., Chu, P., and Rubin, M.K. (1998) *Microporous Mesoporous Mater.*, **23**, 109–117.
67 Chen, N.Y., Maziuk, J., Schwartz, A.B., and Weisz, P.B. (1968) *Oil Gas J.*, **66**, 154–157.
68 Dubbeldam, D. and Smit, B. (2003) *J. Phys. Chem. B*, **107**, 12138–12152.
69 Dubbeldam, D., Calero, S., Maesen, T.L.M., and Smit, B. (2003) *Angew. Chem. Int. Ed.*, **42**, 3623–3626.
70 Dubbeldam, D. and Snurr, R.Q. (2007) *Mol. Sim.*, **33**, 305–325.
71 Chang, C.D. and Silvestri, A.J. (1977) *J. Catal.*, **47**, 249–259.
72 Chang, C.D. (1983) *Catal. Rev. Sci. Eng.*, **25**, 1–183.
73 Stöcker, M. (1999) *Microporous Mesoporous Mater.*, **29**, 3–48.
74 Topp-Jørgensen, J. (1988) *Stud. Surf. Sci. Catal.*, **36**, 293–305.

75 Chen, J.Q., Bozzano, A., Glover, B., Fuglerud, T., and Kvisle, S. (2005) *Catal. Today*, **106**, 103–107.

76 Stöcker, M. (2010) in *Zeolites and Catalysis. Synthesis, Reactions and Applications* (eds J. Cejka, A. Corma, and S. Zones), Wiley-VCH Verlag GmbH, Weinheim, Germany, pp. 687–712.

77 Song, W., Marcus, D.M., Fu, H., Ehresmann, J.O., and Haw, J.F. (2002) *J. Am. Chem. Soc.*, **124**, 3844–3845.

78 Marcus, D.M., McLachlan, K.A., Wildman, M.A., Ehresmann, J.O., Kletnieks, P.W., and Haw, J.F. (2006) *Angew. Chem., Int. Ed.*, **45**, 3133–3136.

79 Lesthaeghe, D., Van Speybroeck, V., Marin, G.B., and Waroquier, M. (2006) *Angew. Chem. Int. Ed.*, **45**, 1–10.

80 Dahl, I.M. and Kolboe, S. (1993) *Catal. Lett.*, **20**, 329–336.

81 Dahl, I.M. and Kolboe, S. (1995) *Stud. Surf. Sci. Catal.*, **98**, 176–177.

82 Dahl, I.M. and Kolboe, S. (1994) *J. Catal.*, **149**, 458–464.

83 Dahl, I.M. and Kolboe, S. (1996) *J. Catal.*, **161**, 304–309.

84 Haw, J.F., Song, W., Marcus, D.M., and Nicholas, J.B. (2003) *Acc. Chem. Res.*, **36**, 317–326.

85 Olsbye, U., Bjørgen, M., Svelle, S., Lillerud, K.-P., and Kolboe, S. (2005) *Catal. Today*, **106**, 108–111.

86 Arstad, B. and Kolboe, S. (2001) *Catal. Lett.*, **71**, 209–212.

87 Arstad, B. and Kolboe, S. (2001) *J. Am. Chem. Soc.*, **123**, 8137–8138.

88 Bjørgen, M., Olsbye, U., and Kolboe, S. (2003) *J. Catal.*, **215**, 30–44.

89 Bjørgen, M., Olsbye, U., Petersen, D., and Kolboe, S. (2004) *J. Catal.*, **221**, 1–10.

90 Bjørgen, M., Svelle, S., Joensen, F., Nerlov, J., Kolboe, S., Bonino, F., Palumbo, L., Bordiga, S., and Olsbye, U. (2007) *J. Catal.*, **249**, 195–207.

91 Svelle, S., Olsbye, U., Joensen, F., and Bjørgen, M. (2007) *J. Phys. Chem. C*, **111**, 17981–17984.

92 Guisnet, M. and Magnoux, P. (1989) *Appl. Catal.*, **54**, 1–27.

93 Lesthaeghe, D., Dee Sterck, B., Van Speybroeck, V., Marin, G.B., and Waroquier, M. (2007) *Angew. Chem. Int. Ed.*, **46**, 1311–1314.

94 Lesthaeghe, D., Van Speybroeck, V., Marin, G.B., and Waroquier, M. (2007) *Stud. Surf. Sci. Catal.*, **170B**, 1668–1676.

95 Song, W., Fu, H., and Haw, J.F. (2001) *J. Am. Chem. Soc.*, **123**, 4749–4754.

96 Matar, S. and Hatch, L.F. (2001) *Chemistry of Petrochemical Processes*, 2nd edn. Gulf Professional Publishing, Houston, TX.

97 Schenk, M., Calero, S., Maesen, T.L.M., Vlugt, T.J.H., van Benthem, L.L., Verbeek, M.G., Schnell, B., and Smit, B. (2003) *J. Catal.*, **214**, 88–99.

98 Jacobs, P.A., Martens, J.A., Weitkamp, J., and Beyer, H.K. (1981) *Faraday Discuss. Chem. Soc.*, **72**, 353–369.

99 Schenk, M., Smit, B., Vlugt, T.J.H., and Maesen, T.L.M. (2001) *Angew. Chem. Int. Ed.*, **40**, 736–739.

100 Beller, M., Cornils, B., Frohning, C.D., and Kohlpaintner, C.W. (1995) *J. Mol. Catal. A*, **104**, 17–85.

101 Jones, J.H. (2000) *Platinum Metals Rev.*, **44**, 94–105.

102 Cheung, P., Bhan, A., Sunley, G.J., and Iglesia, E. (2006) *Angew. Chem. Int. Ed.*, **45**, 1617–1620.

103 Cheung, P., Bhan, A., Sunley, G.J., Law, D.L., and Iglesia, E. (2007) *J. Catal.*, **245**, 110–123.

104 Bhan, A., Allian, A.D., Sunley, G.J., Law, D.J., and Iglesia, E. (2007) *J. Am. Chem. Soc.*, **129**, 4919–4924.

105 Bhan, A. and Iglesia, E. (2008) *Acc. Chem. Res.*, **41**, 559–567.

8
Crystal Engineering of Metal-Organic Frameworks for Heterogeneous Catalysis

Chuan-De Wu

8.1
Introduction

Metal-organic frameworks (MOFs) or metal-organic coordination networks are normally constructed from organic linkers joining up metal nodes in suitable solvents under mild reaction conditions, which can extend into one-, two-, or three-dimensional (1D, 2D, or 3D) polymeric frameworks combined through metal–ligand bondings [1]. Crystal engineering of MOFs have made remarkable progress because of their diverse topologies and fascinating properties, which can provide entries into technologically useful materials for variety applications such as heterogeneous catalysis, gas storage, separation, molecular recognition, fluorescence, magnetics, etc. [2–7].

Connectors (metal ions) and linkers (organic ligands) are the principal components of each MOF. Transition-metal ions are often utilized as versatile connectors in the construction of MOFs. The coordination geometries of transition-metal centers are depending on the metal category and its oxidation state. As the metal coordination numbers can range from 2 to 7, the various geometries can be linear (bend), trigonal, tetrahedral, square-planar, square-pyramidal, octahedral, pentagonal-bipyramidal, etc. [8]. Due to the higher coordination numbers of lanthanide ions (from 7 to 14) and the inherent flexibility of their coordination geometries, using lanthanide ions as connectors for the construction of porous MOFs can generate novel and unusual network topologies [9]. The metal nodes with various geometries have the advantage of offering control of the bond angles and restricting the number of coordination sites, which can subsequently control the topological architectures of MOFs. Linkers must possess electron donor sites, such as lone pair electrons, π electrons, etc., which can enter the electron-defect metal orbitals to generate coordination bonds. If organic linkers have two or more coordination donors, they can link up metal nodes to generate polymeric networks. Such linkers afford a wide variety of linking sites with diversified directionalities to control the steric consequences in the self-organization process.

A crystal structure can in principle be regarded as self-assembly and self-organization of connectors and linkers. The major objective of crystal engineering

Selective Nanocatalysts and Nanoscience, First Edition. Edited by Adriano Zecchina, Silvia Bordiga, Elena Groppo.

is various combinations of connectors and linkers to get specific prescribed functional topological frameworks. The possible topologies of MOFs are determined by the availability of certain coordination geometries of metal nodes and the directionalities of ligands. The combination of metal nodes and organic linkers are also affected by many factors, such as solvent, temperature, reaction time, concentration, anion, and pH-value. Consequently, it is still not possible to precisely predict the overall topological structures from their chemical compositions [10]. Thus, considerable research efforts are dedicated toward the generation of functional materials based on building blocks. By carefully modifying the organic ligands to introduce specialized functional groups, decorated MOFs with predictable topologies can be constructed based on the dominant metal–ligand coordination interactions and tune the physical properties to realize various applications.

Considering many reviews of reported structures and potential applications of MOFs have been addressed [8, 11–19], this perspective of MOFs will give a cross-section of the functionally oriented work involving MOFs for heterogeneous catalysis. The masterly combined metal ions and functional linkers with desired size, shape, and functionalities present the possible catalytic actives of such materials, which follow the use of zeolites [20].

8.2
Volatile Molecules Coordinated Metal Nodes Acted as Catalytic Centers

Generally, the metal coordination sites of MOFs are fully occupied by coordination linkers and/or auxiliary ligands. If the metal coordination sites are fully occupied by strongly coordinated bridging ligands and/or counterions, the metal centers cannot coordinate to organic substrates to promote catalytic process. However, there are a large number of reported MOFs with the metal coordination sites that contain volatile molecules, such as water, alcohols, acetonitrile, DMF, etc., which can be removed/replaced without decomposition of their original frameworks to generate coordinatively unsaturated metal sites for heterogeneous catalysis.

Fujita *et al.* reported a 2D coordination network material $[Cd(NO_3)_2(4,4'\text{-bpy})_2]$ (**1**; 4,4′-bpy = 4,4′-bipyridine) as a Lewis acid catalyst for heterogeneous catalysis [21]. The Cd(II) centers in **1** adopt distorted octahedral environments with four pyridyl groups at the equatorial positions and two monodentate nitrate groups at the apical sites (Figure 8.1). The Cd(II) ions are linked by 4,4′-bpy ligands to extend into a 2D square-grid lamellar framework.

Despite the axial positions of Cd(II) ions are occupied by nitrate counterions, material **1** catalyzed the cyanosilylation of aldehydes effectively. Treatment of benzaldehyde with cyanotrimethylsilane and powdered **1** gave 2-(trimethylsiloxy) phenylacetonitrile in high yield. No reaction took place with $Cd(NO_3)_2$ or 4,4′-bpy or with the supernatant liquid of **1**, suggesting that the reaction is apparently heterogeneous. When benzaldehyde was instead treated by bulk substrates such as 2-tolualdehyde, 3-tolualdehyde, α- and β-naphthaldehyde, 9-anthraldehyde, the

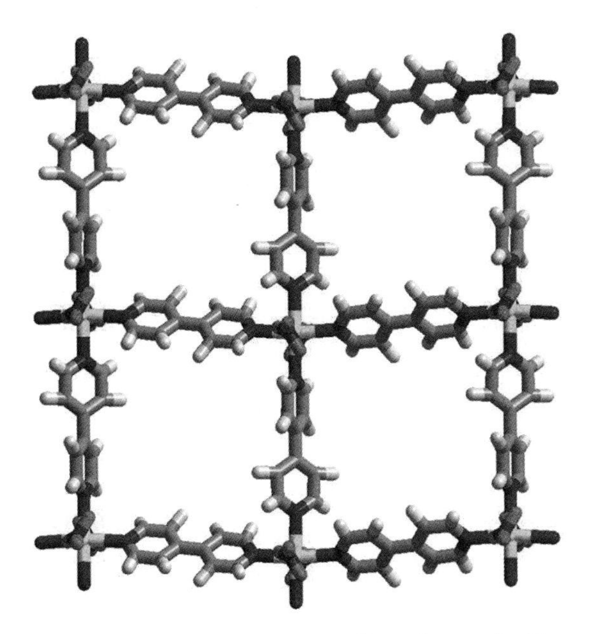

Figure 8.1 Top view of the square-grid layer framework of **1**.

product yields manifest the shape specificities of the cavity size of the network material.

To recover the possible catalytic mechanism, they extended this reaction to the cyanosilylation of imines catalyzed by another similar 2D square-grid coordination network $[Cd(4,4'\text{-bpy})_2(H_2O)_2](NO_3)_2 \cdot 4H_2O$ (**2**) [22]. Each Cd(II) atom coordinates to four pyridines of four 4,4′-bpy ligands and two aqua ligands to propagate into a rhombus lamellar framework of **2** (Figure 8.2). The grid layers stack on each other in such a way that each Cd(II) center is located ca. 4.8 Å above the square cavity of the next layer. The nitrate anions are free within the square-grid cavities.

The cyanosilylation of imine with trimethylsilyl cyanide in the presence of **2** gave the desired product aminonitrile in almost quantitative yield. By examination of the crystal structure, they found that the square cavity-accommodated substrates can easily coordinate to the Cd(II) centers of the next layer. However, they found that the complex took up no detectable amount of imines into the channels when the crystals of **2** were immersed in CH_2Cl_2 solution of the imines. This result suggested that the active sites exist mainly around the surface of **2**, which gave a surface-promoted catalytic reaction. To investigate the possible reaction mechanism, they examined the catalytic activities of $Cd(NO_3)_2 \cdot 4H_2O$ and $Cd(Py_4) \cdot (NO_3)_2$ (Py = pyridine), which have the partial structure of **2** but do not possess infinite frameworks. Surprisingly, they promoted the reaction less effectively, which suggests that the incorporation of catalytic active metal centers in porous network can generate the catalytic properties that are not possessed by the isolated components.

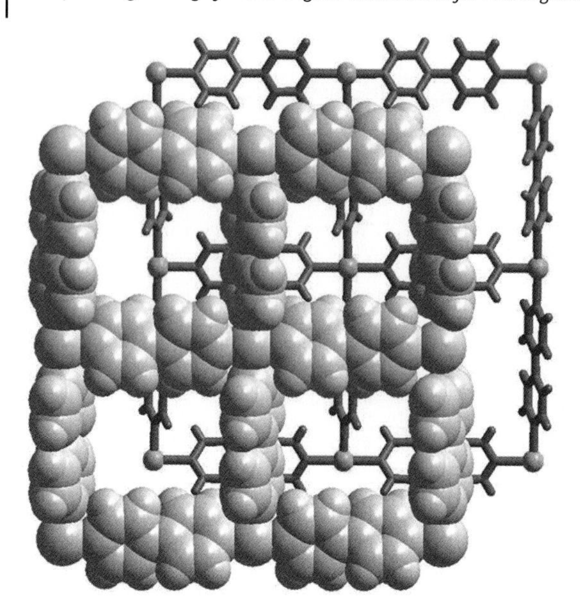

Figure 8.2 Geometrical relationship between the two adjacent layers of **2**. Substrates accommodated in the square cavity of the first layer (space-filling) can interact with Lewis acidic Cd(II) centers of the second layer (ball and stick).

The crystal structure of $Cu_3(BTC)_2(H_2O)_3 \cdot xH_2O$ (**3**, BTC = benzene 1,3,5-tricarboxylate) was first reported by Chui *et al.* [23] In the cubic network, each Cu_2 unit coordinates to four carboxylate groups to give a paddle-wheel unit, which is further linked by BTC ligands to generate a 3D cubic network with open channels filled with water molecular guests (Figure 8.3). The axial coordination sites of Cu(II) ions are occupied by weakly bound water molecules that can be easily removed by heating the compound in vacuum. The dehydration makes the Lewis acid sites to coordinate to organic substrates which results in an activation of the catalytic process.

Kaskel *et al.* have optimized the synthesis conditions to obtain the pure sample **3** for catalytic transformation [24]. The dehydrated Cu_2O-free material **3** catalyzed the selective cyanosilylation of benzaldehyde to give 57% yield and 88.5% selectivity. For comparison, a filtering experiment indicates that the reaction mechanism is heterogeneous. Different reaction solvent experiments showed that the low polarity of solvents is beneficial for the catalytic activity of **3**. They also found that the Cu(II) ions can be reduced by the aldehyde to decompose the catalyst framework at higher temperature, which demonstrates the limitation of the reaction conditions for using the reducible transition metal framework in combination with aldehydes.

More recently, Vos *et al.* have developed the synthesis and washing procedures of **3** [25]. The optimized **3** is identified as a highly selective Lewis acid catalyst for the isomerization of terpene derivatives, such as the rearrangement of α-pinene

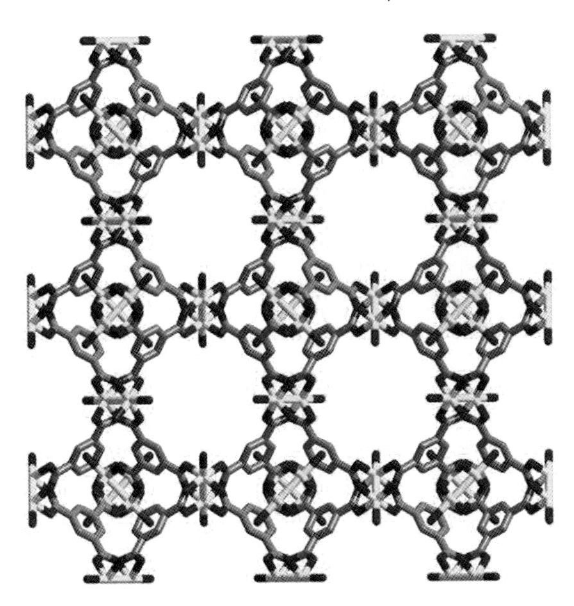

Figure 8.3 A view of the 3D porous framework of **3**, showing the axially bound water molecules point toward the nanochannels.

oxide to campholenic aldehyde and the cyclization of citronellal to isopulegol. By using the ethylene ketal of 2-bromopropiophenone as a test substrate, it was demonstrated that the active sites in **3** are hard Lewis acids. Free H_3BTC ligand has low activity in the isomerization reactions, which suggests that the protonated ligand at the surface or at the defects of **3** could not significantly contribute to the overall acid activity. When **3** is compared with its precursor $Cu(NO_3)_2$, the activity of the Cu(II) ions decreases. The catalytic property investigation of the precursors, H_3BTC ligand and $Cu(NO_3)_2$, suggests that the combined two inactive catalysts can generate a good catalyst **3**.

8.3 Coordinatively Unsaturated Metal Nodes Acted as Catalytic Centers

The higher coordination numbers and the flexible coordination environments of lanthanide nodes can provide coordinatively unsaturated metal sites to act as effective Lewis acid catalytic centers. By modifying BINOL molecule, Lin *et al.* have got a chiral ligand, 2,2′-diethoxy-1,1′-binaphthalene-6,6′-bisphosphonic acid ($\mathbf{L_4}$-H_4), which was used for the construction of a series of homochiral porous lamellar lanthanide bisphosphonates $[Ln(\mathbf{L_4}\text{-}H_2)(\mathbf{L_4}\text{-}H_3)(H_2O)_4]\cdot xH_2O$ (**4**; Ln = La, Ce, Pr, Nd, Sm, Gd, Tb, $x = 9$–14) [26]. These structures are of 2D lamellar frameworks containing four H_2O molecules and four phosphonate groups of four ligand-coordinated square antiprismatic Ln(III) nodes (Figure 8.4). The skewed

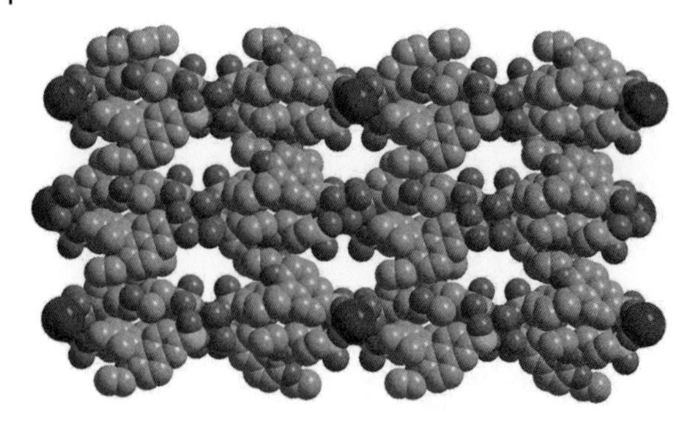

Figure 8.4 A space-filling model of **4** viewed down the *a* axis.

orientation of the binaphthyl subunits together with the phosphonates results in elongated 2D rhombohedral grids with large chiral channels of 12 Å. The PXRD studies suggest that the desolvated frameworks are reversible.

The Gd(III) compound was used as Lewis acid catalyst for the cyanosilylation of aldehydes. Benzaldehyde, 1-naphthaldehyde, and propionaldehyde reacted with trimethylsilyl cyanide to afford the corresponding cyanohydrin products at 69%, 55%, and 61% isolated yields, respectively. Varying sizes of the aldehydes did not affect the catalytic activity, suggesting the expansion of the lamellar lanthanide phosphonates to accommodate the substrates. The Gd(III) compound was also an effective catalyst for ring opening of mesocarboxylic anhydride and Diels–Alder reactions. The supernatants from the reaction mixtures were inactive in these catalyzing reactions, which suggest that all these catalytic reactions are heterogeneous. The lack of enantioselectivity of all these reactions are a direct consequence of the achiral coordination environments around the catalytically active Ln(III) centers.

Nonporous 3D polymer $[In_2(OH)_3(BDC)_{1.5}]$ (**5**; BDC = 1,4-benzendicarboxylate) was hydrothermally synthesized from a mixture of $InCl_3$, H_2BDC, Et_3N, and H_2O [27]. Each In(III) ion octahedrally coordinates to three μ_2-OH groups and three BDC ligands. Each OH group is bridging two In ions to extend into an infinite sheet containing six-member $[In_2(OH)_3]^{3+}$ ring. The BDC ligand coordinates four different In ions of these sheets to give rise to a 3D framework without channel (Figure 8.5).

According to coordinative unsaturation of In(III) ions, the catalytic capability of **5** has been tested in hydrogenation of nitroaromatics and selective oxidation of sulfide reactions. Reduction of nitroaromatics (nitrobenzene and 2-methyl-1-nitronaphthalene) proceeded without induction period and gave high product yields. The catalytic oxidation of sulfides gave sulfoxides as the main product. The catalyst can be recycled in successive runs by a simple filtration without a significant loss of activity and selectivity. PXRD for the recovered material showed no change in the structure of the catalyst. After each cycle, the liquid phase separated from the reaction mixture was used to prove the wholly heterogeneous character

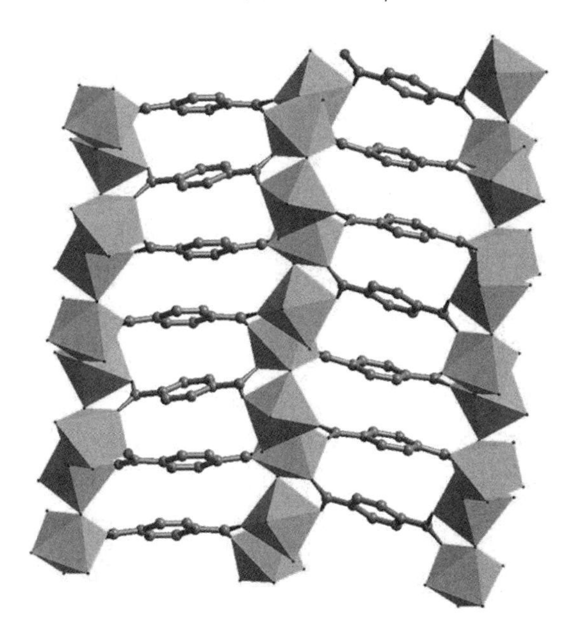

Figure 8.5 A view of the 3D structure of **5**.

of the catalysis. The small size of the pores in this structure avoids the accessibility of the substrates to the metallic centers, which suggests that the catalytic reactions took place on the catalyst surface.

Subsequently, they reported another four polymeric In(III) mixed ligand complexes, $In(BDC)_{1.5}$(2,2′-bpy) (**6**), $In_2(BDC)_2(OH)_2(phen)_2$ (**7**), In(BTC)(H_2O)(2,2′-bpy) (**8**), and In(BTC)(H_2O)(phen) (**9**), constructed from the assembly of the In(III) ions and H_2BDC or H_3BTC in the presence of additional ligands 2,2′-bpy or phen (Figure 8.6) [28]. In compound **6**, each In(III) ion is eight-coordinated to three bidentate chelating carboxylate groups of three BDC units and one 2,2′-bpy molecule. The BDC ligands join the In(III) ions to give (6,3) hexagonal layers, which are interlocked to each other to form a 3D entangled architecture. In compound **7**, each indium ion octahedrally coordinates to two μ_2-OH groups, two BDC carboxyl oxygens, and one phen. The μ_2-OH groups join four In atoms to give tetrameric $[In_4(OH)_4(phen)_4]^{8+}$ units, which are doubly connected by BDC giving rise to an infinite 2D square-type layer. In compounds **8** and **9**, indium ions are six-coordinated to one 2,2′-bpy or phen molecule, three BTC carboxyl groups, and one water molecule to give layer frameworks.

Compounds **7–9** are active and selective catalysts for acetalization of aldehydes. The catalytic reactions of aldehydes with trimethyl orthoformate proceeded with high conversions, and the corresponding products of dimethyl acetals were obtained in good yields. Due to the hindered accessibility of the reactants to the eight-coordinated In atoms, experiment showed lack of catalytic activity for **6**. The heterogeneous Lewis acid catalysts are stable both in water and organic solvents,

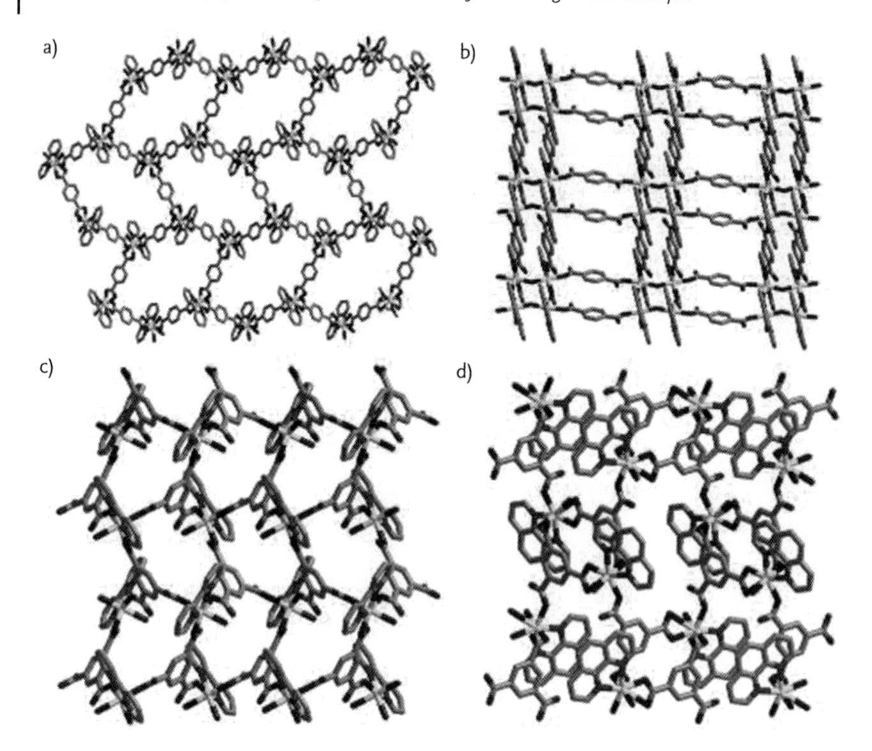

Figure 8.6 The frameworks of **6** (a), **7** (b), **8** (c), and **9** (d).

which can be easily recovered by filtration and reused at least in four cycles without decreasing the product yields or selectivities. PXRD results suggest that the heterogeneous catalysts show no alteration before and after reactions.

Due to the small pore size of above four compounds, the activity took place in the solid surfaces. More recently, they used a bent 4,4′-(hexafluoroisopropylidene) bis(benzoic acid) (H_2hippb) linker to give rise to a porous MOF [In(OH)hippb] (**10**) [29]. Each In(III) ion octahedrally coordinates to two μ_2-OH groups in apical positions and to four different ligands in equatorial positions. The μ_2-OH groups and hippb carboxylate groups bridge In ions to form thick layers with square-shaped channels (Figure 8.7).

Compound **10** is an active and selective catalyst for heterogeneous acetalization of aldehydes with trimethyl orthoformate. The catalytic reaction of benzaldehyde with trimethyl orthoformate gave the corresponding dimethylacetal up to 97% yield. The bulkier *R*-methylbenzeneacetaldehyde was also transformed but needed longer reaction time, showing the hindrance to the diffusion of the reactants inside the catalyst pores. The stability of the solid catalyst was checked by PXRD before and after reaction as aforementioned.

The framework structure of $[Sc_2(BDC)_3]$ (**11**) is a porous 3D framework with each scandium(III) atom octahedrally coordinated to six carboxylic oxygen atoms

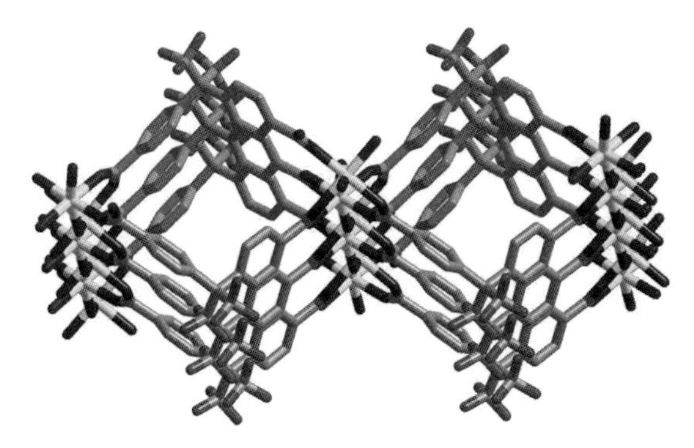

Figure 8.7 Top view of the thick 2D polymeric framework of **10**.

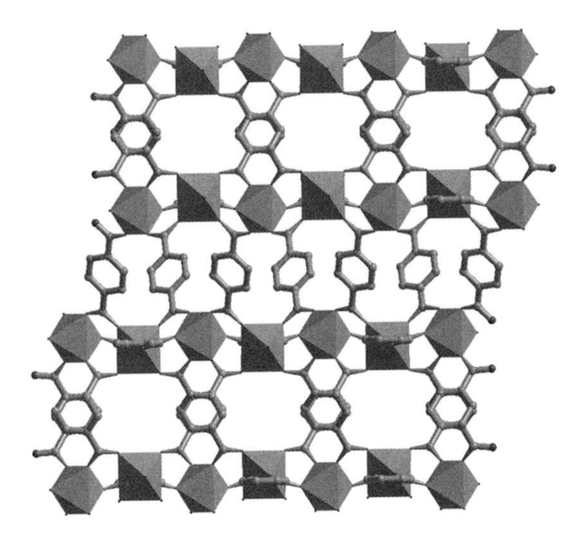

Figure 8.8 A view the 3D structure of **11** with rectangular channels.

of six different terephthalate anions [30]. The carboxylic groups connect the $\{ScO_6\}$ octahedra to form equilateral triangle and rectangular channels (Figure 8.8).

As there are coordinatively unsaturated Sc(III) ions within the framework, compound **11** was tested for the catalytic oxidation of sulfides. The heterogeneous catalytic performance of **11** is better than pure Sc_2O_3, which is a logical consequence of the greater availability of the scandium centers in the pores. While the oxidation of the sulfide continued as the catalyst was present, there was no further significant conversion when the catalyst was removed from the reaction medium. The catalyst could be reused at least for four cycles without loss of catalytic activity and selectivity with catalyst loading as low as 1 mol%. The recovered catalyst is slightly decreased in the reacting rate due to the small amount of the catalyst loss by filtration.

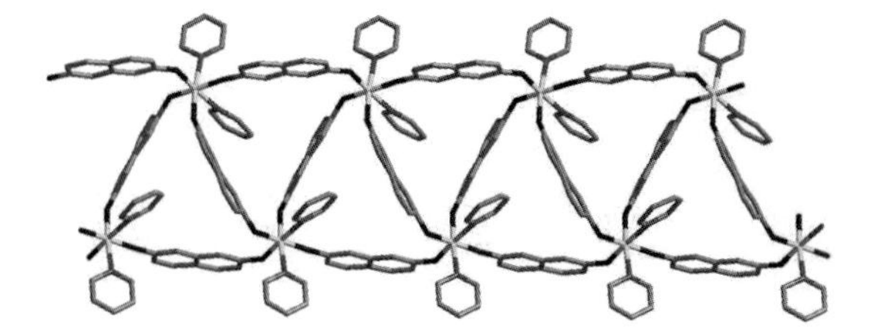

Figure 8.9 Perpendicular view of the zigzag-runged ladder structure of **12**.

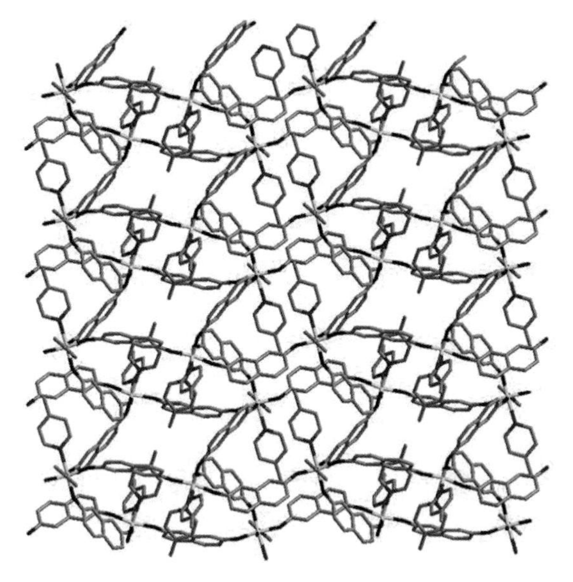

Figure 8.10 The 2D network of **13**.

The reaction of 2,7-dihydroxynaphthalene and (^{i}PrO)$_4$Ti in pyridine formed a 1D zigzag-runged ladder [*cis*-Ti($\mu_{2,7}$-OC$_{10}$H$_6$O)$_2$Py$_2$] (**12**), while the thermolysis of (^{i}PrO)$_4$Ti with 2,7-dihydroxynaphthalene in 4-methylpyridine generated crystals of [Ti($\mu_{2,7}$-OC$_{10}$H$_6$O)$_2$(4-picoline)$_2$ · (4-picoline)$_{0.5}$] (**13**) [31]. A diagonal connectivity of $\mu_{2,7}$-OC$_{10}$H$_6$O bridges was found to link together the strands of *cis*-(ArO)$_4$TiPy$_2$ in **12** (Figure 8.9). When pyridine is replaced by 4-picoline in **13**, alternate "rungs" of the zigzag connections are broken and the ladders are linked together to produce a 2D framework, which consists of stacked ladders (Figure 8.10).

Titanium aryldioxide coordination polymers **12** and **13** were assessed for their Ziegler–Natta polymerization activity toward ethene or propene with methylalumoxane as cocatalyst. The catalysts displayed mediocre activity that was roughly inversely dependent on the dimensionalities of the networks. The polyethylene obtained was polydisperse, although its melting points were within a relatively narrow range indicative of linear material. The data were consistent with the catalyst degradation, which led to numerous sites of variable activity.

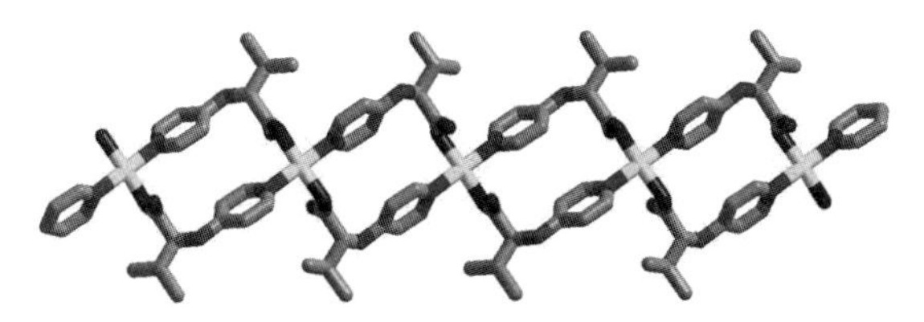

Figure 8.11 The 1D polymeric macrocycle chain in **14**.

Wu *et al.* have modified the amino acid with pyridyl group to generate an effective multiple bridging linker $\mathbf{L_{14}}$ (*N*-(4-pyridyl)-D,L-valine) to join up Cu ions into a 1D coordination polymer [Cu($\mathbf{L_{14}}$)$_2$] (**14**) [32]. Each copper ion coordinates to two carboxyl oxygen atoms and two pyridyl groups of four $\mathbf{L_{14}}$ ligands to form a square-planar geometry. Two $\mathbf{L_{14}}$ ligands thus bidentated *trans*-link two Cu(II) centers to form a 18-membered macrocycle ring, which is further extended into a 1D polymeric chain (Figure 8.11). TGA and PXRD confirmed the stability of **14**, which can tolerate moderate catalytic reaction conditions.

Since the coordination environments of Cu(II) ions are in distorted square-planar geometry, and the axle positions are vacant in **14**, the catalytic cross-coupling reaction of arylboronic acids with imidazole generated the anticipated *N*-phenylimidazoles in good yields. No trace catalytic result observed from the supernate suggests the heterogeneous nature of the present catalyst system. When different solvents such as DMF, THF, CH_2Cl_2, and H_2O were used in the catalytic experiment, almost no product can be observed except trace product for DMF. Obviously, it can be considered that solvent effect is one of the most important factors in the coupling of arylboronic acids with imidazole in the case of compound **14**.

Two MOFs, [Cu(2-pymo)$_2$] (2-pymo = 2-hydroxypyrimidinolate; **15**) and [Co(PhIM)$_2$] (PhIM = phenylimidazolate; **16**), have been successfully used for the aerobic oxidation of tetralin, yielding α-tetralone (T=O) as the main product [33]. The crystal structure of **15** consists of a 3D porous coordination network [34]. The distorted square-planar Cu(II) ions coordinate to four different 2-pymo nitrogen atoms, and each ligand symmetrically bridging two copper centers to give a 3D framework with pores of ~8.1 Å diameter (Figure 8.12). The topological structure of **16** is related to a 3D sodalite-type framework, with tetrahedrally coordinated cobalt(II) ions [35].

Both materials are stable and recyclable under the catalytic reaction conditions. **15** is highly active for the oxidation of tetralin to produce tetralin hydroperoxide (T–OOH), and less efficient in reacting the peroxide. Meanwhile, the use of **16** involves a long induction period for the reaction, but it transforms into T=O rapidly and efficiently. Taking into account the above results, an optimum catalyst was achieved by combining the catalytic advantages of **15** and **16**, that is, minimum induction period, low levels of peroxides, maximum conversion, and high T=O/T–OH ratio, to overcome the drawbacks of the two pure catalysts separately. The combination method gave a convenient strategy for preparing a highly efficient, selective, and reusable catalyst.

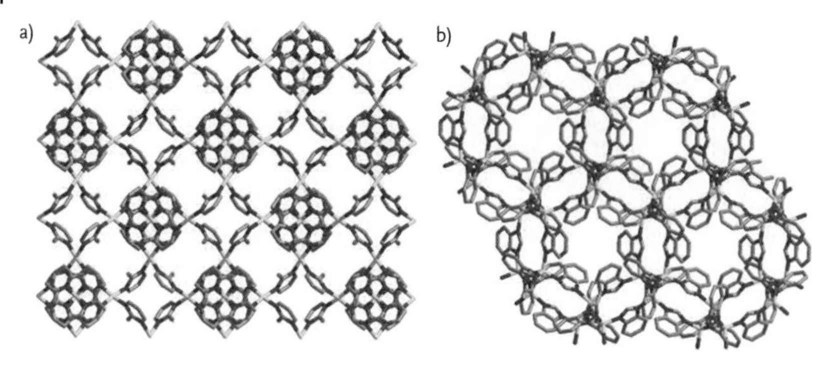

Figure 8.12 The 3D structures of **15** and **16**.

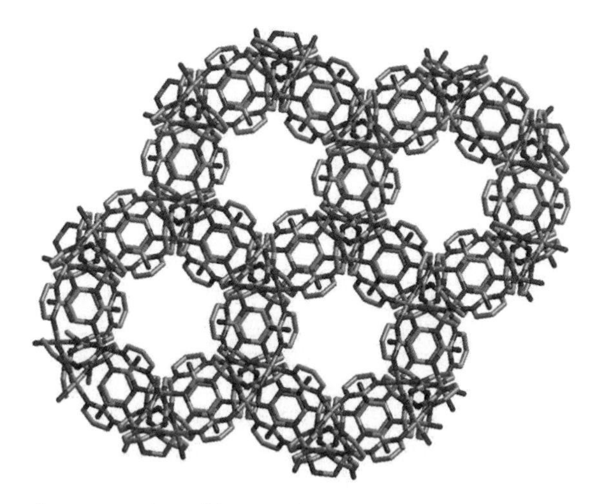

Figure 8.13 The sodalite-type framework of **17**.

Navarro *et al.* reported a porous palladium-containing 3D sodalite-type MOF [Pd(2-pymo)$_2$] (**17**) with BET surface area of 600 m^2 g^{-1} (Figure 8.13) [36]. Garcia *et al.* found that **17** is an active catalyst for alcohol oxidation, Suzuki C–C coupling, and olefin hydrogenation [37]. Suzuki–Miyaura C–C cross-coupling reaction between phenylboronic acid and 4-bromoanisole gave high conversion and selectivity toward the desired product. Alcohol oxidation of 3-phenyl-2-propen-1-ol with **17** gave total conversion of the product with 74% selectivity to cinnamylaldehyde. Shape-selective olefin hydrogenation of 1-octene and cyclododecene showed total conversion of 1-octene after 40 min. In contrast, there was no any disappearance of cyclododecene or formation of cyclododecane even after 5 h at the same reaction condition. PXRD patterns indicate that the structural integrity of the catalyst was maintained during the reaction, and it can be reused without significant diminution of the catalyst activity and selectivity.

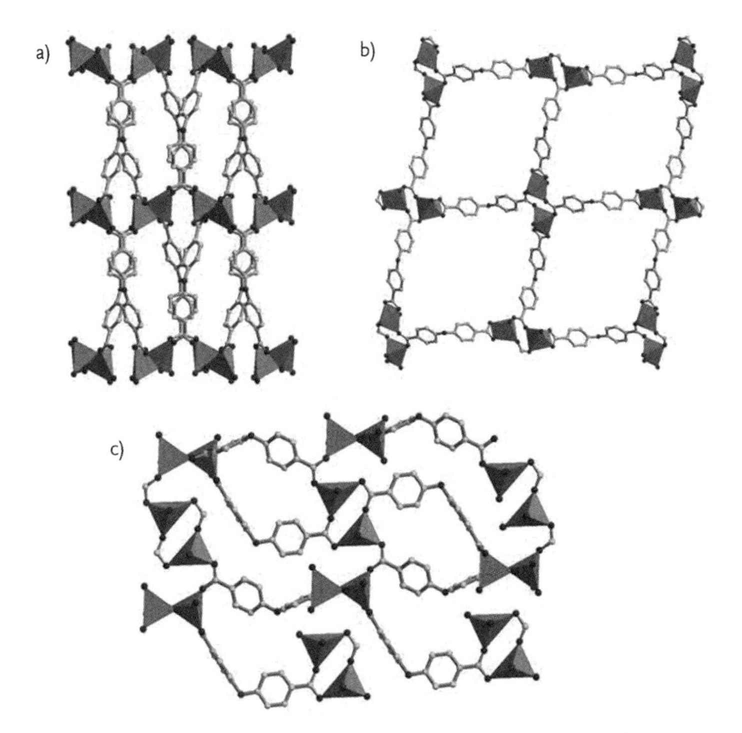

Figure 8.14 Views of the framework structures of **18** (a), **19** (b), and **20** (c).

Natarajan *et al.* reported three 3D MOFs Co_2(4,4′-bpy)$(OBA)_2$, (H_2OBA = 4,4′-oxybis(benzoate) acid; **18**), Ni_2(4,4′-bpy)$_2(OBA)_2 \cdot H_2O$ (**19**), and Zn_2(4,4′-bpy)$(OBA)_2$ (**20**), which were formed by the connectivity of the metal polyhedra ({Co_4N} trigonal bipyramids, {NiO_4N_2} octahedra, {ZnO_4} tetrahedra, and {ZnO_3N_2} trigonal bipyramids), 4,4′-oxybis(benzoate), and 4,4′-bpy (Figure 8.14) [38]. The Co(II) ions are linked by the carboxylate units giving rise to a 1D chain structure, which is further cross-linked by the 4,4′-bpy to form a 3D framework of **18**. Similarly, OBA linking up Ni(II) ions form a 2D porous grid containing eight-membered rings, which are further extended into a 3D framework structure of **19** cross-linked by the 4,4′-bpy. The framework of **20** consists of two kinds of zinc ions, which comprise distorted trigonal-biprismatic geometry and tetrahedral geometry, respectively. The OBA join up Zn(II) ions to form two types of eight-membered rings, which are further connected by the OBA anions to generate a layer structure. The layers are further connected by both the OBA anions and the 4,4′-bpy ligands to form a 3D framework.

The photocatalytic behaviors of the three polymers were tested on four commonly used dyes, orange G, rhodamine B, Remazol Brilliant Blue R, and methylene blue. For the degradation of all four dyes, the order of degradation is **18** > **19** > **20** with activities better than Degussa P-25 TiO_2 catalyst. PXRD patterns clearly indicate that the structures remained the same. As the heterogeneous photocatalytic behaviors of **18**, **19**, and **20** were in a way similar to that in solution, the possible mechanism is through an activated complex involving M(II) sites, which reveals the profitable to investigate metal carboxylates for photocatalytic activity.

Among the reported porous MOFs, the combination of iron(II) or iron(III) atoms in the permanent porosity is scare, despite iron is an environmentally benign and cheap component with nontoxicity and redox properties [39]. Serre *et al.* have reported an iron(III) carboxylate MOF with large accessible and permanent porosity [40]. The solid $[Fe_3O(H_2O)_2F \cdot (BTC)_2] \cdot nH_2O$ ($n = 14.5$; **21**) was hydrothermally synthesized as a polycrystalline powder from a reaction mixture of Fe, 1,3,5-BTC, HF, and HNO_3 in water. The structure of **21** was solved from high-resolution synchrotron PXRD data using chromium BTC compound as the starting model [41]. As shown in Figure 8.15, **21** is built up from trimers of iron(III) octahedra sharing a common vertex μ_3-O. The trimers are then linked by the BTC moieties to the formation of hybrid supertetrahedra which further assemble into an MTN-type zeolitic architecture with mesoporous cages of ca. 25 and 29 Å, and microporous windows of ca. 5.5 and 8.6 Å, respectively. Bond valence calculations indicate a trivalent state of iron, which was confirmed by the transmission

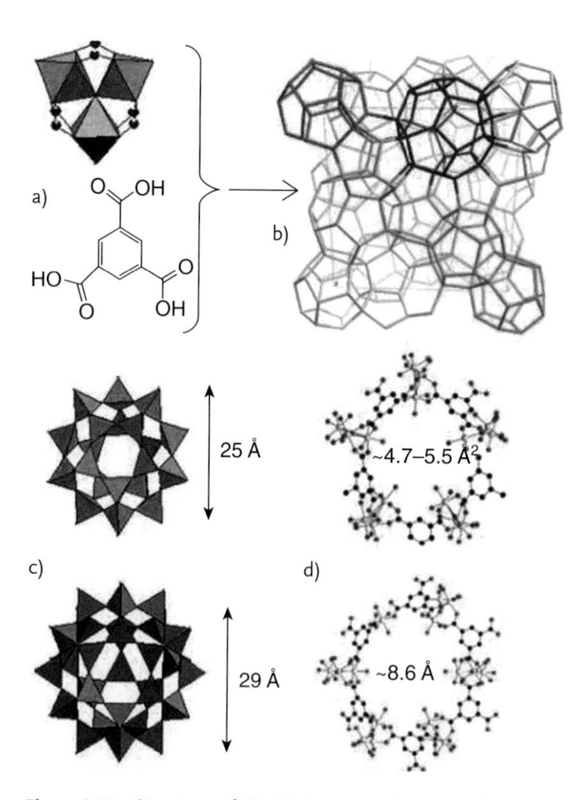

Figure 8.15 Structure of **21**. (a) A trimer of iron octahedra and trimesic acid; (b) schematic view of one unit cell of **21**; (c) the two types of cages in polyhedral mode; (d) pentagonal and hexagonal windows in balls and sticks.

Mössbauer spectra. The Langmuir surface area is estimated to be >2800(100) $m^2 g^{-1}$, while PXRD revealed that the solid is stable up to 270 °C.

Friedel–Crafts benzylation of benzene by benzyl chloride to diphenylmethane was used to confirm the suitability of iron-containing **21** as a porous catalyst. **21** gave high catalytic activity and selectivity, showing complete conversion with nearly 100% selectivity being quickly attained, which is much more efficient than solid acid catalysts such as HBEA and HY zeolites under the same reaction conditions. The observed high benzylation activity of **21** might be attributed to the redox property of trivalent iron species ($Fe^{3+} + e^- \leftrightarrow Fe^{2+}$), which plays a significant role in activating both the reactants.

8.4 Coordinatively Unsaturated Catalytic Metal Ions Exposed in the Pores of MOFs

Multinuclear metal-clustered MOFs can provide an opportunity to create coordinatively unsaturated metal centers in the extended frameworks. Part metal ions occupying the nodes of the network act as connectors and the others bared in the porous channels are accessible for substrates. For high-density hydrogen storage purpose, Long *et al.* have reported a microporous framework, $Mn_3[(Mn_4Cl)_3(BTT)_8(CH_3OH)_{10}]_2$ (**22**; H_3BTT = 1,3,5-benzenetristetrazol-5-yl), baring coordinatively unsaturated metal centers toward the channels [42]. The chloride-centered square-planar $[Mn_4Cl]^{7+}$ units surrounded by eight tetrazolates, together with the trigonal planar nodes presented by the BTT^{3-} ligands to form a 3D 3,8-connected porous cubic network that affords BET surface area of 2100 m^2/g (Figure 8.16). There are two different types of Mn(II) sites that are readily accessible through the porous windows. Site I has one chloride and four bridging tetrazolate ligands, while Site II is bound by just two tetrazolate ligands. Both types of coordinatively unsaturated metal centers are well-positioned to interact with guest molecules that enter the framework pores.

The catalytic experiments demonstrate that the microporous MOF **22**, featuring a high concentration of Lewis acidic Mn(II) sites on its internal surfaces, can catalyze both the cyanosilylation of aromatic aldehydes and the Mukaiyama-aldol reaction [43]. The cyanosilylation of benzaldehyde led to almost quantitative conversion of benzaldehyde. IR of the catalyst confirmed that the activation of the substrates occurred at the unsaturated Mn(II) sites. Wide-dimensional ranged substrates such as 1-naphthaldehyde, 4-phenoxybenzaldehyde, and 4-phenylbenzaldehyde gave a good size-selectivity effect in the cyanosilylation of selected ketones. The Lewis acid catalytic Mukaiyama-aldol reaction of an aldehyde with a silyl enolate is also very active. The observed yield is comparable with the yield obtained for cation-exchanged ZSM-5 or Y-zeolites [44]. The reactions of benzaldehyde with the larger silyl enolates gave an unprecedented size-selective result.

To get permanent porous homochiral MOF, Kim *et al.* have used chiral L-lactic acids (L-H_2lac) to chelate Zn(II) ions to form homochiral secondary building units (SBUs), which are further bridged by BDC spacers to generate a 3D open MOF

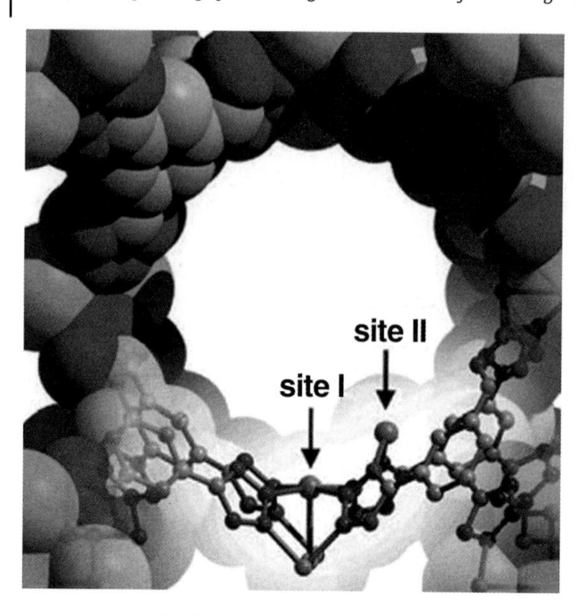

Figure 8.16 Highlight a portion of the crystal structure of **22** showing the two different types of Mn(II) sites exposed within its 3D pore system of 10 Å-wide channel.

$[Zn_2(BDC)(L\text{-lac})(dmf)]\cdot(DMF)$ (**23**·DMF) [45]. Each lactate dianion connects four trigonal-bipyramidaly coordinated Zn(II) ions to extend into 1D chiral chains, which are further interlinked by BDC along the other two directions to form a 3D porous network of **23** (Figure 8.17). The chiral L-lactate moieties are exposed within the voids to result chiral porous environment. TGA and PXRD measurements suggest that the coordinated DMF molecules do not play a direct role in the formation of the MOF architecture, but prevent the porous structure from collapsing.

The porous MOF **23** is capable of mediating highly size- and chemoselective catalytic oxidation of thioethers to sulfoxides by urea hydroperoxide or H_2O_2 (Scheme 8.1). Thioethers with smaller substituents exhibited reasonable conversion and high selectivity, whereas thioethers with bulkier substituents exhibited poor conversion. **23** can perform at least 30 catalytic cycles without loss of oxidation selectivity. Despite the good conversion, no detectable asymmetric induction was found in the catalytic sulfoxidations. However, enantioenriched sulfoxides can be obtained by simultaneously enantioselective sorption of the resulting racemic mixture by the chiral pores of **23**. Thus, the homochiral MOF **23** provides a unique opportunity to produce enantioenriched sulfoxides in a one-pot process, through the size-selective catalytic oxidation of thioethers and the enantioselective separation of the products.

Xu *et al.* have used molecular building block method to constructed a functional MOF $[Na_{20}(Ni_8(\mathbf{L_{24}})_{12})(H_2O)_{28}](H_2O)_{13}(CH_3OH)_2$ ($H_3\mathbf{L_{24}}$ = 4,5-imidazoledicarboxylic acid; **24**) [46]. MOF **24** consists of $[Ni_8(\mathbf{L_{24}})_{12}]^{20-}$ cubic building blocks bridged by Na(I) ions (Figure 8.18). Most of the carboxylate groups in **24** are coordinated in

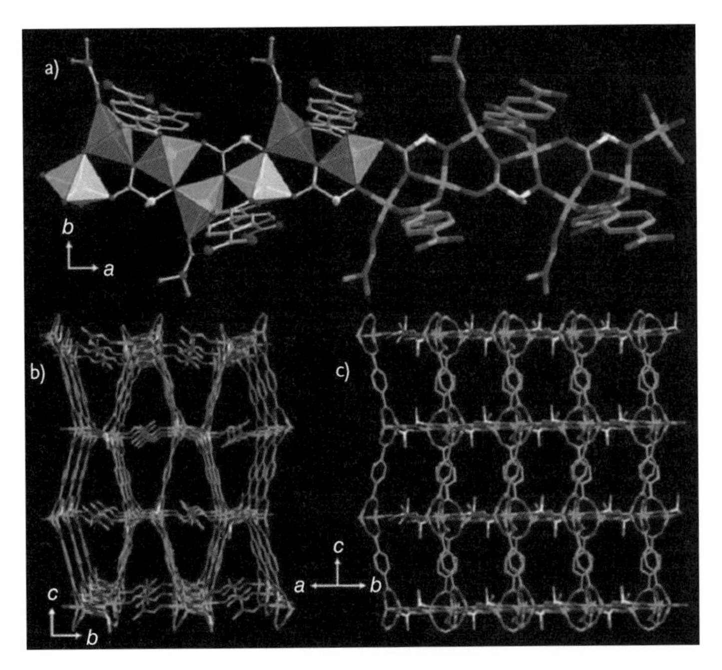

Figure 8.17 (a) A view of the 1D chiral chain in the structure of **23**·DMF; (b) and (c) perspective views of the structure of **23**·DMF.

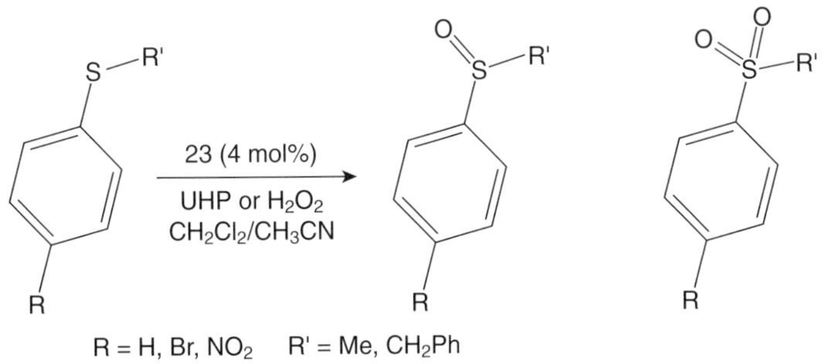

23 (4 mol%)
UHP or H_2O_2
CH_2Cl_2/CH_3CN

R = H, Br, NO_2 R' = Me, CH_2Ph

Scheme 8.1

the μ_2-O mode through Na–O–Na bonds, which further rigidify the porous 3D framework of **24**. TGA and PXRD measurements of **24** indicate that its framework is stable up to 380 °C with BET surface area of 186 $m^2\,g^{-1}$.

24 exhibits stable catalytic activity in the oxidation of CO to CO_2. The reaction rate over **24** was between those of NiO and Ni–Y zeolite at 200 °C, but the activity of **24** is temporally stable while those of Ni–Y zeolite and NiO decreased rapidly at 200 °C. Nearly comparable activity of **24** to NiO is noteworthy because all Ni sites are completely isolated and no oxygen is bound between them. The XRPD patterns of **24** indicate that the framework remains intact after the catalytic reaction.

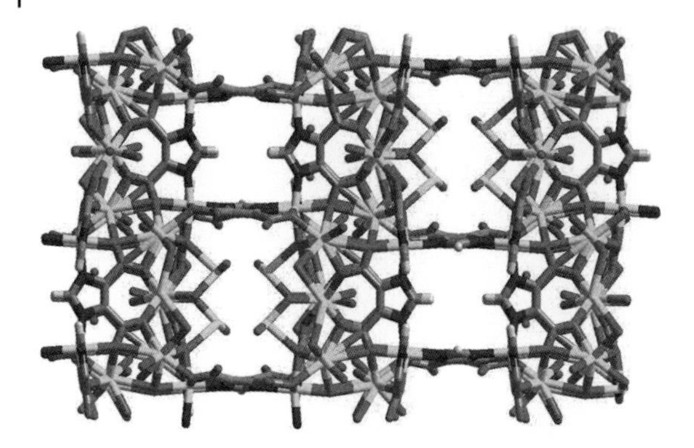

Figure 8.18 The 3D framework of **24** with open channels.

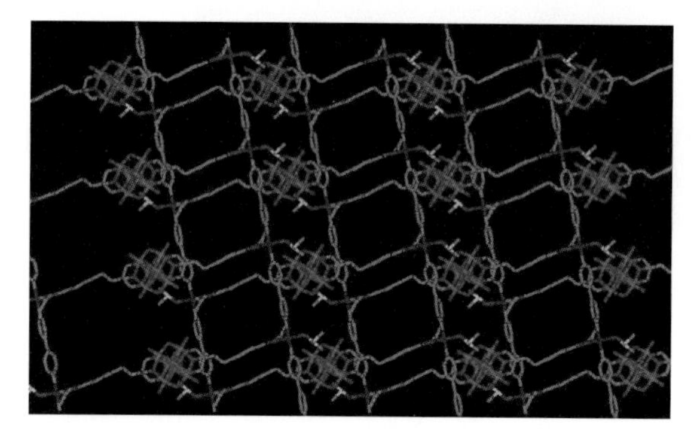

Figure 8.19 A view of the crystal structure of **25**.

The structure of $[Co_2(4,4'\text{-bpy})_2(DMSO)_2(DMF)_2(H_2O)_2(V_6O_{13})\text{-}\{(CH_2O)_3CNHCH_2(4\text{-}C_6H_4COO)\}_2]\cdot 2.1DMF$ (**25**) is a ladder-type coordination network (Figure 8.19) [47]. The Co(II) center octahedrally coordinates to one DMSO, one DMF, one water, one benzoic acid, and two 4,4′-bpy ligands. Co(II) ions are connected by the 4,4′-bpy and $\{V_6O_{13}\}$ units to form a microporous ladder-type coordination network. **25** catalyzed the aerobic oxidation of PrSH to its corresponding disulfide product, PrSSPr, in relative low conversion. In the absence of the catalyst, no PrSSPr was produced, and the supernatant showed no catalytic activity. The FTIR spectra of **25** before and after reaction were identical suggesting that the material is completely stable under the catalytic condition.

8.5
Guest-Accessible Catalytically Functionalized Organic Sites in Porous MOF

To generate guest-accessible functional organic sites into porous MOFs for catalysis, Kitagawa *et al.* have introduced amide groups into a 3D coordination network

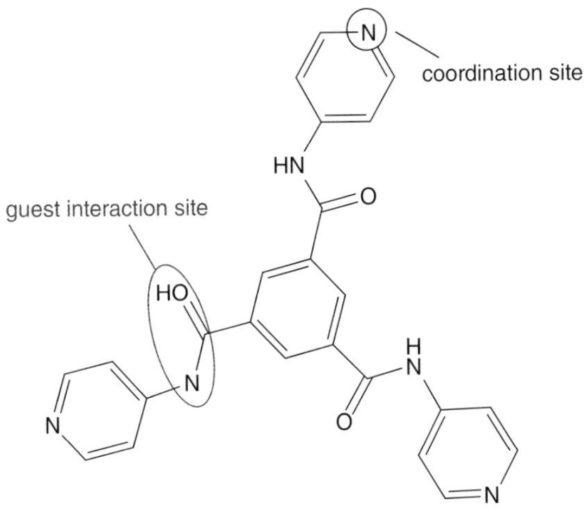

Scheme 8.2

Figure 8.20 Twofold interpenetrating 3D crystal structure of **26** with channels of 4.7×7.3 Å^2 dimensions.

$[Cd(4\text{-btapa})_2(NO_3)_2]\cdot 6H_2O\cdot 2DMF$ (**26**), using a tridentate amide ligand (Scheme 8.2) [48]. The amide group is a fascinating functional group because it possesses two types of hydrogen-bonding sites: the –NH moiety acts as an electron acceptor and the –C=O group acts as an electron donor. The framework of **26** is constructed from the Cd(II) center as a six-connected node using 4-btapa as a three-connected linker (Figure 8.20). Three octahedral Cd(II) moieties are linked together by three 4-btapa units to produce a large six-membered ring that is further extended into an interpenetrated porous 3D network. The amide groups, which are highly ordered on the surfaces of the channels, could interact with guest molecules to generate functional organic sites. TGA and PXRD showed that **26** is stable up to 250 °C and the desolvated amorphous sample can be regenerated into the original crystalline phase immediately after being immersed in methanol or exposed to methanol vapor.

Knoevenagel condensation reaction of benzaldehyde with malononitrile catalyzed by **26** gave 98% conversion, whereas the other bulk substrates reacted negligibly. This guest-selective reaction suggests that the reaction occurs in the channels of **26**. Only as-synthesized **26** promoted the reaction with a good yield compared with desolvated sample or the 4-btapa ligand or $Cd(NO_3)_2 \cdot 4H_2O$. The catalyst **26** is easily isolated from the reaction suspension by filtration and can be reused without loss of activity. Under solvent-free condition, total conversion was observed *via* Knovenagel condensation of benzaldehyde with malononitrile.

8.6
Nanochannel-Promoted Polymerization of Organic Substrates in Porous MOFs

Recent studies have showed that the functionalized porous MOFs bearing specific interaction sites on the nanochannels can be used for catalytic organic polymer synthesis to form well-ordered single polymer chains [49]. Kitagawa *et al.* have pioneered the radical polymerization of styrene in porous MOFs of $[M_2(1,4\text{-bdc})_2(\text{triethylenediamine})]$ ($M = Zn^{2+}, Cu^{2+}$)[50] and $[Cu(pzdc)_2(4,4'\text{-bpy})]$ [51], showing a specific space effect of the host frameworks on the monomer reactivity [49].

Subsequently, they carried out the polymerization of methyl propiolate (MP) in the nanochannels of $[Cu_2(pzdc)_2(4,4'\text{-bpy})]$ (**27**) [52]. The narrow nanochannel structure is directly controlling selective polymerizations of substituted acetylenes in the 1D specific nanochannels of **27** with carboxylate oxygen atoms as catalytic interaction sites on the pore walls. In this experiment, the reaction of neat MP with the sky-blue complex **27** for 12 h at room temperature provided a dark-green powder composite (**27**⊃polyMP). The PXRD patterns of **27**⊃polyMP showed that the channel structure of **27** was preserved during polymerization. Accommodated polyMP was released from the host framework **27** by extraction with DMF at 80 °C. The overall process is a character of the nanochannel-promoted polymerization in **27** (Figure 8.21). Only acidic monosubstituted acetylenes with **27** yield the corre-

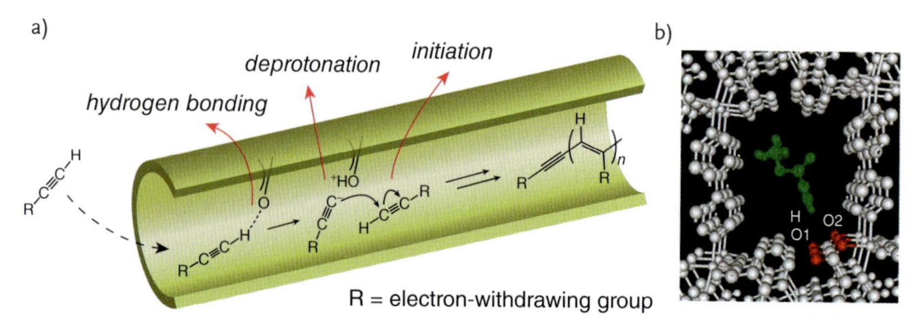

Figure 8.21 (a) Mechanism of the nanochannel-promoted polymerization of acidic acetylenes in porous **27**; (b) optimized structure of MP (green) incorporated in a nanochannel of **27**.

sponding polyacetylenes, which indicates that the strong hydrogen-bonding interaction (acid–base) between the monomers and the channel surface is a key factor to this unique spontaneous polymerization. The polymerization system with MOF **27** features highly accelerated, stereocontrolled, and monomerselective polymerization of substituted acetylenes. In a control experiment, MP was treated with sodium benzoate as a discrete model catalyst that gave only a trace amount of the product.

8.7
Homochiral MOFs Used as Enantioselective Catalysts

Homochiral MOFs are normally constructed from chiral organic ligands to link up metal centers or SBUs. Despite the great progress has been made in the synthesis of homochiral MOFs [12], the homochiral MOFs with permanent porosity for enantioselective catalysis are extremely rare. Kimoon *et al.* reported the first example of asymmetric catalysis using a homochiral MOF [53]. Chiral $\mathbf{L_{28}}$ ligands (Scheme 8.3) bridge Zn(II) atoms to give porous homochiral MOF $[Zn_3(\mu_3\text{-}O)(\mathbf{L_{28}}\text{-}H)_6]\cdot 2H_3O\cdot 12H_2O$ (**28**). In **28**, three zinc ions are held together with six carboxylate groups of $\mathbf{L_{28}}$ and bridging oxygens to form trinuclear units, which are interconnected through three pyridyl groups of $\mathbf{L_{28}}$ to generate a 2D infinite layer consisting of hexagons with trinuclear units at the corners (Figure 8.22). The 2D layers stack together to form large chiral 1D channels to accommodate water molecule guests. Compound **28** lost crystallinity upon evacuation of the solvate molecules, but the PXRD patterns of **28** can be regenerated upon exposure to ethanol or water vapor.

The remaining uncoordinated dangling pyridyl groups in the 1D chiral channels of **28** can catalyze transesterification reaction of **29** (Scheme 8.4) and ethanol to produce ethyl ester in 77% yield. No or little transesterification occurs without **28** or with the *N*-methylated **28**, respectively. Transesterification of **29** with bulkier

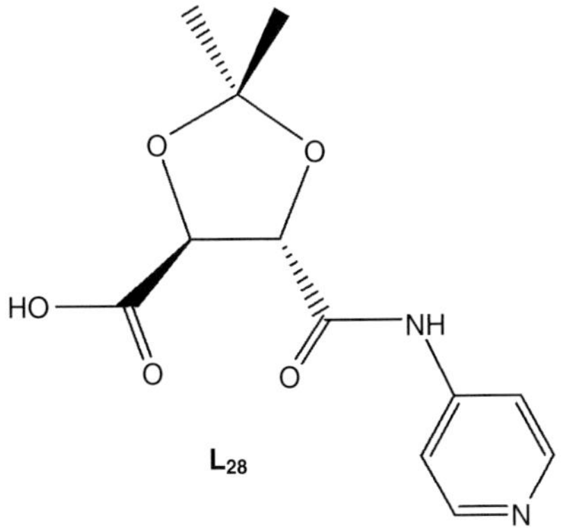

Scheme 8.3

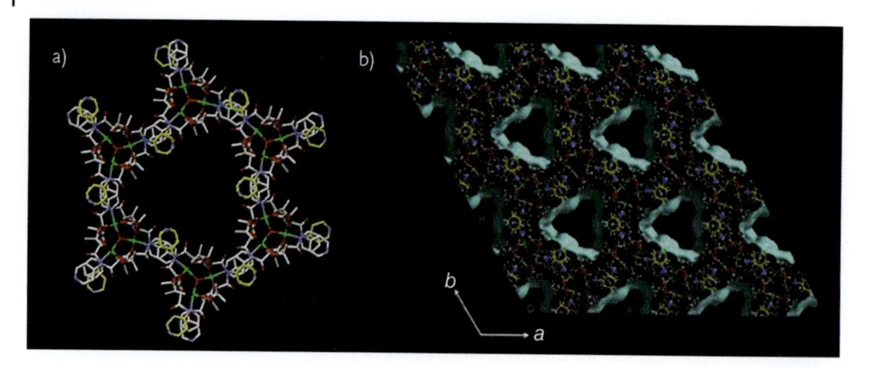

Figure 8.22 (a) The hexagonal framework that is formed with the trinuclear SBUs; (b) a view of the chiral trigonal channels of **28**.

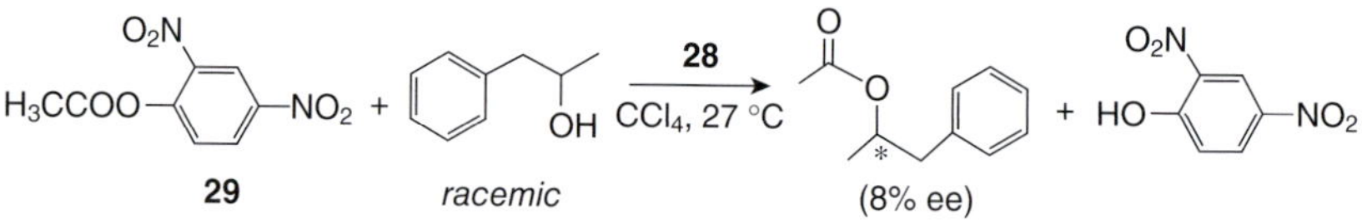

H_3CCOO–(O_2N)–C$_6$H$_3$–NO_2 (**29**) + 1-phenyl-2-propanol (OH, *racemic*) $\xrightarrow[CCl_4,\ 27\ °C]{\mathbf{28}}$ ester (8% ee) + HO–(O_2N)–C$_6$H$_3$–NO_2

Scheme 8.4

L30

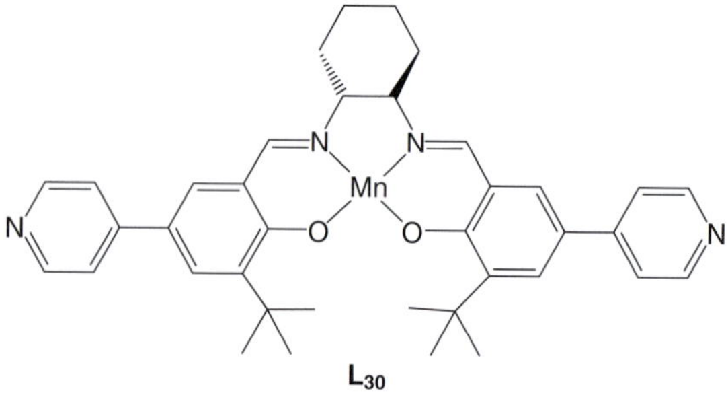

Scheme 8.5

alcohols such as isobutanol, neopentanol, and 3,3,3-triphenyl-1-propanol occurs with a much slower rate under otherwise identical reaction conditions. Such size selectivity suggests that the catalytic process mainly occurred in the channels. **28** was also used to catalyze kinetic resolution of *rac*-1-phenyl-2-propanol *via* transesterification of **29**. The reaction of **29** with a large excess of *rac*-1-phenyl-2-propanol in the presence of **28** produces the corresponding esters with ~8% *ee* in favor of *S* enantiomer.

A homochiral microporous MOF $Zn_2(bpdc)_2\mathbf{L_{30}}\cdot 10DMF\cdot 8H_2O$ (H_2bpdc = biphenyldicarboxylic acid; **30**), containing chiral (salen)Mn SBUs (Scheme 8.5; $\mathbf{L_{30}}$) as a highly effective asymmetric catalyst for olefin epoxidation, was reported by

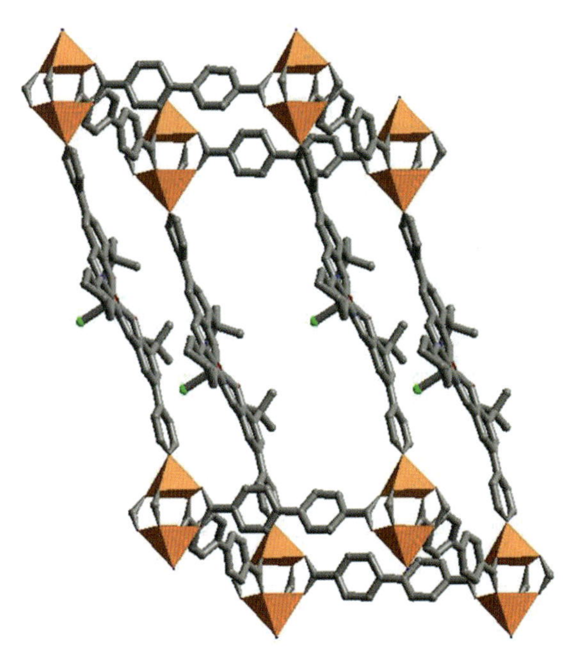

Figure 8.23 The 3D framework structure of **30**.

Cho *et al.* [54]. The structure of **30** is constructed from paddle wheel type Zn(II) dimers and the bridging ligands $\mathbf{L_{30}}$, forming a square network leaving distorted-rectangular and rhombic channels (Figure 8.23). The active species of the Mn(III) sites for oxidation catalysis are incorporated into the bridging ligands, which are accessible to the channels.

The heterogeneous catalytic property of **30** tested the oxidation of 2,2-dimethyl-2H-chorome by 2-(*tert*-butylsulfonyl)iodosylbenzene to give the epoxide with an *ee* value of 82%. Experiments showed that locking $\mathbf{L_{30}}$ into the MOF structure can effectively prevent the encounter of the catalyst with oxidants. The catalyst did not lose enantioselectivity after three cycles and only a small loss of activity. The plasma spectroscopy of **30** showed the small quantity of dissolved manganese but did not catalyze the reaction.

In order to enhance the enantioselective properties of homochiral MOFs, Lin *et al.* have rationally designed a BINOL-based multifunctional chiral bridging ligand ((*R*)-6,6′-dichloro-2,2′-dihydroxy-1,1′-binaphthyl-4,4′-bipyridine, $\mathbf{L_{31}}$) for the construction of a homochiral porous MOF $[Cd_3Cl_6(\mathbf{L_{31}})_3]\cdot 4DMF\cdot 6MeOH\cdot 3H_2O$ (**31**) [55]. The Cd(II) ions in **31** are doubly bridged by the chlorides to form 1D zigzag $[CdCl_2]_n$ chain SBUs, which are further linked by $\mathbf{L_{31}}$ to form a noninterpenetrating 3D network with very large chiral channels of $\sim 1.6 \times 1.8\,nm^2$ (Figure 8.24). PXRD patterns demonstrate that the framework of **31** was maintained upon the removal of all the solvent molecules with specific surface area of $601\,m^2/g$.

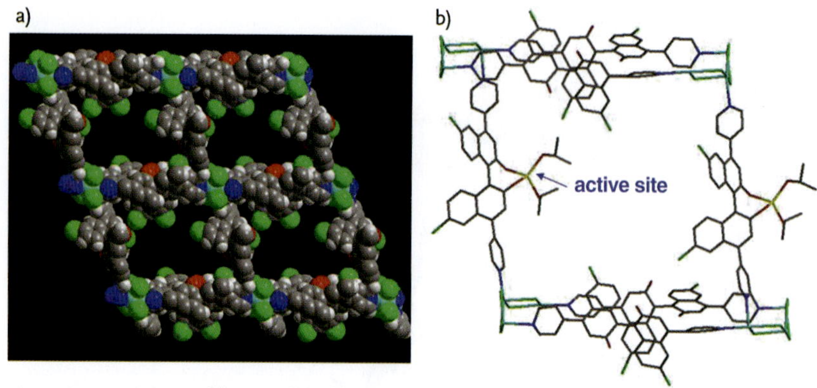

Figure 8.24 (a) Space-filling model of **31**, showing the large chiral 1D channels; (b) a schematic representation of the active (BINOLate)Ti(OiPr)$_2$ catalytic sites in the open channels of **31·Ti**.

Homochiral MOF **31** was utilized in heterogeneous asymmetric catalysis by taking advantage of the chiral dihydroxy groups that are readily accessible through the large open channels. Treatment of the dried sample **31** with excess $Ti(O^iPr)_4$ led to an active catalyst (designated as **31·Ti**) for the diethylzinc addition reaction. **31·Ti** catalyzed the addition of diethylzinc to a range of aromatic aldehydes with complete conversions and *ee* values rival that of the homogeneous analogs under similar conditions. Control experiments with a series of dendriticaldehydes of varying sizes (from 0.8 to 2.0 nm) demonstrate that **31·Ti** is a true heterogeneous asymmetric catalyst, and both diethylzinc and aromatic aldehydes are accessing the (BINOLate)$Ti(O^iPr)_2$ sites *via* the open channels to generate the chiral secondary alcohol products.

The single-crystalline nature of MOFs allows to examine the structure–catalytic activity relationship of porous MOFs in great details. To understand the relationship of catalytic activities and the framework structures, Lin *et al.* reported two homochiral MOFs built from the same chiral connectors and metal linkers [56]. Homochiral MOF $[Cd_3(\mathbf{L}_{31})_4(NO_3)_6]\cdot 7MeOH\cdot 5H_2O$ (**32**) was formed by two polymeric subunits, which are $\mathbf{L}_{31}$ ligands linked 2D square-grid layer and zigzag polymeric chain. The 2D grids and the 1D zigzag chains are joined to each other by the bridging nitrate groups to form a twofold interpenetrated open 3D framework structure (Figure 8.25). PLATON calculation, PXRD, and CO_2 adsorption measurements confirmed the porosity and stability of **32**. Another 2D rhombic grid homochiral MOF $[Cd\mathbf{L}_2(H_2O)_2][ClO_4]_2\cdot DMF\cdot 4MeOH\cdot 3H_2O$ (**33**) was constructed from Cd(II) ions coordinating to $\mathbf{L}_{31}$ pyridyl groups in the equatorial plane (Figure 8.26). Two sets of the 2D rhombic grids interpenetrate to each other to form a 3D network.

Similar to **31**, the generated active catalyst (**32**·Ti) is highly effective for the diethylzinc addition to aromatic aldehydes in very high yields and high enantioselectivities. However, under identical conditions, a mixture of **33** and $Ti(O^iPr)_4$ was

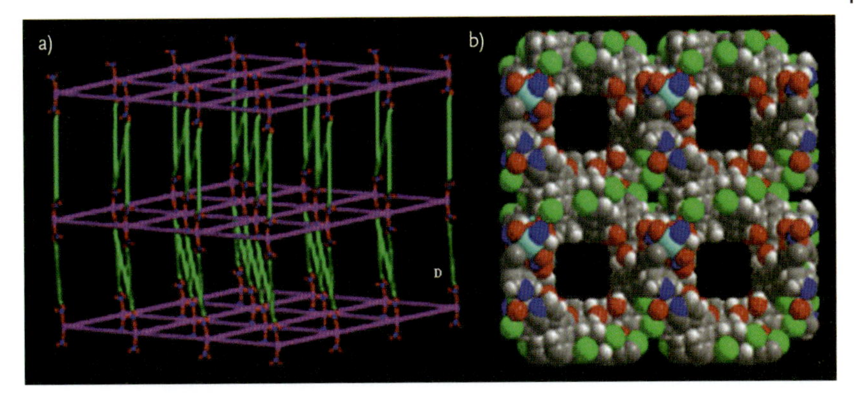

Figure 8.25 (a) Schematic representation of the 3D framework of **32**; (b) space-filling model of **32** showing the chiral 1D channels of $13.5 \times 13.5\,\text{Å}^2$.

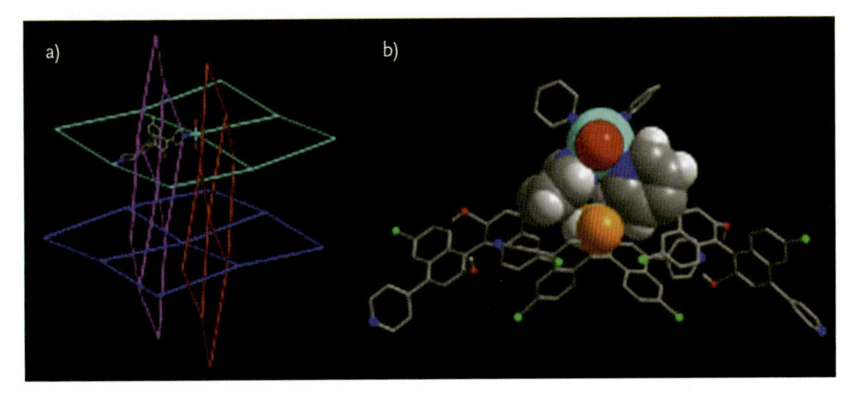

Figure 8.26 (a) Schematic showing the interpenetration of mutually perpendicular 2D rhombic grids in **33**; (b) a drawing showing the steric congestion around chiral dihydroxy groups of $\mathbf{L_{31}}$ ligand (orange balls).

not active in this reaction. A closer examination of the structure of **33** reveals that the lack of catalytic activity of the **33**/Ti(O^{i}Pr)$_4$ system is a result of the steric congestion around the chiral dihydroxy groups which prevents the substitution of two isopropoxide groups by the BINOLate functionality. The drastically different catalytic activities observed for the **32**/Ti(O^{i}Pr)$_4$ and **33**/Ti(O^{i}Pr)$_4$ systems highlight the important role of the frameworks in determining the catalytic performances.

8.8 Conclusions and Outlook

This chapter highlights recent advances in the area of MOFs for heterogeneous catalysis. It is evident from the examples illustrated in this chapter that catalytic functionalities of MOFs can be designed rationally. The topologies of MOFs are defined by the arrangements of the metal coordination sites with the orientation and number of coordinating donor groups in the multifunctional linkers. However, most structural arrangements are not desirable for the application of MOFs in heterogeneous catalysis, because connecting the coordination sites to the linkers often block active sites. Precise selection of the structural SBUs and the ways in which they are connected allows systematic modification of the pore structures that are responsible for the high catalytic activities. The active sites should be freely accessible to reagent molecules; preferably the sites must be oriented toward the interior pores that are accessible for molecules. There are some precedents with accessible metal sites, opening the possibility of their use in catalysis. Most of them containing the metal sites coordinate to volatile guests, which can be removed by suitable techniques to generate catalytic centers. However, the dissociation of coordinating block molecules inevitably destroys the frameworks in most cases, which hampered the applications in heterogeneously catalytic processes.

A promising approach to the rational assembly of networks is the use of molecular building blocks to fix the catalytically active sites in rigid lattices as basic units to be self-existent apart from the whole frameworks. The facile tunability should allow precise engineering of a multitude of catalytic active properties *via* systematic tuning of the molecular modules. Many interesting functional materials for heterogeneous catalysis can be expected from this burgeoning field of the solid-state supramolecular chemistry.

Acknowledgments

Funding support from the NSF of China (Grant No. 21073158), Zhejiang Provincial Natural Science Foundation of China (Grant No. Z4100038), and the Fundamental Research Funds for the Central Universities (Grant No. 2010QNA3013) made the writing of this chapter possible.

References

1 Moulton, B. and Zaworotko, M.J. (2001) *Chem. Rev.*, **101**, 1629.
2 Batten, S.R. and Robson, R. (1998) *Angew. Chem. Int. Ed. Engl.*, **37**, 1461.
3 Munakata, M., Wu, L.P., and Kuroda-Sowa, T. (1999) *Adv. Inorg. Chem.*, **46**, 173.
4 Eddaoudi, M., Moler, D.B., Li, H., Chen, B., Reineke, T.M., O'Keeffe, M., and Yaghi, O.M. (2001) *Acc. Chem. Res.*, **34**, 319.
5 Evans, O.R. and Lin, W. (2002) *Acc. Chem. Res.*, **35**, 511.
6 Lee, S., Mallik, A.B., Xu, Z., Lobkovsky, E.B., and Tran, L. (2005) *Acc. Chem. Res.*, **38**, 251.
7 Suslick, K.S., Bhyrappa, P., Chou, J.-H., Kosal, M.E., Nakagaki, S., Smithenry, D.W., and Wilson, S.R. (2005) *Acc. Chem. Res.*, **38**, 283.
8 Kitagawa, S., Kitaura, R., and Noro, S.-I. (2004) *Angew. Chem. Int. Ed.*, **43**, 2334.
9 Morris, R.E. and Wheatley, P.S. (2008) *Angew. Chem. Int. Ed.*, **47**, 4966.
10 Gavezzotti, A. (1994) *Acc. Chem. Res.*, **27**, 309.

11 Hagrman, P.J., Hagrman, D., and Zubieta, J. (1999) *Angew. Chem. Int. Ed.*, **38**, 2638.
12 Kesanli, B. and Lin, W. (2003) *Coord. Chem. Rev.*, **246**, 305.
13 Cheetham, A.K., Ferey, G., and Loiseau, T. (1999) *Angew. Chem. Int. Ed.*, **38**, 3268.
14 Wells, A.F. (1977) *Three Dimensional Nets and Polyhedra*, John Wiley & Sons, Inc., New York.
15 O'Keeffe, M. and Hyde, B.G. (1996) *Crystal Structure I: Patterns and Symmetry*, American Mineralogical Association, Washington.
16 Yaghi, O.M., Davis, C., H.L., Richardson, D., Groy, and T.L. (1998) *Acc. Chem. Res.*, **31**, 474.
17 Munakata, M., L.P. Wu, and T. Kuroda-Sowa (1999) *Adv. Inorg. Chem.*, **46**, 173.
18 Janiak, C. (1997) *Angew. Chem. Int. Ed. Engl.*, **36**, 1431.
19 Batten, S.R. and Robson, R. (1998) *Angew. Chem. Int. Ed. Eng.*, **37**, 1460.
20 Breck, D.W. (1974) *Zeolite Molecular Sieves, Structure, Chemistry, and Use*, John Wiley & Sons, Inc., New York.
21 Fujita, M., Kwon, Y.J., Washizu, S., and Ogura, K. (1994) *J. Am. Chem. Soc.*, **116**, 1151.
22 Ohmori, O. and Fujita, M. (2004) *Chem. Commun.*, 1586.
23 Chui, S.S.Y., Lo, S.M.F., Charmant, J.P.H., Orpen, A.G., and Williams, I.D. (1999) *Science*, **283**, 1148.
24 Schlichte, K., Kratzke, T., and Kaskel, S. (2004) *Microporous Mesoporous Mater.*, **73**, 81.
25 Alaerts, L., Séguin, E., Poelman, H., Thibault-Starzyk, F., Jacobs, P.A., and Vos, D.E.D. (2006) *Chem. Eur. J.*, **12**, 7353.
26 Evans, O.R., Ngo, H.L., and Lin, W.B. (2001) *J. Am. Chem. Soc.*, **123**, 10395.
27 Gomez-Lor, B., Gutiérrez-Puebla, E., Iglesias, M., Monge, M.A., Ruiz-Valero, C., and Snejko, N. (2002) *Inorg. Chem.*, **41**, 2429.
28 Gomez-Lor, B., Gutierrez-Puebla, E., Iglesias, M., Monge, M.A., Ruiz-Valero, C., and Snejko, N. (2005) *Chem. Mater.*, **17**, 2568.
29 Gándara, F., Gomez-Lor, B., Gutiérrez-Puebla, E., Iglesias, M., Monge, M.A., Proserpio, D.M., and Snejko, N. (2008) *Chem. Mater.*, **20**, 72.
30 Perles, J., Iglesias, M., Martín-Luengo, M.-A., Monge, M.A., Ruiz-Valero, C., and Snejko, N. (2005) *Chem. Mater.*, **17**, 5837.
31 Tanski, J.M. and Wolczanski, P.T. (2001) *Inorg. Chem.*, **40**, 2026.
32 Wu, C.-D., Li, L., and Shi, L.-X. (2009) *Dalton Trans.*, 6790–6794.
33 Xamena, F.X.L.i., Casanova, O., Tailleur, R.G., Garcia, H., and Corma, A. (2008) *J. Cat.*, **255**, 220.
34 Tabares, L.C., Navarro, J.A.R., and Salas, J.M. (2001) *J. Am. Chem. Soc.*, **123**, 383.
35 Park, K.S., Ni, Z., Cote, A.P., Choi, J.Y., Huang, R.D., Uribe-Romo, F.J., Chae, H.K., O, M., Keeffe, O.M., and Yaghi (2006) *Proc. Natl. Acad. Sci.*, **103**, 10186.
36 Navarro, J.A.R., Barea, E., Salas, J.M., Masciocchi, N., Galli, S., Sironi, A., Ania, C.O., and Parra, J.B. (2006) *Inorg. Chem.*, **45**, 2397.
37 Xamena, F.X.L.i., Abad, A., Corma, A., and Garcia, H. (2007) *J. Cata.*, **250**, 294.
38 Mahata, P., Madras, G., and Natarajan, S. (2006) *J. Phys. Chem. B*, **110**, 13759.
39 (a) Sudik, A.C., Côté, A.P., Wong-Foy, A.G., O'Keefe, M., and Yaghi, O.M. (2005) *Angew. Chem. Int. Ed.*, **118**, 2590; (b) Jia, J., Lin, X., Wilson, C., Blake, A. J., Champness, N. R., Hubberstey, P., Walker, G., Cussena, E. J., and Schröder, M. (2007) *Chem. Commun.*, 840.
40 Horcajada, P., Surblé, S., Serre, C., Hong, D.-Y., Seo, Y.-K., Chang, J.-S., Grenèche, J.-M., Margiolakid, I., and Féreya, G. (2007) *Chem. Commun.*, 2820.
41 Férey, G., Serre, C., Mellot-Draznieks, C., Millange, F., Surblé, S., Dutour, J., and Margiolaki, I. (2004) *Angew. Chem. Int. Ed.*, **43**, 6296.
42 Dincă, M., Dailly, A., Liu, Y., Brown, C.M., Neumann, D.A., and Long, J.R. (2006) *J. Am. Chem. Soc.*, **128**, 16876.
43 Horike, S., Dincă, M., Tamaki, K., and Long, J.R. (2008) *J. Am. Chem. Soc.*, **130**, 5854.
44 (a) Sasidharan, M. and Kumar, R. (2003) *J. Catal.*, **220**, 326; (b) Garro, R., Navarro, M. T., Primo, J., and Corma, A. (2005) *J. Catal.*, **233**, 342.
45 Dybtsev, D.N., Nuzhdin, A.L., Chun, H., Bryliakov, K.P., Talsi, E.P., Fedin, V.P., and Kim, K. (2006) *Angew. Chem. Int. Ed.*, **45**, 916.

46 Zou, R.-Q., Sakurai, H., and Xu, Q. (2006) *Angew. Chem. Int. Ed.*, **45**, 2542.

47 Hill, C.L., Anderson, T.M., Han, J.W., Hillesheim, D.A., Geletii, Y.V., Okun, N.M., Cao, R., Botar, B., Musaev, D.G., and Morokuma, K. (2006) *J. Mol. Catal., A Chem.*, **251**, 234.

48 Hasegawa, S., Horike, S., Matsuda, R., Furukawa, S., Mochizuki, K., Kinoshita, Y., and Kitagawa, S. (2007) *J. Am. Chem. Soc.*, **129**, 2607.

49 Uemura, T., Kitagawa, K., Horike, S., Kawamura, T., Kitagawa, S., Mizuno, M., and Endo, K. (2005) *Chem. Commun.*, 5968.

50 (a) Dybtsev, D.N., Chun, H., and Kim, K. (2004) *Angew. Chem. Int. Ed.*, **43**, 5033; (b) Chun, H., Dybtsev, D. N., Kimand, H., and Kim, K. (2005) *Chem. Eur. J.*, **11**, 3521; (c) Seki, K. and Mori, Y. (2002) *J. Phys. Chem. B*, **106**, 1380; (d) Seki, K. (2002) *Langmuir*, **18**, 2441; (e) Kitaura, R., Iwahori, F., Matsuda, M., Kitagawa, S., Kubota, Y., Takata, M., and Kobayashi, T. C. (2004) *Inorg. Chem.*, **43**, 6522.

51 (a) Kondo, M., Okubo, T., Asami, A., Noro, S., Yoshitomi, T., Kitagawa, S., Ishii, T., Matsuzaka, H., and Seki, K. (1999) *Angew. Chem. Int. Ed.*, **38**, 140; (b) Matsuda, R., Kitaura, R., Kubota, Y., Kobayashi, T. C., Horike, S., and Takata, M. (2004) *J. Am. Chem. Soc.*, **126**, 14063.

52 Uemura, T., Kitaura, R., Ohta, Y., Nagaoka, M., and Kitagawa, S. (2006) *Angew. Chem. Int. Ed.*, **45**, 4112.

53 Seo, J.S., Whang, D., Lee, H., Jun, S.I., Oh, J., Jeon, Y.J., and Kimoon, K. (2000) *Nature*, **404**, 982.

54 Cho, S.-H., Ma, B., Nguyen, S.T., Hupp, J.T., and Albrecht-Schmitt, T.E. (2006) *Chem. Commun.*, 2563.

55 Wu, C.-D., Hu, A., Zhang, L., and Lin, W. (2005) *J. Am. Chem. Soc.*, **127**, 8940.

56 Wu, C.-D. and Lin, W. (2007) *Angew. Chem., Int. Ed.*, **46**, 1075.

9
Mechanism of Stereospecific Propene Polymerization Promoted by Metallocene and Nonmetallocene Catalysts

Andrea Correa and Luigi Cavallo

9.1
Introduction

The discovery of stereoregular polymers and of the catalysts needed to synthesize them not only has broken a monopoly of Nature [1] but provided us also with new and extremely versatile materials, that is, thermoplastic polyolefins. Indeed, few discoveries have had so much impact on the everyday life as the one made by Ziegler and Natta in the 1950s. Today, roughly 50 000 000 tons/year of isotactic polypropylene (see Scheme 9.1) are produced worldwide, and the large majority of this impressive amount is produced by heterogeneous $MgCl_2/TiCl_4$-based catalytic systems, which are a straight derivation of the $TiCl_3$-based catalytic systems discovered by Ziegler for ethene polymerization [2], and subsequently used by Natta for the first synthesis of isotactic polypropylene [3]. However, despite of the wonderful performances of the heterogeneous catalysts, a detailed knowledge of these systems at atomistic level is still rather scarce [4].

Thus, it is no surprise that immediately after the sparkling discovery many attempts have been made to synthesize catalysts with better defined chemistry, and $Cp_2TiCl_2/AlRCl_2$ or AlR_3 (Cp = cyclopentadienyl, R = alkyl group) was used as early as 1957 [5–7]. However, these early homogeneous catalysts had a quite low activity in ethene polymerization and failed to homopolymerize 1-olefins. Analogous research with Cp_2ZrCl_2/AlR_3 met with limited success [8]. In short, for almost 30 years, little steps forward were made. Probably, the only exception are the vanadium-based catalytic systems [9–11] that lead to syndiotactic polypropylene (see Scheme 9.1) and are industrially used for the synthesis of elastomeric ethene–propene copolymers, but are even less defined than the classical heterogeneous Ziegler–Natta catalytic systems.

The giant step forward was made in the 1980s, with the serendipitous discovery of the activating effect of small amounts of water [12] on the system $Cp_2MtX_2/AlMe_3$ (X = Cl or alkyl group) [13]. The subsequent controlled synthesis of methylalumoxane (known as MAO) by the group of Sinn and Kaminsky [14] has provided organometallic and polymer chemists with a potent cocatalyst able to activate group 4 metallocenes (and subsequently a large number of other transition metal

Selective Nanocatalysts and Nanoscience, First Edition. Edited by Adriano Zecchina, Silvia Bordiga, Elena Groppo.

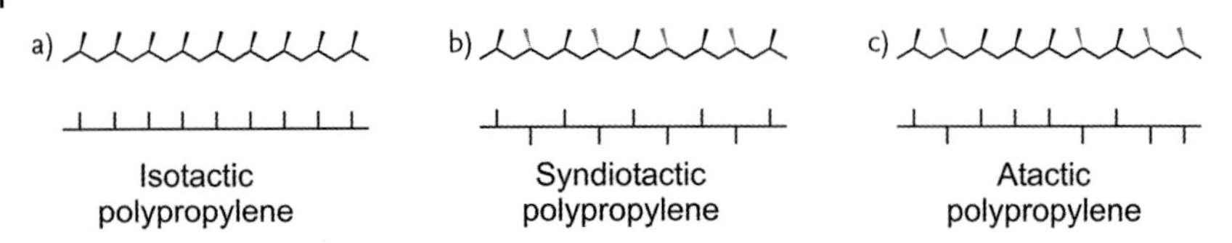

Scheme 9.1

complexes) toward the polymerization of virtually any 1-olefins, including several cyclic olefins.

This ability to activate organometallic complexes to effective polymerization catalysts initiated 20 years of impressive academic and industrial efforts, and as a result, a great number of effective precatalysts and activators with very different structures have been designed. Besides a huge number of group 4 metallocenes and half-metallocenes (the *ansa*-monocyclopentadienyl amido complexes also known as constrained-geometry catalysts) [15, 16], nonmetallocenes catalysts ranging from tetrahedral to octahedral geometries have been synthesized. The variety of complexes and of activators is too large to be listed here, and we refer the reader to specialized and comprehensive reviews [11, 17–29].

During the last 20 years, the very well characterized chemistry of metallocene and nonmetallocene catalysts offered the exceptional opportunity to investigate in details the mechanisms that control their behavior, and the way information is transferred from the catalyst structure to the microstructure of the produced polymers. Indeed, the understanding we have at molecular level of all these systems is very detailed. We know the factors that control stereospecifity, regiospecificity, molecular masses, and, in some cases, even activity. This knowledge has also been used to acquire insight into the more complicate and less-characterized heterogenous Ziegler–Natta systems [30–38]. Further, the same kind of reasoning has been applied to rationalize the $(E)/(Z)$ selectivity in the copolymerization of ethene with 2-butene [39, 40], the stereospecificity in the Ziegler–Natta 1,2, and *cis*-1,4 polymerizations of conjugated dienes [41, 42], and the syndiospecific polymerization of styrene [43, 44].

In this chapter, we will discuss the chemical mechanics of a series of catalysts that, in our opinion, represent milestones in this wonderful story. These systems are shown in Scheme 9.2.

To this end, we organized this chapter as follows: in Section 9.2.1, we will discuss the mechanism of chain growth, while in Section 9.2.2, we will discuss the regioselective behavior of the systems shown in Scheme 9.2. In Section 9.3, we will introduce the elements of chirality that are needed to rationalize the mechanism of stereoselectivity that will be discussed in Sections 9.4–9.6. In Section 9.4.1, we will introduce the mechanism of stereocontrol using the well-defined $Me_2Si(1\text{-Ind})_2ZrCl_2$ (Ind = indenyl) system [45, 46] as an example of the C_2-symmetric metallocenes introduced by Ewen and Brintzinger around 1985 for the synthesis of isotactic polypropylene [47, 48]. In Section 9.4.2, we will show that the same mechanism can be applied to rationalize the isotacticity of the

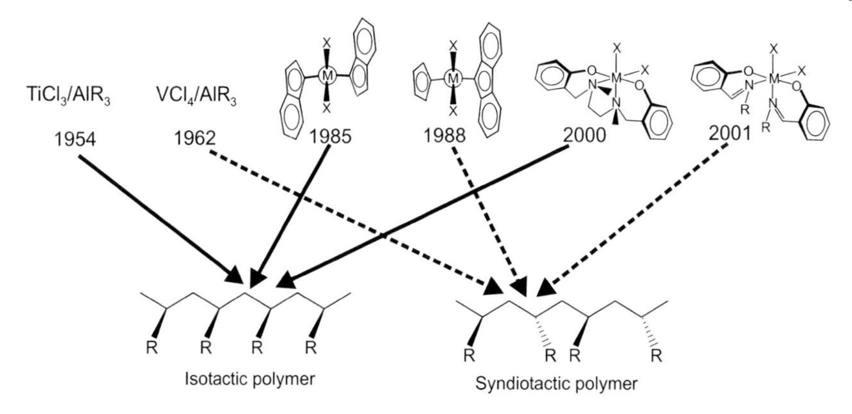

Scheme 9.2

well-defined nonmetallocene octahedral catalysts developed by Kol in 2000, and based on bridged bis(phenoxy-amine) catalysts [49, 50], and that this mechanism can be extended to rationalize the isospecificity of poorly defined heterogeneous Ziegler–Natta systems. In Section 9.5, we will show that the mechanism developed to rationalize the behavior of isospecific metallocenes can be applied *tout court* to rationalize the behavior of the well-defined $Me_2C(Cp)(9\text{-}Flu)ZrCl_2$ (Flu = fluorenyl) metallocene system introduced by Ewen in 1998 for the synthesis of syndiotactic polypropylene [51].

Conversely, in Section 9.6.1, we will show that a somewhat different mechanism has to be applied to rationalize the syndiospecific behavior of the well-defined nonmetallocene octahedral catalysts based on unbridged bis(phenoxy-imine) complexes [52, 53] developed by Fujita at Mitsui Chemicals in 1999 for the polymerization of ethene and 1-olefins [54–56], and proposed as catalyst for the syndiospecific polymerization of propene around 2001 [57, 58]. Finally, in Section 9.6.2, we will show that the mechanism developed to rationalize the behavior of syndiospecific unbridged octahedral catalysts can be extended to rationalize the syndiospecificity of the poorly defined homogeneous V-based Ziegler–Natta catalysts [9, 10].

9.2 Mechanism of Polymerization

9.2.1 The Chain Growth Reaction

Although a few neutral group 4 catalysts have been synthesized [59–62], almost all effective group 4 complexes are inactive in polymerization if they are not activated by a suitable cocatalyst. Activation and formation of the cationic species is accomplished through a suitable activating species, the cocatalyst. The activator

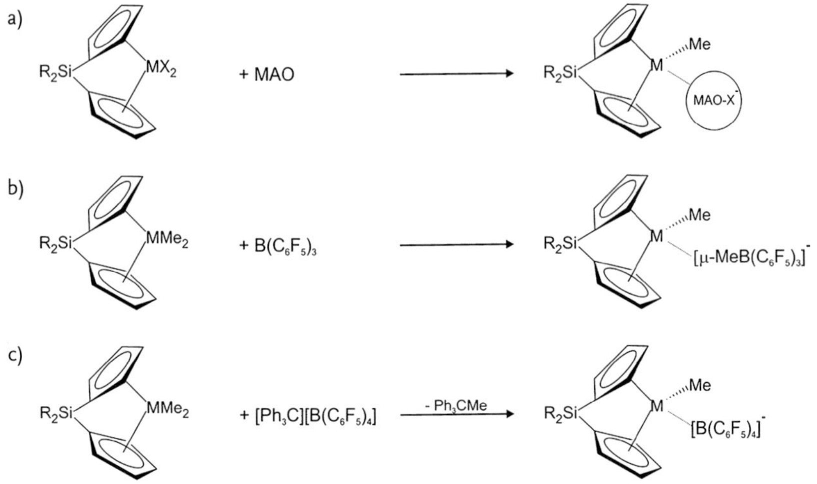

Chart 1

becomes an anion after the activation process, see Chart 1, forming a cation–anion pair, which is now accepted to be the real catalytically active polymerization species. Consequently, it is not a surprise that the ion pair has been quite investigated both experimentally [28, 63–73] and theoretically [74–82], but probably is one of the few points of this catalysis that are still not understood in full details, due to its intrinsically flexible and complicated structure.

Different activators can result in dramatic differences in the activity of a given precatalyst structure [27, 28]. Indeed, the metallocene revolution was started by the discovery of MAO [14], and the development of new catalytic systems has been sided by the development of new activators. Indubitably, beside MAO that exists in solution as an equilibrium of species with different aggregation numbers and structures [83–86], the most famous activators are noncoordinating borane and borate salts such as $B(C_6F_5)_3$ [87–90] and $[Ph_3C^+][B(C_6F_5)^-_4]$ [91, 92]. Excellent reviews on the subject have appeared [11, 27–29].

The propagating active site in olefin polymerizations mediated by group 4 catalysts is the M–C (polymer) bond of a metal–alkyl complex [93–97], and the most accepted mechanism for the chain growth reaction is based on that proposed by Cossee in 1964 for the heterogeneous $TiCl_3$-based catalytic systems [98]. This mechanism is monometallic and the active center is a transition metal–carbon bond (see Figure 9.1). It basically occurs in two steps: (i) olefin coordination to a vacant site and (ii) olefin insertion into the metal-growing chain bond through a *cis*-opening of the olefin double bond. Green *et al* slightly modified this mechanism, introducing an α-agostic interaction that would facilitate the insertion reaction [99]. Of course, in the case of homogeneous cationic catalysts, the original Cossee mechanism was modified to include a more or less tightly coordinating counterion.

It is important to note that at the end of the chain growth step, the growing chain migrates to the coordination position previously occupied by the monomer.

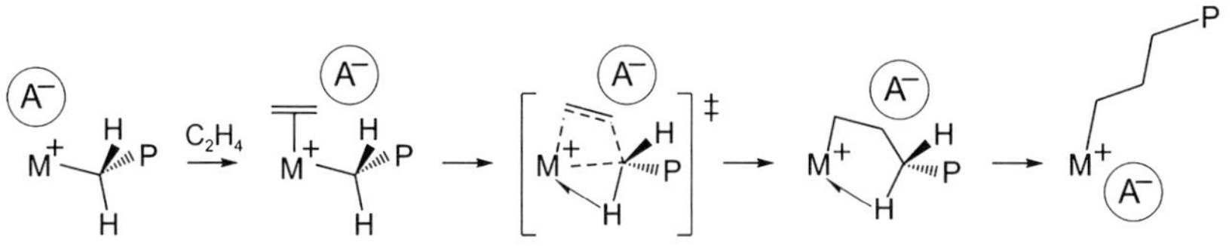

Figure 9.1 Modified Cossee mechanism. A^- is the counterion.

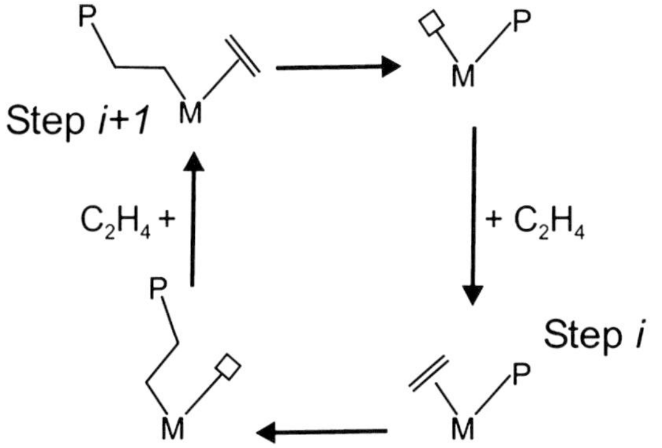

Scheme 9.3

That is, at successive chain growth steps, the relative coordination positions of the monomer and the growing chain are exchanged (chain migratory mechanism, see Scheme 9.3).

9.2.2
Regioselectivity of Propene Insertion

Fundamental to understand 1-olefin polymerizations are the concepts of regio- and stereochemistry of monomer insertion. Regiochemistry of 1-olefins polymerization can occur via 1,2 (or primary) as well as via 2,1 (or secondary) insertion, which lead to growing chains with a primary or a secondary C atom bonded to the metal atom, respectively (Scheme 9.4).

Primary insertion is largely dominant with the traditional heterogeneous catalysts, with group 4 metallocenes and the bridged bis(phenoxy-amine) Zr-based catalysts (regiomistakes roughly 1–2%) [19, 20, 49, 100, 101], whereas secondary insertion is dominant with the traditional V-based homogeneous catalysts and the unbridged bis(phenoxy-imine) Ti-based catalysts [11, 102–105]. Several studies have focused on the origin of the regioselective behavior of different systems, and the main conclusion is that in the case of metallocenes and constrained-geometry-based catalysts, the regiochemistry of monomer insertion is essentially dominated by steric effects because the transition state corresponding to secondary insertion of propene is destabilized by steric interactions between the incoming monomer and the ligand framework (see Figure 9.2) [106–108].

Primary or (1,2) insertion

Secondary or (2,1) insertion

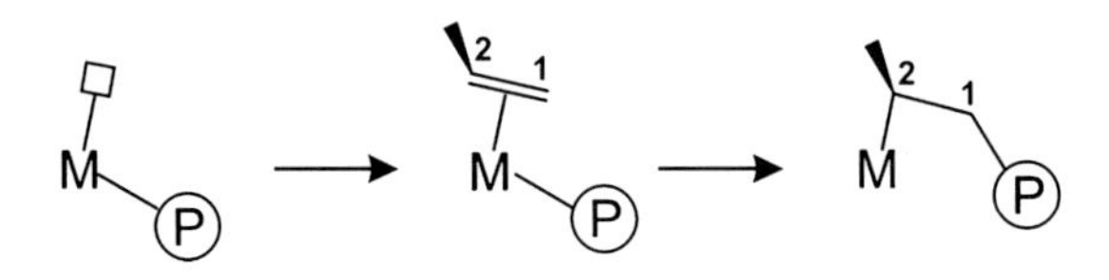

Scheme.9.4

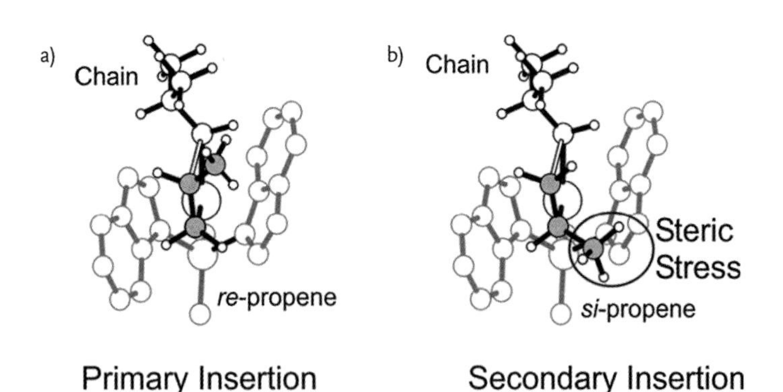

Figure 9.2 Primary (a) and secondary (b) transition states for propene insertion into the M–^{i}Bu bond of the metallocene based on the $Me_2Si(1\text{-}Ind)_2$ ligand.

Differently, in the case of the phenoxy-amine- and phenoxy-imine-based octahedral catalysts, the regiochemistry of monomer insertion is determined by a balance between steric and electronic effects [108, 109]. In the case of the bis(phenoxy-amine)-based catalysts developed by Kol, there is a steric interaction between the secondary inserting propene molecule and the ligand, whereas this interaction, due to the flat geometry around the sp^2-hybridized imine N atom, is less relevant in the case of the bis(phenoxy-imine)-based catalysts developed at Mitsui Chemicals (see Figure 9.3).

Further, electronic effects seem to contribute synergically to this difference. In fact, the higher electron density at the metal of the bis(phenoxy-amine)-based catalysts, due to the electron-donating amine N atom, favors primary propene insertion, whereas reduced electron density at the metal of the bis(phenoxy-amine)-based catalysts, due to the less electron-donating imine N atom, makes secondary

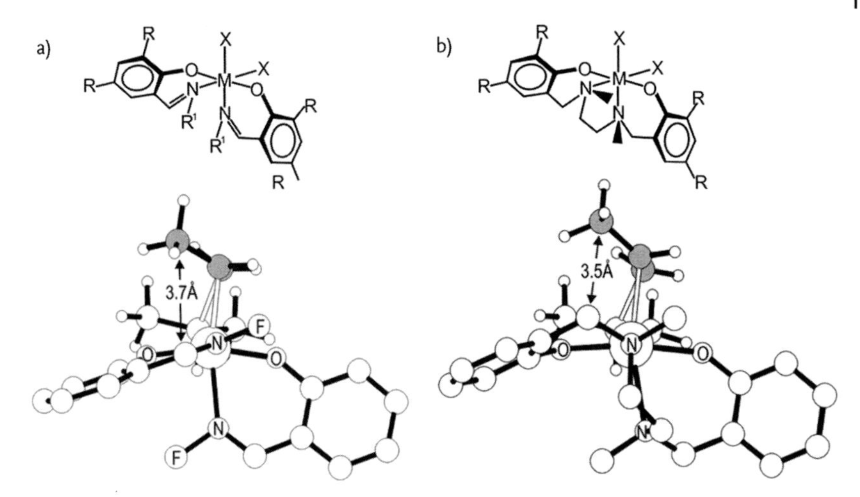

Figure 9.3 Rear view of the transition states for secondary propene insertion into the M–^{i}Pr bond of the bis(phenoxy-imine) Ti-based catalyst (a) and of the bis(phenoxy-amine) Zr-based catalyst (b). Short distances between the propene CH_3 group and C atoms of the ligand are indicated.

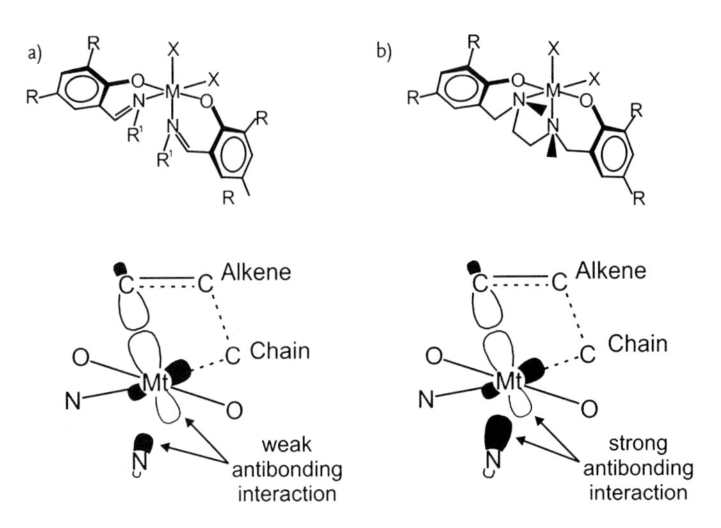

Figure 9.4 Schematic representation of the most relevant part of the molecular orbital corresponding to the incipient M–C bond in bis(phenoxy-imine) (a) and bis(phenoxy-amine) (b) catalyst.

insertion more likely. This electronic effect is due to destabilizing interaction between electron densities at the metal and the C atom that will form the new M–C bond. In this framework, secondary propene insertion is disfavored relative to primary insertion by the electron-donating Me group, which increases electron density on the C_2 atom of the propene molecule (see Figure 9.4) [109].

9.3
Elements of Chirality

Stereochemistry of monomer insertion is connected to the 1-olefin enantioface that inserts into the M-growing chain bond. Scheme 9.5 shows how coordination of the two enantiofaces of a prochiral 1-olefin, such as propene, gives rise to chiral *si* and *re* 1-olefin coordinations [110]. Considering the definition of iso- and syndiotactic polymers, isotactic polymers are generated by multiple insertion of 1-olefin molecules with the same enantioface (either *re* or *si*), while syndiotactic polymers are generated by a regular alternance of chain growth reactions corresponding to insertion of *re*- and *si*-coordinated 1-olefin molecules.

Stereoselection between the two monomer enantiofaces requires chiral active species, and the elements of chirality usually found in these catalysts are described below.

If the chirality is connected to ligand coordination, the following possibilities arise: (i) In the case of pseudotetrahedral complexes with prochiral ligands, it is possible to use the notation (*R*) or (*S*), in parenthesis, according to the Cahn–Ingold–Prelog rules [111] as extended by Schlögl to metallocenes [112]. For instance, in the case of a C_2-symmetric metallocene, the (*R,R*) and (*S,S*) chiralities of coordination of the $H_2C(1\text{-Ind})_2$ ligand, labeled according to the absolute configuration of the bridgehead carbon atoms (marked by arrows), are shown in Figure 9.5.

In the case of octahedral complexes with two bidentate ligands like those proposed for the traditional Ziegler–Natta catalytic systems, as well as for the new century catalysts of Scheme 9.2, the relative orientation of the two bidentate ligands is chiral and can be labeled with the Λ or Δ nomenclature (see Figure 9.6) [113]. If the two bidentate ligands are not bridged, the complex is stereoflexible and interconversion between opposite configurations is relatively easy, whereas the presence of chemical bonds that connect the two ligands confers stereorigidity, freezing the configuration at the metal atom (at least in the timescale of a polymerization reaction).

The other possibilities are:

(i) An intrinsic chirality at the central metal atom that for tetrahedral or assimilable to tetrahedral situations can be labeled with the notation *R* or *S*, by the extension of the Cahn–Ingold–Prelog rules, as proposed by Stanley and Baird [114]. For instance, the two enantiomers with intrinsic chirality at the central

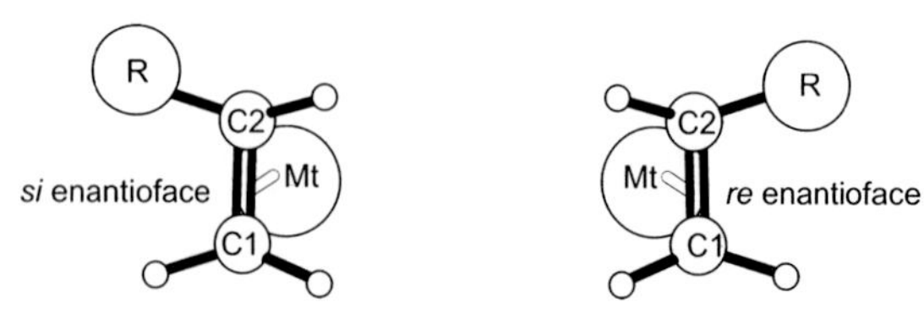

Scheme 9.5

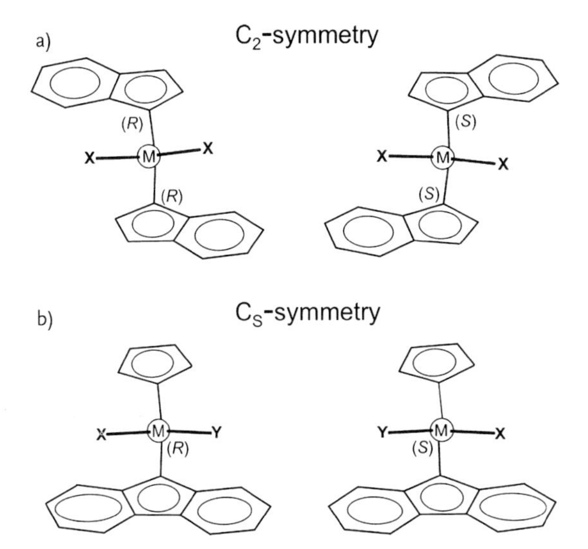

Figure 9.5 (a) C_2-symmetric model catalytic complexes comprising an $H_2C(1\text{-Ind})$ ligand. (b) C_S-symmetric model catalytic complexes comprising an $H_2C(Cp)(9\text{-Flu})$ ligand. The chirality at the metal atom is determined by a difference in the X and Y ligands, as occurs during chain growth, when X and Y are the growing chain and/or the monomer. The main difference between the two systems is that configuration does not change during polymerization for the C_2-symmetric catalyst, whereas (in the framework of the chain migratory mechanism) it changes at each insertion step for the C_S-symmetric catalyst.

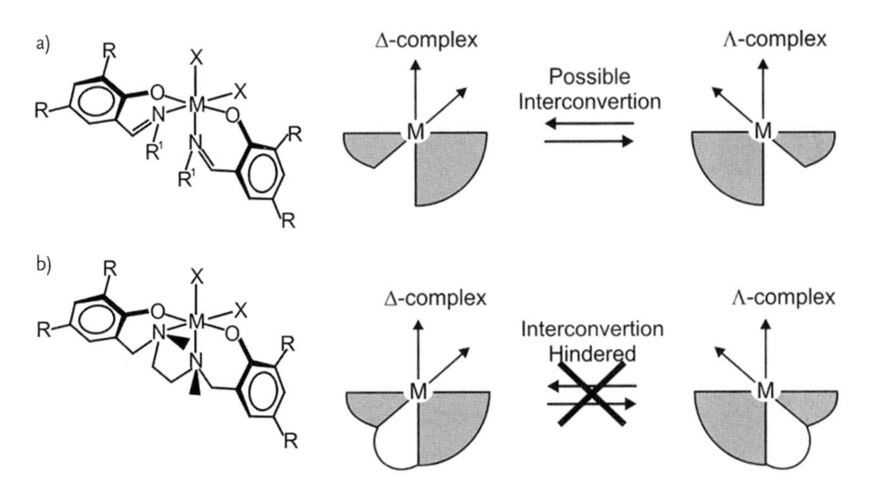

Figure 9.6 (a) C_2-symmetric model catalytic complexes comprising a nonbridged octahedral nonmetallocene complex. (b) C_2-symmetric model catalytic complexes comprising a bridged octahedral nonmetallocene complex. The main difference between the two systems is that configuration does not change during polymerization for the catalyst with a bridged ligand, whereas it can change at each insertion step in the case of the nonbridged catalyst.

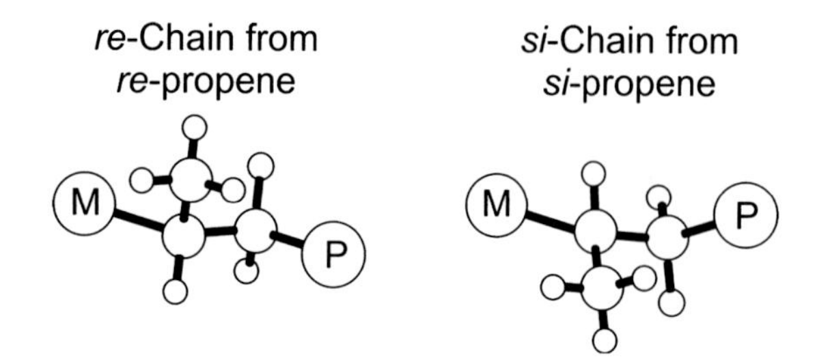

Figure 9.7 Examples of chiral chain-end in the case of secondary propene insertion. This kind of chirality is of relevance only in the case of prevailing secondary propagation, while it is of scarce (or almost none) relevance in the case of prevailing primary propagation.

metal atom *R* and *S* are shown in Figure 9.5, for the case of a C_S-symmetric metallocene with a $H_2C(Cp)(9\text{-Flu})$ ligand.

(ii) The chirality of the last tertiary C atom of the growing chain, which is determined by the chirality of monomer coordination in the last insertion step (see Figure 9.7). Although the *R*/*S* CIP nomenclature could be used, for clarity, we label the two configurations as *si*- or *re*-ending growing chains. It is then straightforward that insertion of a *re*-coordinated monomer on a *re*-ending or on a *si*-ending growing chain would lead to an *m* (iso) or to an *r* (syndio) diad, respectively. This chirality depends only on the chirality of the last inserted monomer, and can change during polymerization if there is a switch in the chirality of monomer insertion.

Both the above elements of chirality can be at the origin of stereoselectivity. If discrimination between the faces of the inserting prochiral monomer is dictated by the chirality of the catalytic site, we are in presence of *chiral-site stereocontrol*, while we are in presence of *chain-end stereocontrol* if discrimination is dictated by the chirality of the last inserted monomer unit. The relative amounts of stereodefects in the polymer chain, easily determined with standard NMR techniques [115, 116], indicates which kind of stereocontrol is operative. Bernoullian statistics are consistent with chain-end stereocontrol [117], while non-Bernoullian distributions originate from chiral-site stereocontrol [118]. The difference can be readily realized for isotactic propagation. Intuitively, for chain-end stereocontrol, a stereomistake is propagated, while for chiral-site stereocontrol, the same stereomistake, having no effect on the site chirality, remains isolated (see Scheme 9.6). It is generally assumed that the stereoselectivity of stereospecific olefin polymerizations is connected with the energy differences between diastereoisomeric situations that originate from combination of two or more of the above elements of chirality.

Finally, we define a conformational chirality connected to the placement in space of the growing chain. The chiral orientation of the growing chain is determined by the sign of the dihedral angle X_{Ol}-M-C-P (X_{Ol} is the midpoint of the C=C double bond) defined in Figure 9.8. Following IUPAC recommendations for stereochemistry [119], we label the two chiral orientations of the growing chain of Figure 9.8

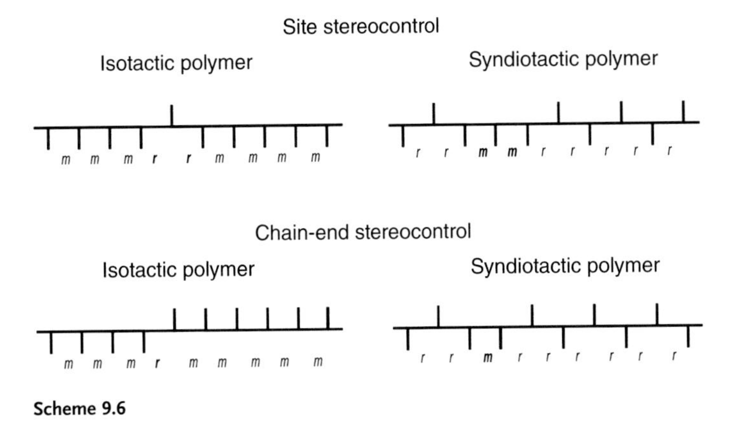

Scheme 9.6

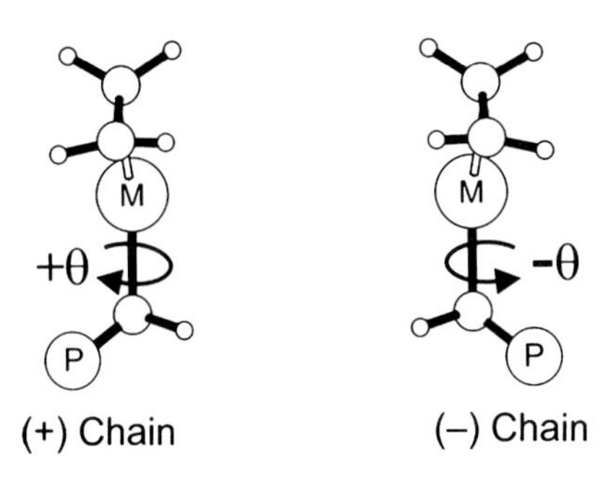

Figure 9.8 Schematic representation of chirally oriented growing chains.

as (–) or (+) growing chains. This chirality is extremely flexible since it depends only on the conformation assumed by the growing chain before/during the formation of the new C–C bond.

9.4
Chiral-Site Stereocontrol: Isotactic Polypropylene by Primary Propene Insertion

9.4.1
Well-Defined C_2-Symmetric Metallocene Catalysts

The discovery that properly activated group 4 metallocenes are excellent catalysts for propene polymerization allowed to investigate in details the origin of stereoselectivity operative in site-controlled propene polymerization. This understanding has been possible because of the extremely well-defined chemistry of group 4 metallocenes. Structural and energetic analysis of the geometries of the transition

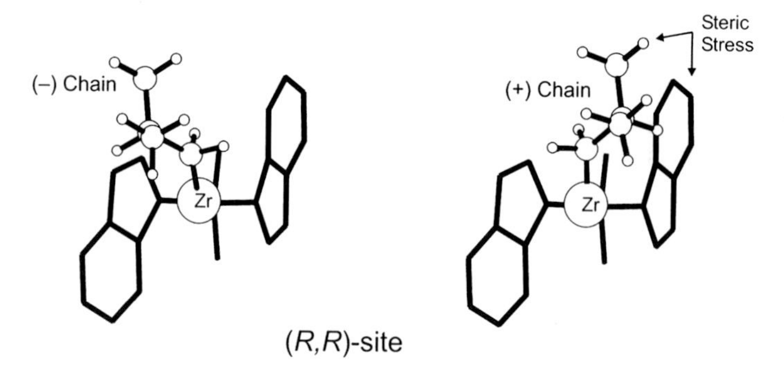

Figure 9.9 Models of the C_2-symmetric metallocene catalyst based on the $Me_2C(1\text{-}Ind)_2$ ligand, with an *i*-Bu group (to simulate the chirally oriented growing chain) bonded to the Zr atom.

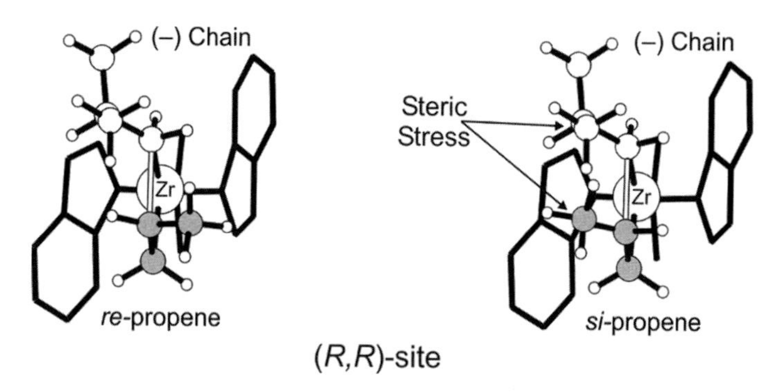

Figure 9.10 Models of the transition states for propene insertion into the Zr–^{i}Bu bond, of the C_2-symmetric metallocene catalyst based on the $Me_2C(1\text{-}Ind)_2$ ligand.

states for primary propene insertion clearly evidenced that in the case of C_2-symmetric metallocenes, the orientation in space of the growing chain is determined by steric interactions with the chiral ligand [120]. For example, in the case of the C_2-symmetric $Me_2C(1\text{-}Ind)_2Zr$ metallocene, the growing chain is forced on the less encumbered side to avoid steric interaction with the nearby six-membered aromatic ring of the ligand (see Figure 9.9). In the case of an (*R*,*R*)-coordinated ligand, this implies that a growing chain with a (−) chiral orientation is favored over a chain with a (+) chiral orientation.

The next step considered was the interaction between the two prochiral faces of the incoming monomer molecule and the system composed by the metallocene with a chirally oriented growing chain bonded to the metal. Analysis of the steric interactions at the transition state for propene insertion immediately revealed that in the case of an (*R*,*R*)-coordinated $Me_2C(1\text{-}Ind)_2$ ligand, which imposes a (−) chiral orientation to the growing chain, insertion of the *si* enantioface of propene was disfavored by steric interactions between the methyl group of the inserting propene molecule and atoms of the chirally oriented growing chain (see Figure 9.10).

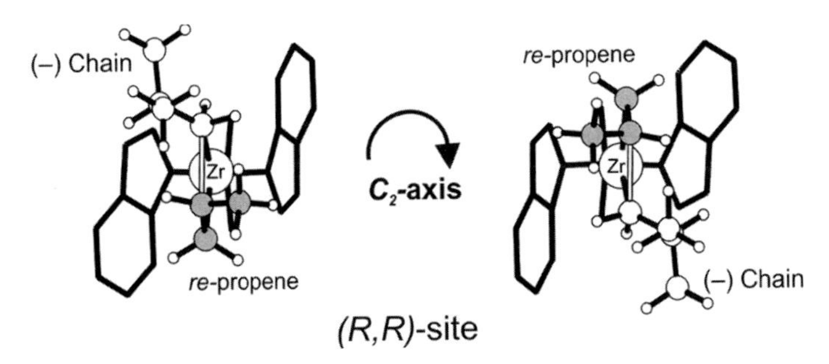

Figure 9.11 Transition states for propene insertion in successive chain growth steps, with a C_2-symmetric metallocene catalyst based on the $Me_2C(1\text{-Ind})_2$ ligand.

Differently, insertion of the *re* enantioface of propene proceeds smoothly because the methyl group of the inserting propene molecule is far away from the growing chain. Thus, enantioselectivity in propene insertion is determined by steric interactions between the monomer and the chirally oriented growing chain, which, in turn, is imposed by the chirality of the catalyst.

However, to fully understand the stereoregularity of the resulting polypropylene, it has to be considered what happens in the following step. In fact, in the framework of the chain migratory insertion mechanism (see Scheme 9.3), the stereospecific behavior of a given catalyst depends on the relationship between the two situations obtained by exchanging, at each chain growth step, the relative positions of the growing chain and the incoming monomer. In the case of a C_2-symmetric catalyst, the two available coordination positions are identical, that is, they are homotopic, since they are related by a local C_2-symmetry axis. Consequently, if the insertion step is enantioselective, as in the case of the systems of Figure 9.11, the polymerization process is isospecific.

This mechanism of stereoselectivity is in agreement with several available experimental data, and has been used to rationalize the isospecific behavior of a large number of C_2 and also, with some modifications, C_1-symmetric metallocenes [33, 121–126]. Further, it is mandatory to recall that a relevant and elegant experimental proof of the importance of the role played by the growing chain in determining the steric course of the chain growth reaction was given by Zambelli, who investigated the stereospecificity in the first step of polymerization. Analysis by ^{13}C NMR techniques of the polymer end groups showed that in the first step of polymerization, when the alkyl group bonded to the metal is a methyl group, propene insertion is essentially nonenantioselective. Conversely, when the alkyl group is an ethyl group, the first insertion is enantioselective as the successive insertions, both for heterogeneous [127] and homogeneous Ziegler–Natta catalysts [128].

It is of interest that the two groups of atoms that are responsible for the stereoselectivity of propene insertion (atoms 4 and 5 of the indenyl ligands, see Figure 9.11) are located roughly 0.8–1.0 nm away from each other. Thus, even if the geometry of the four-center transition state is rather compact, the stereoregularity

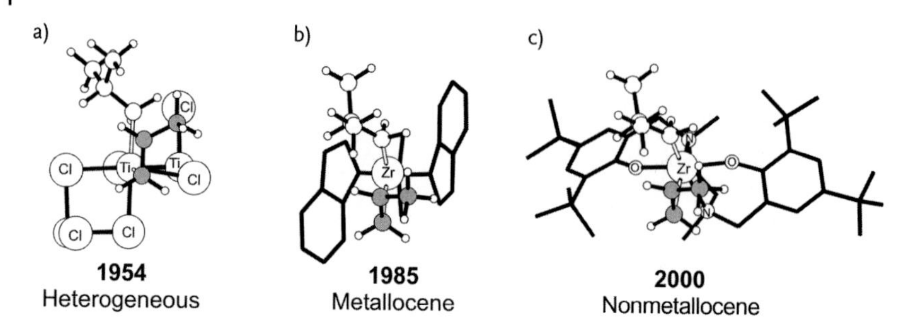

Figure 9.12 Models of the transition states for chain growth with a possible active species in the case of heterogeneous Ziegler–Natta catalysts (a), with a C_2-symmetric metallocene based on the $Me_2C(1\text{-}Ind)_2$ ligand (b), and with a C_2-symmetric nonmetallocene catalyst based on the bis(phenoxy-amine) bridged ligand (c).

of the resulting polymer is dictated by the relative position of groups of atoms at a larger scale.

9.4.2
Well-Defined Bis(Phenoxy-Amine)-Based Octahedral Catalysts and Poorly Defined Heterogeneous Ziegler–Natta Catalytic Systems

The concepts developed in Section 9.4.1 can be used to rationalize easily the isospecificity exhibited by the recently developed C_2-symmetric nonmetallocene catalysts based on bridged octahedral bis(phenoxy-amine) Zr-complexes (see Scheme 9.2), as well as to possible catalytic sites of the classical and poorly characterized heterogeneous Ziegler–Natta catalytic systems [129, 130]. The most favored transition states for propene insertion into the M-chain bond for the three classes of catalysts are reported in Figure 9.12. In the middle is the same C_2-symmetric metallocene discussed before.

Comparison of the most favored transition state for the metallocene with that for the primary propene insertion into the Zr-chain bond of the nonmetallocene and octahedral system, and into the Ti-chain bond of a possible active site in the case of the classical Ziegler–Natta systems, reveals strong similarities. Indeed, all the structures of Figure 9.12 show the classical features that characterize the *mechanism of the chiral orientation of the growing chain*. These are: (i) the growing chain assumes a chiral orientation to minimize steric interactions with the chiral ligand; (ii) the monomer inserts with the methyl group trans-oriented relative to the growing chain, to minimize its steric interaction with the growing chain itself.

Further, it is clear that the information that has to be programmed into the catalyst in order to achieve an isotactic polymer is the placement of two bulky groups of atoms at roughly 1 nm of distance from each other, and that the two groups must be correlated by a C_2-symmetry axis. This result was spectacularly achieved in the case of the heterogeneous catalysts based on $TiCl_4$ supported on $MgCl_2$, which are stereoselective only in the presence of Lewis bases that act as donors.

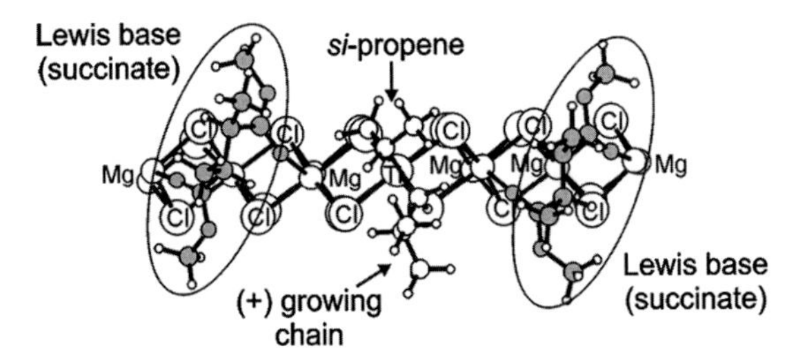

Figure 9.13 Favored transition state leading to primary propene insertion into the Ti–^{i}Bu bond of an $MgCl_2$-supported Ziegler–Natta catalysts. The Ti-active site is flanked by two *rac*-1,4-dimethoxy-2,3-dimethyl succinate molecules coordinated to the $MgCl_2$-support.

In fact, the most reasonable model proposed so far to explain the role of the Lewis bases requires coordination of the Lewis bases at both sides of the active Ti atom (see Figure 9.13). The coordinated Lewis bases restrict the conformational freedom of the growing chain, which is forced to assume a chiral orientation. The mean distance between the two coordinated Lewis bases is roughly 1.0–1.5 nm. In short, the most reasonable model for the last generation of heterogeneous Ziegler–Natta catalysts is a complex structure that is obtained by the interaction between different entities, which is the support, the $TiCl_4$, and the Lewis base.

9.5 Chiral-Site Stereocontrol: Syndiotactic Polypropylene by Primary Propene Insertion

The concepts developed in the previous sections can be used to rationalize also the syndiosospecificity exhibited by some C_S-symmetric metallocenes, such as the catalyst based on the $Me_2C(Cp)(9\text{-Flu})$ ligand (see Scheme 9.2) [51, 131]. Indeed, although the overall stereospecificity is different from that exhibited by some C_2-symmetric metallocenes, the single enantioselective event obeys the same rules both in the C_2- and C_S-symmetric metallocenes. In fact, the most favored transition states for propene insertion into the M-chain bond for a C_S-catalyst are reported in Figure 9.14. As in the case of the C_2-symmetric metallocene discussed before, the growing chain is chirally oriented in an open part of space, to avoid steric interactions with the ligand skeleton, and the propene molecule inserts with the enantioface that places the methyl group of the propene away from the chirally oriented growing chain.

The fundamental difference with respect to the C_2-symmetric metallocenes is in the relationship between the preferred propene enantiofaces in successive insertion steps. In fact, in the case of the C_S-symmetric metallocene, due to the presence of a symmetry plane, the two coordination positions available to the growing chain and the monomer are enantiotopic (see Figure 9.14). Thus, with a C_S-symmetric metallocene, opposite propene enantiofaces are enchained at successive chain growth steps, which explains the syndiotacticity of the produced polypropylenes.

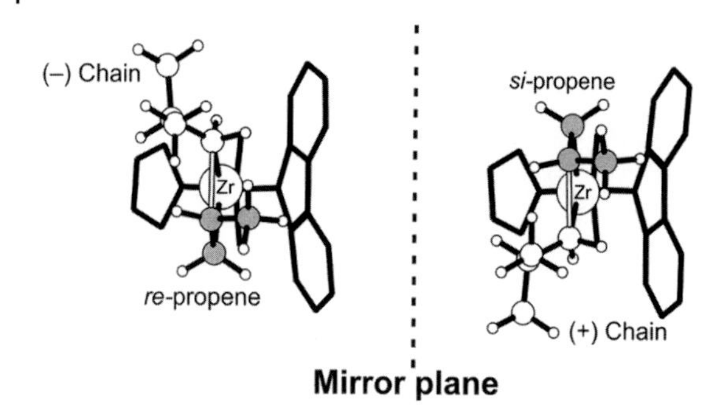

Figure 9.14 Transition states for chain growth in successive chain growth steps, with a C_S-symmetric metallocene catalyst based on the $Me_2C(Cp)(Flu)$ ligand.

In summary, also in the case of the syndiospecific catalysts, stereoselectivity is determined by bulky groups that are roughly 0.7 nm away from each other, although they are correlated by a local symmetry plane rather than from a symmetry axis, as in the case of the C_2-symmetric catalysts.

9.6 Chain-End Stereocontrol: Syndiotactic Polypropylene by Secondary Propene Insertion

9.6.1 Well-Defined Bis(Phenoxy-Imine)-Based Octahedral Catalysts

In this section we discuss the origin of stereospecificity operative in the case of the nonmetallocene catalysts based on unbridged bis(phenoxy-imine) Ti-complexes. There are two main differences between these catalysts and those based on the bridged bis(phenoxy-amine) Zr-complexes, discussed previously. The first is the absence of a bridge between the N atoms of the ligand, which allows for possible interconversion between opposite configurations at the metal atom (see Figure 9.6). The second is connected to the regiochemistry of propene insertion, which is primary in the case of the bridged bis(phenoxy-amine) Zr-based catalysts, whereas it is secondary in the case of unbridged bis(phenoxy-imine) Ti-based catalysts. This means that with the latter catalysts, the first chiral C atom of the growing chain is directly bonded to the metal atom (see Figure 9.7), which has strong influences on the origin of stereoselectivity.

The most favored transition states for secondary propene insertion into the Ti-(growing chain) bond (the secondary growing chain is simulated by the chiral $-CH(CH_3)CH_2CH(CH_3)_2$ group) are reported in Figure 9.15. Both structures present a *re*-ending chain, which is a chain originated from the secondary insertion of a *re*-coordinated propene. The configuration of the chiral complexes is Λ in the structure on the left, while it is Δ in the one on the right. This change in

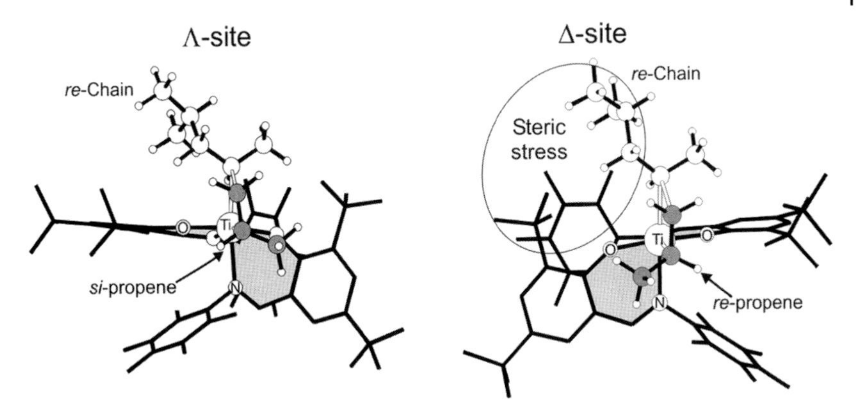

Figure 9.15 Models of the transition states for chain growth with a with a C_2-symmetric nonmetallocene catalysts based on an unbridged bis(phenoxy-imine) Ti-complex.

the configuration at the metal atom is possible because of the stereoflexible nature of the complex (see Figure 9.6). In order to avoid steric interactions with the ligand, the enantioface of the inserting propene is *si* on the structure on the left, while it is *re* in the one on the right. Finally, the secondary *re*-growing chain is forced to assume always a (–) orientation to form a stabilizing α-agostic interaction.

The structure on the right, with a *re*-chain and a Δ site, is disfavored by steric interactions between the growing chain and the nearby $-C_6F_5$ and tBu groups. Thus, a *re*-chain imposes a Λ configuration to the catalytic complex that, in turn, selects the *si* enantioface of the propene molecule. The favored transition state is characterized by the following chiralities, *re*-chain/Λ-site/*si*-propene, and leads to formation of an *r* (syndio) diad. Since insertion of a *si*-propene leads to a *si*-chain, in the following step will be favored the *si*-chain/Δ-site/*re*-propene transition state, leading again to the formation of an *r* (syndio) diad. The regular and alternate succession of transition states of this kind explains the syndiospecific behavior of nonmetallocene catalysts based on unbridged bis(phenoxy-imine) Ti-complexes.

In this case, the selectivity of the reaction depends on the bulky groups *t*-Bu groups ortho to the O atoms of the phenoxy ligands that are, as in the case of the metallocenes, around 1 nm away from each other. What makes this catalyst very peculiar is the fact that the bulky groups responsible for the stereoselectivity do not occupy a fixed position in the coordination sphere of the metal, but can fluctuate in order to minimize steric interactions with the growing chain and the monomer molecule.

9.6.2
Poorly Defined V-Based Catalytic Systems

The mechanism just described was first proposed [132] to explain the origin of stereoselectivity in secondary and syndiospecific chain-end-controlled

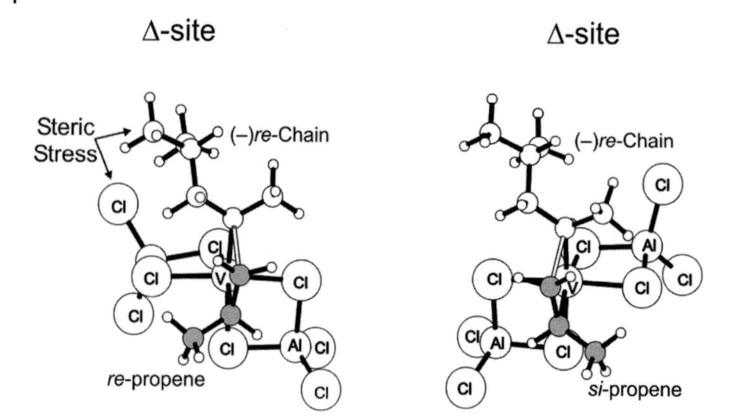

Figure 9.16 Models of the transition states for chain growth with a possible active species in the case of V-based Ziegler–Natta catalysts.

polymerization of propene with the homogeneous V-based catalytic systems [9, 11], still used for the industrial production of ethene–propene copolymers. Also in this case, the origin of syndiospecificity was supposed to be based on the formation of a fluxional chiral site. The assumed model consisted of a hexacoordinated V-atom surrounded by four chlorine atoms assumed to be bridge bonded to other metal atoms (e.g., the Al atoms of the AlR_3 cocatalysts). Thus, the catalytic site is chiral and, similarly to the homogeneous bis(phenoxy-imine) Ti-based catalyst of Section 9.4.1, interconversion between enantiomeric complexes is assumed to be possible. In Figure 9.16, we report the most stable transition states for secondary propene insertion into the V-(growing chain) bond (the secondary growing chain is again simulated by the chiral $-CH(CH_3)CH_2CH(CH_3)_2$ group). Both structures present a *re*-ending chain. As in the case of the unbridged non-metallocene catalysts just described, the chirality of the chain-end imposes a specific configuration at the metal atom that, in turn, selects between the two enantiofaces of the inserting monomer. Specifically, insertion of a *si*-propene is favored in the case of a *re*-ending chain, which corresponds to formation of a syndiotactic diad.

9.7 Conclusions

In this chapter, we discussed the origin of stereoselectivity in catalytic systems that can be considered milestones in the coordinative polymerization of propene. The power of these models lies in the fact that extremely different catalysts can be described with simple and elegant concepts, and in the introduction of the concept of the chiral orientation of the growing chain as key to understand these catalysts.

More in detail, the mechanisms operative in the case of primary and secondary propene insertion (Sections 9.5 and 9.6, respectively) have some similarities, since in both cases, the "crucial" element of chirality at the origin of the enantioselectivity (site chirality for the isospecific and stereorigid systems, and chain-end chirality for the syndiospecific and unbridged systems) are some bulky groups of atoms that are strategically located at 0.7–1.5 nm from each other. These atoms can be part of the ligands, as in the metallocene and nonmetallocene catalysts; a coordinated molecule, as in the $MgCl_2$-supported Ziegler–Natta catalytic systems; and can also fluctuate, as in the case of the unbridged bis(phenoxy-imine) catalysts. These groups do not select the enantioface of the inserting monomer directly. In all cases, there is a messenger of information between the "crucial" element of chirality and the monomer. However, the flow of information that originates stereoselectivity is rather different in the case of primary and secondary propene insertion. In fact, Scheme 9.7 shows that for primary propene insertion with stereorigid catalysts the chirally oriented growing chain acts as messenger of information between the chiral active site and the inserting prochiral monomer.

Instead, in the case of secondary propene insertion with stereoflexible catalysts, the chirally oriented growing chain acts as a messenger between the chiral growing chain-end and the configuration at the metal atom (see Scheme 9.8). The so determined chirality of the catalytic site selects between the two enantiofaces of the prochiral monomer.

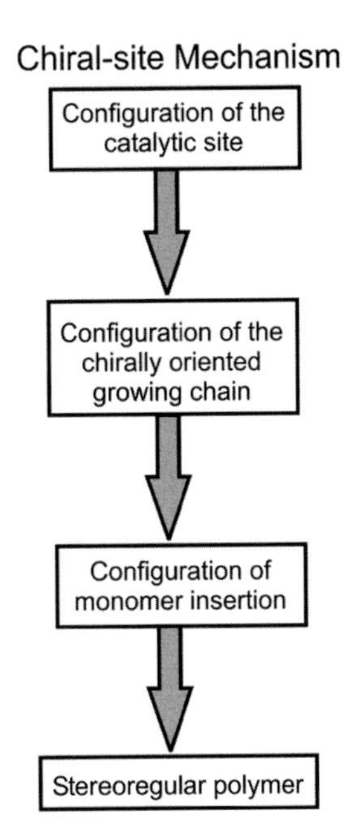

Scheme 9.7

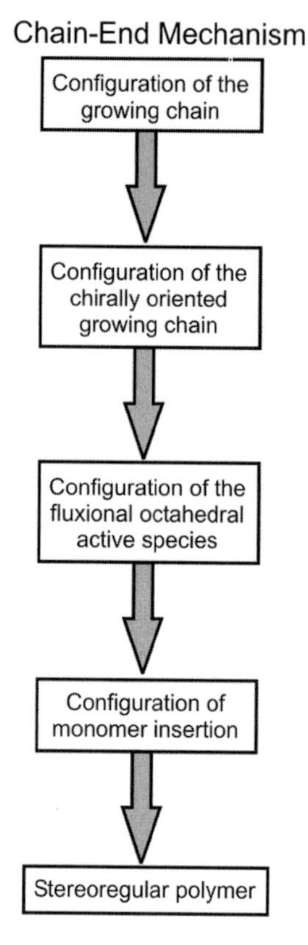

Scheme 9.8

References

1 Nobel e-Museum. http://www.nobel.se/chemistry/laureates/1963 (accessed March 2011).
2 Ziegler, K. (1972) Nobel lectures in chemistry, 1963–1970, in *Nobel Lectures in Chemistry, 1963–1970* (ed. Elsevier), Elsevier, Amsterdam, The Netherlands, p. 6.
3 Natta, G. (1972) Nobel lectures in chemistry, 1963–1970, in *Nobel Lectures in Chemistry, 1963–1970* (ed. Elsevier), Elsevier, Amsterdam, The Netherlands, p. 27.
4 Moore, E.P. (2005) *Polypropylene Handbook*, 2nd edn. Munich, Hanser.
5 Wilkinson, J. and Birmingham, J.M. (1954) *J. Am. Chem. Soc.*, **76**, 4281.
6 Natta, G., Pino, P., Corradini, P., Danusso, F., Mantica, E., Mazzanti, G., and Moraglio, G. (1955) *J. Am. Chem. Soc.*, **77**, 1708.
7 Breslow, D.S. and Newburg, N.R. (1957) *J. Am. Chem. Soc.*, **79**, 5072.
8 Breslow, D.S. (Hercules) US 2924593, 1960.
9 Natta, G., Pasquon, I., and Zambelli, A. (1962) *J. Am. Chem. Soc.*, **84**, 1488.
10 Pasquon, I., Zambelli, A., Signorini, R., and Natta, G. (1968) *Makromol. Chem.*, **112**, 160.
11 Zambelli, A., Sessa, I., Grisi, F., Fusco, R., and Accomazzi, P. (2001) *Macromol. Rapid Commun.*, **22**, 297.
12 Breslow, D.S. and Long, W.P. (1975) *Liebigs Ann. Chem.*, 463.

13 Anderson, A., Cordes, H.G., Herwig, J., Kaminsky, W., Merk, A., Motweiler, R., Sinn, J.H., and Vollmer, H.J. (1976) *Angew. Chem. Int. Ed. Engl.*, **88**, 689.
14 Sinn, H.J. and Kaminsky, W. (1980) *Adv. Organomet. Chem.*, **18**, 99.
15 Stevens, J.C. (1994) *Stud. Surf. Sci. Catal.*, **89**, 277.
16 Stevens, J.C. (1996) *Stud. Surf. Sci. Catal.*, **101**, 11.
17 Cecchin, G., Morini, G., and Piemontesi, F. (2003) Ziegler-Natta catalysts, in *Encyclopedia of Chemical Technology*, John Wiley & Sons, Inc., New York, pp. 502–554.
18 Chadwick, J.C. (2003) Ziegler-Natta catalysis, in *Encyclopedia of Polymer Science and Technology*, John Wiley & Sons, Inc., New York, pp. 517–531.
19 Brintzinger, H.-H., Fischer, D., Mülhaupt, R., Rieger, B., and Waymouth, R.M. (1995) *Angew. Chem. Int. Ed.*, **34**, 1143.
20 Resconi, L., Cavallo, L., Fait, A., and Piemontesi, F. (2000) *Chem. Rev.*, **100**, 1253.
21 Coates, G.W. (2000) *Chem. Rev.*, **100**, 1223.
22 Lin, S. and Waymouth, R.M. (2002) *Acc. Chem. Res.*, **35**, 765.
23 Park, S.J., Han, Y., Kim, S.K., Lee, J.Y., Kim, H.K., and Do, Y. (2004) *J. Organomet. Chem.*, **689**, 4263.
24 Britovsek, G.J.P., Gibson, V.C., and Wass, D.F. (1999) *Angew. Chem. Int. Ed.*, **38**, 428.
25 Gibson, V. and Spitzmesser, S.K. (2003) *Chem. Rev.*, **103**, 283.
26 Resconi, L., Chadwick, J.C., and Cavallo, L. (2006) Olefin polymerizations with group IV metal catalysts, in *Comprehensive Organometallic Chemistry III*, vol. 4, 3rd edn (eds D.M.P. Mingos and R. Crabtree), Elsevier, New York, 2007, pp. 1005–1166.
27 Chen, E.Y.-X. and Marks, T.J. (2000) *Chem. Rev.*, **100**, 1391.
28 Bochmann, M. (2004) *J. Organomet. Chem.*, **689**, 3982.
29 Macchioni, A. (2005) *Chem Rev.*, **105**, 2039.
30 Corradini, P., Barone, V., Fusco, R., and Guerra, G. (1979) *Eur. Polym. J.*, **15**, 1133.
31 Corradini, P., Guerra, G., Fusco, R., and Barone, V. (1980) *Eur. Polym. J.*, **16**, 835.
32 Corradini, P., Barone, V., Fusco, R., and Guerra, G. (1982) *J. Catal.*, **77**, 32.
33 Cavallo, L., Guerra, G., Corradini, P., and Vacatello, M. (1991) *Chirality*, **3**, 299.
34 Corradini, P., Busico, V., Cavallo, L., Guerra, G., Vacatello, M., and Venditto, V. (1992) *J. Mol. Catal.*, **74**, 433.
35 Guerra, G., Cavallo, L., Venditto, V., Vacatello, M., and Corradini, P. (1993) *Makromol. Chem. Makromol. Symp.*, **69**, 237.
36 Corradini, P., Guerra, G., Cavallo, L., Moscardi, G., and Vacatello, M. (1995) Models for the explanation of the stereospecific behavior of Ziegler Natta catalysts, in *Ziegler Catalysts* (eds G. Fink, R. Mülhaupt, and H.-H. Brintzinger), Springer, Berlin, p. 237.
37 Guerra, G., Corradini, P., Cavallo, L., and Vacatello, M. (1995) *Makromol. Chem. Makromol. Symp.*, **89**, 307.
38 Cavallo, L., Guerra, G., and Corradini, P. (1998) *J. Am. Chem. Soc.*, **120**, 2428.
39 Guerra, G., Longo, P., Corradini, P., and Cavallo, L. (1999) *J. Am. Chem. Soc.*, **121**, 8651.
40 Longo, P., Grisi, F., Guerra, G., and Cavallo, L. (2000) *Macromolecules*, **33**, 4647.
41 Guerra, G., Cavallo, L., Corradini, P., and Fusco, R. (1997) *Macromolecules*, **30**, 677.
42 Costabile, C., Milano, G., Cavallo, L., and Guerra, G. (2001) *Macromolecules*, **34**, 7952.
43 Minieri, G., Corradini, P., Zambelli, A., Guerra, G., and Cavallo, L. (2001) *Macromolecules*, **34**, 2459.
44 Minieri, G., Corradini, P., Guerra, G., Zambelli, A., and Cavallo, L. (2001) *Macromolecules*, **34**, 5379.
45 Spaleck, W., Antberg, M., Rohrmann, J., Winter, A., Bachmann, B., Kiprof, P., Behm, J., and Herrmann, W.A. (1992) *Angew. Chem., Int. Ed. Engl.*, **31**, 1347.
46 Resconi, L., Piemontesi, F., Camurati, I., Sudmeijer, O., Nifant'ev, I.E., Ivchenko, P.V., and Kuz'mina, L.G. (1998) *J. Am. Chem. Soc.*, **120**, 2308.
47 Ewen, J.A. (1984) *J. Am. Chem. Soc.*, **106**, 6355.

48 Kaminsky, W., Külper, K., Brintzinger, H.H., and Wild, F.R.W.P. (1985) *Angew. Chem. Int. Ed. Engl.*, **24**, 507.

49 Tshuva, E.Y., Goldberg, I., and Kol, M. (2000) *J. Am. Chem. Soc.*, **122**, 10706.

50 Segal, S., Goldberg, I., and Kol, M. (2005) *Organometallics*, **24**, 200.

51 Ewen, J.A., Jones, R.L., Razavi, A., and Ferrara, J.D. (1988) *J. Am. Chem. Soc.*, **110**, 6255.

52 Saito, J., Mitani, M., Onda, M., Mohri, J.-I., Ishii, S.-I., Yoshida, Y., Nakano, T., Tanaka, H., Matsugi, T., Kojoh, S.-I., Kashiwa, N., and Fujita, T. (2001) *Macromol. Rapid Commun.*, **22**, 1072.

53 Coates, G.W. and Tian, J. (2000) *Angew. Chem. Int. Ed.*, **39**, 3626.

54 Fujita, T., Tohi, Y., Mitani, M., Matsui, S., Saito, J., Nitabaru, M., Sugi, K., and Makio, H. Tsutsui (Mitsui Chemicals) EP 874005, 1998.

55 Matsui, S., Tohi, Y., Mitani, M., Saito, J., Makio, H., Tanaka, H., Nitabaru, M., Nakano, T., and Fujita, T. (1999) *Chem. Lett.*, **10**, 1065.

56 Matsui, S., Mitani, M., Saito, J., Tohi, Y., Makio, H., Tanaka, H., and Fujita, T. (1999) *Chem. Lett.*, **12**, 1263.

57 Mitani, M., Yoshida, Y., Mohri, J., Tsuru, K., Ishii, S., Kojoh, S.-I., Matsugi, T., Saito, J., Matsukawa, N., Matsui, S., Nakano, T., Tanaka, H., Kashiwa, N., and Fujita, T. (Mitsui Chemicals) WO 0155231, 2001.

58 Coates, G.W., Tian, J., and Hustad, P.D. (Cornell Res. Foundation Inc.) US 6562930, 2001.

59 Crowther, D.J., Baenziger, N.C., and Jordan, R.F. (1991) *J. Am. Chem. Soc.*, **113**, 1455.

60 Bochmann, M., Lancaster, S.J., and Robinson, O.B. (1995) *J. Chem. Soc. Chem. Commun.*, 2081.

61 Ruwwe, J., Erker, G., and Fröhlich, R. (1996) *Angew. Chem. Int. Ed. Engl.*, **35**, 80.

62 Sun, Y., Spence, R.E.v.H., Piers, W.E., Parvez, M., and Yap, G.P.A. (1997) *J. Am. Chem. Soc.*, **119**, 5132.

63 Song, F., Lancaster, S.J., Cannon, R.D., Schormann, M., Humphrey, M.B., Zuccaccia, C., Macchioni, A., and Bochmann, M. (2005) *Organometallics*, **24**, 1315.

64 Song, F., Cannon, R.D., Lancaster, S.J., and Bochmann, M. (2004) *J. Mol. Catal., A Chem.*, **218**, 21.

65 Rodriguez-Delgado, A., Hannant, M.D., Lancaster, S.J., and Bochmann, M. (2004) *Macromol. Chem. Phys.*, **205**, 334.

66 Li, H., Li, L., and Marks, T.J. (2004) *Angew. Chem., Int. Ed.*, **43**, 4937.

67 Chen, M.-C., Roberts, J.A.S., and Marks, T.J. (2004) *J. Am. Chem. Soc.*, **126**, 4605.

68 Li, H., Li, L., Marks, T.J., Liable-Sands, L.M., and Rheingold, A.L. (2003) *J. Am. Chem. Soc.*, **125**, 10788.

69 Chen, M.-C. and Marks, T.J. (2001) *J. Am. Chem. Soc.*, **123**, 11803.

70 Sillars, D.R. and Landis, C.R. (2003) *J. Am. Chem. Soc.*, **125**, 9894.

71 Landis, C.R., Rosaaen, K.A., and Sillars, D.R. (2003) *J. Am. Chem. Soc.*, **125**, 1710.

72 Landis, C.R., Rosaaen, K.A., and Uddin, J. (2002) *J. Am. Chem. Soc.*, **124**, 12062.

73 Liu, Z., Somsook, E., White, C.B., Rosaaen, K.A., and Landis, C.R. (2001) *J. Am. Chem. Soc.*, **123**, 11193.

74 Fusco, R., Longo, L., Masi, F., and Garbassi, F. (1997) *Macromolecules*, **30**, 7673.

75 Chan, M.S.W., Vanka, K., Pye, C.C., and Ziegler, T. (1999) *Organometallics*, **18**, 4624.

76 Vanka, K. and Ziegler, T. (2001) *Organometallics*, **20**, 905.

77 Xu, Z., Vanka, K., and Ziegler, T. (2004) *Organometallics*, **23**, 104.

78 Lanza, G., Fragalà, I.L., and Marks, T.J. (2000) *J. Am. Chem. Soc.*, **122**, 12764.

79 Lanza, G., Fragalà, I.L., and Marks, T.J. (2001) *Organometallics*, **20**, 4006.

80 Lanza, G., Fragalà, I.L., and Marks, T.J. (2002) *Organometallics*, **21**, 5594.

81 Correa, A. and Cavallo, L. (2006) *J. Am. Chem. Soc.*, **128**, 10952.

82 Ducéré, J.-M. and Cavallo, L. (2006) *Organometallics*, **25**, 1431.

83 Resconi, L., Bossi, S., and Abis, L. (1990) *Macromolecules*, **23**, 4489.

84 Sinn, H., Kaminsky, W., and Hoker, H. (1995) *Alumoxanes*, Huthig & Wepf, Heidelberg, Germany.

85 Siedle, A.R., Lamanna, W.M., Newmark, R.A., and Schroepfer, J.N. (1998) *J. Mol. Catal., A Chem.*, **128**, 257.
86 Zurek, E. and Ziegler, T. (2004) *Prog. Polym. Sci.*, **29**, 107.
87 Yang, X., Stern, C., and Marks, T.J. (1994) *J. Am. Chem. Soc.*, **116**, 10015.
88 Yang, X., Stern, C., and Marks, T.J. (1991) *J. Am. Chem. Soc.*, **113**, 3623.
89 Ewen, J.A. and Elder, M.J. (Fina Technology). EP 427697, 1991.
90 Ewen, J.A. and Elder, M.J. (Fina Technology). US 5561092, 1996.
91 Ewen, J.A. and Elder, M.J. (Fina Technology). EP 0426637, 1991.
92 Elder, M.J. and Ewen, J.A. (Fina Technology). EP 0573403, 1993.
93 Dyachkovskii, F.S., Shilova, F.S., and Shilov, A.K. (1967) *J. Polym. Sci. C*, **16**, 2333.
94 Eisch, J.J., Piotrowski, A.M., Brownstein, S.K., Gabe, E.J., and Lee, F.L. (1985) *J. Am. Chem. Soc.*, **107**, 7219.
95 Bochmann, M. and Wilson, L.M. (1986) *J. Chem. Soc. Chem. Commun.*, 1610.
96 Jordan, R.F., Bajgur, C.S., Willet, R., and Scott, B. (1986) *J. Am. Chem. Soc.*, **108**, 7410.
97 Turner, H.W., Hlatky, G.G., and Eckman, R.R. (Exxon Chemical) US 5384299, 1993.
98 Cossee, P. (1964) *J. Catal.*, **3**, 80.
99 Brookhart, M., Green, M.L.H., and Wong, L.L. (1988) *Prog. Inorg. Chem.*, **36**, 1.
100 Corradini, P., Busico, V., and Guerra, G. (1988) *Comprehensive Polymer Science*, vol. 4 (eds G. Allen and J.C. Bevington), Pergamon Press, Oxford, p. 29.
101 McKnight, A.L. and Waymouth, R.M. (1998) *Chem. Rev.*, **98**, 2587.
102 Zambelli, A. and Allegra, G. (1980) *Macromolecules*, **13**, 42.
103 Makio, H., Kashiwa, N., and Fujita, T. (2002) *Adv. Synth. Catal.*, **344**, 477.
104 Coates, G.W., Hustad, P.D., and Reinartz, S. (2002) *Angew. Chem. Int. Ed.*, **41**, 2236.
105 Lamberti, M., Pappalardo, D., Zambelli, A., and Pellecchia, C. (2002) *Macromolecules*, **35**, 658.
106 Guerra, G., Cavallo, L., Moscardi, G., Vacatello, M., and Corradini, P. (1994) *J. Am. Chem. Soc.*, **116**, 2988.
107 Guerra, G., Longo, P., Cavallo, L., Corradini, P., and Resconi, L. (1997) *J. Am. Chem. Soc.*, **119**, 4394.
108 Correa, A., Talarico, G., and Cavallo, L. (2007) *J. Organomet. Chem.*, **692**, 4519.
109 Talarico, G., Busico, V., and Cavallo, L. (2003) *J. Am. Chem. Soc.*, **125**, 7172.
110 Hanson, K.R. (1966) *J. Am. Chem. Soc.*, **88**, 2731.
111 Cahn, R.S., Ingold, C., and Prelog, V. (1966) *Angew. Chem. Int. Ed. Engl.*, **5**, 385.
112 Schlögl, K. (1966) *Top. Stereochem.*, **1**, 39.
113 IUPAC (1971) Nomenclature of inorganic chemistry. *Pure Appl. Chem.*, **18**, 77.
114 Stanley, K. and Baird, M.C. (1975) *J. Am. Chem. Soc.*, **97**, 6598.
115 Zambelli, A. and Ammendola, P. (1991) *Prog. Polym. Sci.*, 16.
116 Busico, V. and Cipullo, R. (2001) *Prog. Polym. Sci.*, **26**, 443.
117 Bovey, F.A. and Tiers, G.V.D. (1960) *J. Polym. Sci.*, **44**, 173.
118 Shelden, R.A., Fueno, T., Tsunetsugu, T., and Furukawa, J. (1965) *J. Pol. Sci. Pol. Lett. Ed.*, **3**, 23.
119 Moss, G.P. (1996) *Pure Appl. Chem.*, **68**, 2193.
120 Corradini, P., Guerra, G., Vacatello, M., and Villani, V. (1988) *Gazz. Chim. Ital.*, **118**, 173.
121 Cavallo, L., Corradini, P., Guerra, G., and Vacatello, M. (1991) *Polymer*, **32**, 1329.
122 Cavallo, L., Corradini, P., Guerra, G., and Resconi, L. (1996) *Organometallics*, **15**, 2254.
123 Guerra, G., Cavallo, L., Moscardi, G., Vacatello, M., and Corradini, P. (1996) *Macromolecules*, **29**, 4834.
124 Corradini, P., Cavallo, L., and Guerra, G. (2000) Molecular modeling studies on stereospecificity and regiospecificity of propene polymerization by metallocenes, in *Metallocene-Based Polyolefins: Preparation, Properties and Technology*, vol. 2 (eds J. Scheirs and W. Kaminsky), John Wiley & Sons, New York, 2000, pp. 3–28.
125 Moscardi, G., Resconi, L., and Cavallo, L. (2001) *Organometallics*, **20**, 1918.

126 Corradini, P., Guerra, G., and Cavallo, L. (2003) *Top. Stereochem.*, **24**, 1.

127 Zambelli, A., Sacchi, M.C., Locatelli, P., and Zannoni, G. (1982) *Macromolecules*, **15**, 211.

128 Longo, P., Grassi, A., Pellecchia, C., and Zambelli, A. (1987) *Macromolecules*, **20**, 1015.

129 Corradini, P., Guerra, G., and Cavallo, L. (2004) *Acc. Chem. Res.*, **37**, 231.

130 Correa, A., Piemontesi, F., Morini, G., and Cavallo, L. (2007) *Macromolecules*, **40**, 9181.

131 Cavallo, L., Guerra, G., Vacatello, M., and Corradini, P. (1991) *Macromolecules*, **24**, 1784.

132 Corradini, P., Guerra, G., and Pucciariello, R. (1985) *Macromolecules*, **18**, 2030.

Index

Page numbers in *italics* refer to figures

Selective Nanocatalysts and Nanoscience, First Edition. Edited by Adriano Zecchina, Silvia Bordiga, Elena Groppo.

h

i

t